Ecophysiology of Vascular Halophytes

Author

Irwin A. Ungar, Ph.D.

Professor
Department of Botany
Ohio University
Athens, Ohio

CRC Press is an imprint of the
Taylor & Francis Group, an **informa** business

CRC Press
Taylor & Francis Group
6000 Broken Sound Parkway NW, Suite 300
Boca Raton, FL 33487-2742

© 1991 by Taylor & Francis Group, LLC
CRC Press is an imprint of Taylor & Francis Group, an Informa business

First issued in paperback 2019

No claim to original U.S. Government works

ISBN 13: 978-0-367-45053-3 (pbk)
ISBN 13: 978-0-8493-6217-0 (hbk)

This book contains information obtained from authentic and highly regarded sources. Reasonable efforts have been made to publish reliable data and information, but the author and publisher cannot assume responsibility for the validity of all materials or the consequences of their use. The authors and publishers have attempted to trace the copyright holders of all material reproduced in this publication and apologize to copyright holders if permission to publish in this form has not been obtained. If any copyright material has not been acknowledged please write and let us know so we may rectify in any future reprint.

Except as permitted under U.S. Copyright Law, no part of this book may be reprinted, reproduced, transmitted, or utilized in any form by any electronic, mechanical, or other means, now known or hereafter invented, including photocopying, microfilming, and recording, or in any information storage or retrieval system, without written permission from the publishers.

For permission to photocopy or use material electronically from this work, please access www.copyright.com (http://www.copyright.com/) or contact the Copyright Clearance Center, Inc. (CCC), 222 Rosewood Drive, Danvers, MA 01923, 978-750-8400. CCC is a not-for-profit organization that provides licenses and registration for a variety of users. For organizations that have been granted a photocopy license by the CCC, a separate system of payment has been arranged.

Trademark Notice: Product or corporate names may be trademarks or registered trademarks, and are used only for identification and explanation without intent to infringe.

Visit the Taylor & Francis Web site at
http://www.taylorandfrancis.com

and the CRC Press Web site at
http://www.crcpress.com

PREFACE

The present volume represents the results of my research interests in the study of halophytes over the last thirty years. Much of my personal research has been devoted to the study of the responses of inland halophytes of North America to environmental stress. A number of researchers throughout the world have made important contributions to our understanding of the physiological ecology of coastal halophytes and their research efforts have contributed significantly in the preparation of this manuscript.

In recent years there has been an increase in research activity concerning the physiological ecology of halophytes, which has been stimulated by a number of significant problems that have developed because of the current usage of our land resources. The natural coastal and inland salt marsh habitats have been strongly impacted by human activity and efforts are being made to conserve existing salt marshes and in the creation of new salt marshes. Knowledge of the physiology and ecology of marsh species is essential to carry out these revegetation projects and for better management of salt marsh habitats. Irrigation of farmland in drier regions of the world has caused the salinization of large areas, gas and oil well drilling has produced large amounts of brine, and salting of roads in cooler climates has led to the development of either saline soils or loss of vegetation. The activities of people living in coastal areas of the world have caused major damage to salt marsh habitats and resulted in the disappearance of salt marshes in many regions of the world.

The need for more food to feed our increasing population has stimulated a concerted effort to find halophytes which are suitable for direct use as good for the human population or as feed for livestock. Efforts have also been directed toward determining the physiological and genetic mechanisms responsible for salt tolerance in plants. The more salt tolerant halophytes could serve as donors for genetic material that would be utilized in the development of new salt tolerant races for our most important agricultural plants. Although coastal and inland salt marshes and salt deserts are currently used for grazing animals and some harvesting of species is done for food, greater agricultural use may need to be made of these lands in the future to meet the growing demands on currently used agricultural land to feed the world's increasing population of people.

The main objective of this book is to review the literature which has been published over the last two decades concerning the ecophysiology of halopytes. The topics covered in this book range from the physiological responses of plants to salt stress and other edaphic factors at different stages of development to the effect of competition on the growth and distribution of halophytes. Other reviews in the past have either contained a very general discussion of halophyte biology or emphasized one limited portion of the problem that halophytes must respond to in adjusting to environmental conditions in saline habitats. In this survey of the literature an attempt is made to discuss how the physical and biotic environment may be influencing the physiological and ecological responses of halophytes.

I am sincerely indebted to my colleagues and graduate students who have worked diligently in the area of halophyte ecology and provided much of the data reported on in this book. My special thanks to Ohio University for providing me with a Faculty Fellowship and John Baker research award to complete and revise the manuscript for this publication. A number of researchers have provided me with encouragement and inspiration to continue my research on the ecology of halophytes, including Paul Binet, Jean Boucaud, Jacques Langlois, Olubukanla Okusanya, Frank Drysdale, Stanley Woodell, Yoav Waisel, and Velentine Chapman. I wish to thank my former graduate students for their exchange of ideas and the dedicated work they have carried out to make this publication more complete. A number of the graduate students that I have worked with, including William Hogan, Allan Macke, Karen McMahon, David Benner, David McGraw, David Loveland, Willa Ewing, Barbara Wertis, Jeyarany Philipupillai, Donald Drake, Terrence Riehl, Hamid Karimi, Ajmal

Khan, Marlis Rahman, Kemuel Badger, Tonya Selby, Tira Katembe, and Barbara Ballard, have stimulated my continuing interest in the ecology of halophytes and made significant contributions to our understanding of the physiological ecology of halophytes.

The manuscript was completed while the author was a Visiting Fellow at Wolfson College and Visiting Professor in the Laboratory of Plant Ecology, Department of Plant Sciences, Oxford University. I thank Roger Hall for providing me space in his laboratory.

I am most sincerely appreciative of the support that my parents, wife and children have given me through the years in my pursuit of research and during the preparation of the manuscript for this book.

AUTHOR

Irwin A. Ungar, Ph.D., is Professor of Botany in the Department of Botany, College of Arts and Sciences, at Ohio University, Athens, Ohio. He received his Bachelor of Science degree from the City College of New York in 1955 and his M.A. and Ph.D. degrees from the University of Kansas in 1957 and 1961, respectively. In 1961 he was an Instructor in the Department of Botany, University of Rhode Island and from 1962 to 1966 an Assistant Professor in the Department of Biology at Quincy College. Since 1966, Dr. Ungar has held successive appointments as Assistant Professor of Botany, Associate Professor of Botany, and Professor of Botany in the Department of Botany at Ohio University. He was Chairman of the Department of Botany from 1984 to 1989.

For the 1972 academic year, Dr. Ungar was a Visiting Professor and Chercheur Associe in the Laboratoire de Physiologie Vegetale at the University of Caen and in the 1990 academic year he was a Visiting Fellow at Wolfson College and a Visiting Professor in the Department of Plant Sciences at Oxford University. Dr. Ungar is a fellow of the Ohio Academy of Science and a member of the Ecological Society of America, Botanical Society of America, and the American Association for the Advancement of Science. He is the author of over 80 publications and has presented his research findings at numerous national and international meetings. His major research interest is in the effect of stress on the seed germination, growth, and distribution of plants, with special emphasis in the ecology of halophytes. In addition to his research interests, he is very active in both undergraduate and graduate student teaching, Director of the Dysart Woods Laboratory since 1985, and other committee activities at Ohio University.

TABLE OF CONTENTS

Chapter 11

Chapter 1

INTRODUCTION

I. SIGNIFICANCE OF SALINE HABITATS

The salt marsh habitat represents a very significant natural resource. Inland and coastal salt marshes contain specific flora and fauna that are not found in other freshwater wetlands or terrestrial plant communities.[1,2] Coastal salt marshes play a significant role in food chains of marine animals, and their disturbance could severely alter animal communities. Anthropogenic activities have severely impacted these salt marsh ecosystems with pollutants and damage from construction. Because of developmental activities, many natural salt marshes have been disturbed or completely destroyed due to industrial development or other activities of people living in coastal areas of the world. A number of efforts have been made to preserve and reconstruct salt marsh communities, using the dominant halophytic species for their revegetation.[3,4]

Because of the need for continually increasing our food supply to feed the expanding world population, significant efforts have been made to develop species of plants that can be grown for agricultural purposes in saline lands.[5-8] First, attempts were made to select cultivars of common crop species that demonstrated some degree of salt tolerance.[9,10] Second, attempts have been made to use native halophytes either as a direct source of food or as feed for farm animals. Several reviews have been written describing the multiple uses that we are currenty making of halophytes for agricultural purposes.[11-13] Many inland and coastal salt marsh and salt desert habitats have been utilized for the grazing of cattle and sheep, and the vegetation of some of these marshes has been greatly modified by the effects of long-term grazing.[14-16] In arid regions of the world, the continuous irrigation of agricultural fields has led to the development of saline soils. Evaporation of irrigation water has produced an accumulation of salts on the soil surface and made large areas which previously did not have a soil salinity problem useless for farming.[5-7]

II. LITERATURE REVIEW

My primary goal for writing this book is to present a review of the current data about the contributions of physiological ecology to our understanding of the ecology of halophytes. A significant research effort, which has been made in recent years in the area of environmental plant physiology, has contributed strongly to our understanding of how plants are responding to saline environments. Both field research and laboratory investigations have been carried out to ascertain the factors which are significantly affecting the seed germination, growth, and reproductive responses of halophytes under environmental conditions that are prevalent in coastal and inland salt marshes.

Earlier literature relating to the biology of halophytes has been described in considerable detail by several researchers.[17-21] A number of recent reviews are also available concerning mangrove vegetation,[22-27] and no attempt was made to review the literature concerning these broadly distributed tropical coastal forest species in this book. Many publications have described the distinct zonational patterns of plant species that occur in salt marsh habitats.[2,28-40] Efforts have been made to determine specifically which environmental factors are limiting the growth and distribution of halophytes along soil salinity-flooding gradients.[41-45] The physiological and biochemical responses of plants to salinity stress have been reviewed by a number of researchers in recent years.[8,46-62]

Several earlier comprehensive reviews have been written concerning the biology of halophytes.[2,15,28] In recent years a number of reviews have concentrated on the functional responses of plants to some aspect of salt stress.[59,60,63-66] The emphasis in the latter publications was on the physiological and biochemical responses of plants to salt stress. In my analysis of physiological and ecological data, the main thrust will be to describe and clarify some of the adaptations that species of plants have developed, which permit them to become established and to grow and reproduce in saline environments. Successful establishment of plants in salt marsh habits, which are subjected to a number of deleterious factors in their physical environment such as flooding, salt stress, tidal abrasion and deposition, nutritional ion imbalance, oxygen deficiencies, reducing environments, specific ion toxicities, and low water potentials, is dependent on a number of physiological, morphological, and phenological adaptations that make halophytic species successful in these stressful and unpredictable habitats. Other abiotic and biotic factors, as well as combinations of the above-mentioned factors, may also compound the problem that plant species must adjust to in order to become successfully integrated in the salt marsh habitat.

The floristic composition of most plant communities that occur in salt marshes is low when compared with adjacent vegetation from nonsaline habitats. Reduced diversity of plant species in salt marsh and salt desert communities is correlated with the fact that many terrestrial and aquatic vascular plant species do not have the necessary adaptations that would provide them with a significant degree of salt tolerance.[6-8] When we discuss the salt tolerance of halophytes, we usually recognize that a group of biological characteristics have apparently evolved in diverse groups of vascular plants that are necessary for their successful establishment in saline environments. No species may have developed all of the following characteristics, but at least some of them are necessary adaptations for plants that are found growing in these highly saline and unpredictable habitats: osmotic adjustment at the cellular level, ion compartmentation, succulence, salt glands, salt hairs, ionic exclusion at the root, tolerance to the dominant specific ions as well as tolerance to heavy metals, poorly aerated soils, reducing environments, and low nitrogen availability — common problems that can also be found in other nonsaline wetland ecosystems.

In this book I will use an ecophysiological approach to attempt to elucidate the factors affecting growth at the different developmental stages and distribution of halophytes in saline environments. It is clear that salt tolerance is one of the primary factors determining whether or not plants can grow in salt marsh habitats. Intolerant glycophytes are not present in saline environments, and most salt marsh and salt desert habitats generally contain a low diversity of vascular plant species. Salt tolerance involves both cellular tolerance to salt stress and whole organism phenomena, such as plant water balance and ion transport.[2,50] The results of investigations with salt tolerance of cell cultures often do not correlate well with the data for salt tolerance of mature plants and their growth and distribution in field situations.[67,68]

A holistic approach is necessary to determine the nature of the responses of plants to saline habitats. Besides the tolerance of species of plants to salinity, which is necessary for the successful establishment of halophytes in saline habitats, a complex of biotic[69-72] and abiotic factors[2,28,30,73-76] interacts to determine the level of biomass production, reproductive output, and distribution of species along environmental gradients in salt marshes. Biotic factors that are involved in limiting the growth of halophytes include herbivory, parasitism, and both intraspecific and interspecific competition.[69-72] The degree to which these biotic factors are affecting the growth and distribution of halophytes varies in different zonal communities and from one year to the next on any one marsh. Climatic and edaphic factors also play a significant role in determining the success of halophytes in a particular habitat. During a year of drought, mortality of plants is generally high because of the low soil-water potentials that develop in marsh soils.[72,73] Seasonal changes in precipitation pattern must also be adjusted to by those plants which are successful invaders of salt marsh environments. Low soil-water potentials are produced in saline soils because of low levels of precipitation and high evaporation rates in areas of the world that are dominated by grassland and desert plant communities. Unpredictable precipitation patterns from one year to the next can cause dramatic changes in the zonational pattern of plants growing in inland salt marsh habitats.[74,75]

Edaphic conditions, such as soil-water potential, soil salinity, flooding, aeration, sulfide accumulation, specific ion effects, and nutritional deficiencies, all play a role in determining the physiological status of halophytes. The productivity and distribution of plants that are found growing in saline environments are closely related to soil edaphic conditions. One of the main tasks in the study of the ecophysiology of halophytes in the future will be to sort out the relative significance of these solitary and interacting factors and the role they play in determining the growth responses of halophytes at different stages of development. In this book, I will attempt to describe from data available in the current literature some of the general responses of halophytes to the various factors prevalent in saline habitats and how they interact to determine the production and distribution patterns that currently exist in salt marsh habitats.

III. OUTLINE OF CHAPTERS INCLUDED IN THIS BOOK

The chapters in this book contain a review of the recent literature concerning the ecophysiology of vascular halophytes. Chapter 2 contains a discussion of the responses of seeds to salinity and the variety of germination characteristics that halophytes have evolved to survive the problem of salinity stress. Recent investigations with seed germination indicate that the seeds of halophytes have developed a number of adaptations that distinguish them from glycophytes. These include a general tolerance of seeds to low water potentials which is greater than that for most glycophytes, various dormancy mechanisms which may serve to avoid salt stress, and the capacity of seeds to resist extended periods of exposure to hypersaline conditions and then continue to germinate when the salinity stress is removed. Germination of halophytes often takes place when salinity levels are reduced under field conditions. A delay of germination to periods of reduced soil salinity is of adaptive value because it assures that at least some of the seedlings produced by a seed cohort will become reproducing plants. Germination for most of the annual species of halophytes occurs over an extended period of time, and some species form a persistent seed bank. Multiple germination periods play a significant role in preventing the local extinction of species in saline environments, since they permit several periods for the establishment of plants in the highly stressful and unpredictable salt marsh habitats.

Chapter 3 provides some details concerning the growth responses of halophytes to salinity stress under both field and experimental laboratory conditions. Most halophytic species examined vary in their degree of salt tolerance at different stages of development. Increases in salinity from that in control nutrient solutions caused a reduction in growth for a number of species of halophytes. However, the biomass production of some species is enhanced by micronutrient levels of NaCl and other species have their growth stimulated by up to 200 mM NaCl. It is sometimes difficult to predict the distribution of halophytes based on laboratory experiments that are used to determine their physiological limits of tolerance. Other biotic and abiotic environmental factors may play a significant role in determining the biomass production, fecundity, and survival of species in salt marsh habitats. Tolerance to salinity may either represent the ability of a species to adapt to the general effects of low water potentials or to the adjustment of plants to the presence of a specific ion. The occurrence of specific ion toxicities, which are caused by some combination of specific ions, have been reported in experimental investigations with halophytes.

Chapter 4 represents a discussion of the effect of salinity on the photosynthetic responses of halophytes. Much of the early work concerning the carboxylation rates of halophytes was done under laboratory conditions. However, in recent years with the development of portable infrared gas analyzers, it has been possible to make diurnal and seasonal measurements of the net photosynthetic rates of individual leaves, plants, and communities under field conditions. Halophytes can generally tolerate higher levels of salinity than glycophytes, and for some of the most salt tolerant species, there is apparently no significant reduction in their net photosynthetic rates, even at the highest salinities examined. Several factors contribute to the reduction in net photosynthetic rates of halophytes that are exposed to salinity stress. These include reduced stomatal conductance, increased mesophyll resistance caused by a decrease in carboxylating enzyme or its substrate, and inhibition of thylakoid membranes. Induction of the CAM pathway has been reported for some species of succulents when they are exposed to saline conditions, while other species are apparently not stimulated to change their photosynthetic pathway when exposed to salt stress.

Chapter 5 contains a discussion of our current knowledge regarding the effect of salinity on the growth regulator activity of halophytes. Several researchers have determined that exogenous applications of gibberellic acid could overcome the inhibitory effects of salinity on the growth of plants. However, growth regulator treatments were not found to be effective at the highest salinities to which plants were exposed. Cytokinin levels have been reported to decrease when plants were exposed to salt stress. Reductions in cytokinin levels could negatively affect the rate of cell division and also cause closure of the stomatal apparatus, inhibiting both the growth and net primary production of plants. Recently, it has been reported that ethylene production is enhanced when plants are exposed to increases in salt stress and that this effect is in response to low water potentials rather than to a specific ion toxicity. Increases in the water potential of the growth medium caused a reversal in the level of ethylene production.

Chapter 6 contains a review of the literature regarding the adjustment of plant water status to salinity stress. Halophytes must be capable of adjusting the water potentials of their tissues to that of the fluctuating soil water potentials in order to become successfully established in saline environments. The development of the Scholander bomb and hygrometric techniques for analyzing plant water status has allowed researchers to make field as well as laboratory measurements of plant water potentials and osmotic potentials. Halophytes are able to adjust osmotically to increasing levels of soil salinity, and the capacity to reduce their tissue water potential is a determining factor in where they can grow and survive along a soil salinity gradient. The two major mechanisms for the adjustment of water potentials of plants are represented either by a reduction in tissue water content or by the accumulation of inorganic ions. Other adaptations which control the level of salts in tissues include succulence, salt glands, salt hairs, exclusion from roots, and abscission of organs.

Chapter 7 provides a review of the literature pertaining to field and laboratory investigations regarding the ionic content of plants under a variety of soil salinity and experimental conditions. Ash content of some species of halophytes reaches more than 50% of their dry mass yields, indicating that a number of highly salt-tolerant species are able to accumulate high ionic concentrations. A number of investigations have been made to determine the specific ion content of plant organs. Some species of halophytes accumulate high concentrations of Na and Cl in their tissues, while other species are able to restrict the uptake of these ions. Experimental data indicate that most species of halophytic-dicotyledons respond with an increase in tissue ion content when the soil salinity level increases. X-ray probe microanalysis has been used to determine the relative content of elements in different tissue regions and for different parts of plant cells. Data using these techniques indicate that plants can compartmentalize ions among organs and tissues and within cells. Uptake of Na is effected by the salinity of the medium, but it is also affected by the presence of Ca and other cations in the soil solution. Salt tolerance may be in part regulated by the tolerance level of species to high concentrations of salt, compartmentation of ions, and production of compatible solutes to regulate the water potential of different cell compartments.

Chapter 8 contains a discussion of the tolerance of halophytes to flooding conditions in salt marshes. Tidal action effects plants by two major mechanisms: first, there is a mechanical effect, where deposition and removal of silt by tides causes mechanical damage to plants and uproots individuals. Second, there is the problem of tidal flooding at periodic intervals. In inland salt marshes, plants are exposed to high water tables and periods of flooding after heavy rainfall. Several soil factors are correlated with flooding, including low soil-oxygen concentration, reducing environments, reduced nutrient availability, high sulfide content, and heavy metal accumulation. Variation in the level of aeration of soils along a salt marsh environmental gradient may play a significant role in determining the biomass production, survival, and distribution of halophytes. A number of adaptations are necessary for plants to carry out aerobic root respiration under flooding conditions. Some species of plants produce large amounts of aerenchyma tissue, which allows oxygen to be transported from above-ground organs to the roots. Transfer of oxygen from the roots to the soil may actually increase the aeration level of soils and improve the availability of soil nutrients to plants.

Chapter 9 provides a review of the recent literature relating to the effect of nitrogen availability on plant growth in saline habitats. Several researchers have reported that the availability of nitrogen is one of the primary factors limiting the growth of halophytes in salt marshes. It has been determined that the growth of halophytes could vary in different areas of a salt marsh, even though the nitrogen content of the soil was similar in the two areas investigated. A number of other soil factors, including high soil salinity, lack of aeration, sulfide accumulation, and the presence of heavy metals, could inhibit the nitrogen available to plants in salt marsh habitats, even when the soil nitrogen content did not vary. Field fertilization experiments indicated that the fertilization of salt marsh soils with nitrogen would increase the biomass production of plants growing in salt marsh habitats. Laboratory studies have demonstrated that nitrogen fertilization could overcome some of the inhibitory effects of salinity on the production of plant biomass.

Chapter 10 discusses the prevalence of ecotypes in salt marsh and salt desert habitats. The occurrence of genetically distinct races has been reported for plants growing in saline habitats. Greater salinity tolerance has been demonstrated for populations found in the more saline part of a marsh gradient than for populations growing in areas of lower salinity. Two major types of ecotypic differentiation appear to occur. First, broadly distributed species of halophytes in coastal and inland habitats apparently develop genetically distinct ecotypes along climatic gradients, in which the plants demonstrate different patterns of phenological development. Second, one finds ecotypic differentiation and the development of microspecies along salinity gradients in local marshes. Reciprocal transplant and uniform garden experiments both indicate that genetic selection occurs in isolated plant populations on salt marshes. In some cases, cleistogamy has developed as the primary breeding system, which further increases inbreeding and the reproductive isolation of populations. Because of the reproductive or geographic isolation of these populations from other populations on a salt marsh, selection for microspecies will occur.

Chapter 11 contains a review of the available literature concerning the effect of competition on the growth and distribution of halophytes. Both intraspecific and interspecific competition have been reported to play a significant role in determining the biomass production of plants, population size, and distribution of halophytes in saline habitats. Interspecific competition for space, nutrients, and light has been suggested to be the primary factor limiting the growth of halophytes in nonsaline habitats. Competition also could restrict species from zones on a salt marsh that are less saline than the habitat in which a species dominates. Removal and transplant experiments in the field indicate that biomass production of the target species increased when other species were removed from a plot. Intraspecific competition has been shown to reduce plant biomass production in salt marsh habitats. In saline environments, competition has been demonstrated to cause a decrease in yields and to inhibit the reproductive capacity of species, but it has not been shown to effect survival. This is probably because salt stress induced very high mortality in extreme habitats, and competition plays little or no role in determining the level of survival of species in the most stressful environments where salt stress is the prevailing factor controlling the mortality of plants.

In this book I will review the literature on the ecophysiology of vascular halophytes. Emphasis will be placed on the data available from publications concerning the inland and coastal salt marsh vegetation over the past 20 years.

REFERENCES

1. **Chapman, V. J.**, *Ecosystems of the World 1. West Coastal Ecosystems*, Elsevier, Amsterdam, 1977.
2. **Waisel, Y.**, *Biology of Halophytes*, Academic Press, New York, 1972.
3. **Cole, D. P.**, The restoration of coastal vegetation in Florida, *Proc. 5th Annu. Conf.*, Tampa, Florida, 1978.
4. **Niesen, T. and Josselyn, M.**, The Hayward regional shoreline marsh restoration: biological succession during the first year following dike removal, *Tiburon Center Environ. Stud. Tech. Rep.*, 1, 1, 1981.
5. **Boyko, H.**, *Salinity and Aridity*, Junk Publishers, The Hague, 1966.
6. **Hollaender, A., Aller, J. C., Epstein, E., San Pietro, A., and Zaborsky, O. R.**, *The Biosaline Concept*, Plenum Press, New York, 1979.
7. **San Pietro, A.**, *Biosaline Research, A Look to The Future*, Plenum Press, New York, 1982.
8. **Staples, R. C. and Toenniessen, G. H.**, *Salinity Tolerance in Plants*, John Wiley & Sons, New York, 1984.
9. **Shannon, M. C.**, Breeding, selection, and the genetics of salt tolerance, in *Salinity Tolerance in Plants*, Staples, R. C. and Toenniessen, G. H., Eds., John Wiley & Sons, New York, 1984, Chap. 13.
10. **Ponnamperuma, F. N.**, Role of cultivar tolerance in increasing rice production on saline lands, in *Salinity Tolerance in Plants*, Staples, R. C. and Toenniessen, G. H., Eds., John Wiley & Sons, New York, 1984, Chap. 14.
11. **Mudie, P. J.**, The potential economic uses of halophytes, in *Ecology of Halophytes*, Reimold, R. J. and Queen, W. H., Eds., Academic Press, New York, 1974, 465.
12. **Somers, G. F.**, Food and economic plants: a review, in *Biosaline Research, A Look to the Future*, San Pietro, A., Ed., Plenum Press, New York, 1982, 127.
13. **O'Leary, J. W.**, The role of halophytes in irrigated agriculture, in *Salinity Tolerance in Plants*, Staples, R. C. and Toenniessen, G. H., Eds., John Wiley & Sons, New York, 1984, Chap. 16.
14. **Zahran, M. A. and Abdel Wahid, A. A.**, Halophytes and human welfare, in *Contributions to the Ecology of Halophytes*, Sen, D. N. and Rajpurohit, K. S., Eds., Junk Publishers, The Hague, 1982, 233.
15. **Chapman, V. J.**, *Salt Marshes and Salt Deserts of the World*, Interscience, New York, 1960, Chap. 12.
16. **Tiedemann, A. R., McArthur, E. D., Stutz, H. C., Stevens, R., and Johnson, K. L.**, Eds., Symposium on the Biology of Atriplex and Related Chenopods, Tech. Rep. INT-172, Forest Service, U.S. Department of Agriculture, Ogden, UT, 1984.
17. **Uphoff, J. C. T.**, Halophytes, *Bot. Rev.*, 7, 1, 1941.
18. **Stocker, O.**, Das Halophytenproblem, *Ergebnisse Biol.*, 3, 265, 1928.
19. **Chapman, V. J.**, The new perspective in the halophytes, *Q. Rev. Biol.*, 17, 291, 1942.
20. **Adriani, M. J.**, Der Wasserhaushalt der Halophyten, in *Handbuch der Pflanzenphysiologie, Band III. Pflanze und Wasser*, Ruhland, W., Ed., Springer-Verlag, Berlin, 1956, 902.
21. **Adriani, M. J.**, Halophyten, in *Encyclopedia of Plant Physiology*, Vol. 4, Ruhland, W., Ed., Springer-Verlag, Berlin, 1958, 709.
22. **Chapman, V. J.**, *Mangrove Vegetation*, J. Cramer, Vaduz, Liechtenstein, 1976.
23. **Teas, H. J.**, *Physiology and Management of Mangroves*, Junk Publishers, The Hague, 1984.
24. **Tomlinson, P. B.**, *The Botany of Mangroves*, Cambridge University Press, Cambridge, 1986.
25. **Hutchings, P. and Saenger, P.**, *Ecology of Mangroves*, University of Queensland Press, St. Lucia, 1987.
26. **Walsh, G. E.**, Mangroves: a review, in *Ecology of Halophytes*, Reimold, R. J. and Queen, W. H., Eds., Academic Press, New York, 1974, 51.
27. **Barth, H.**, The biogeography of mangroves, in *Contributions to the Ecology of Halophytes*, Sen, D. N. and Rajpurohit, K. S., Eds., Junk Publishers, The Hague, 1982, Chap. 3.
28. **Chapman, V. J.**, *Salt Marshes and Salt Deserts of the World*, J. Cramer, Lehre, 1974.

29. **Ward, J. M.**, Studies in ecology on a shell barrier beach. I. Physiography and vegetation of shell barrier beaches, *Vegetation*, 14, 240, 1967.
30. **Ranwell, D. S.**, *Ecology of Salt Marshes and Sand Dunes*, Chapman and Hall, London, 1972.
31. **Ungar, I. A.**, Inland halophytes of the United States, in *Ecology of Halophytes*, Reimold, R. J. and Queen, W. H., Eds., Academic Press, New York, 1974, 235.
32. **Macdonald, K. and Barbour, M. G.**, Beach and salt marsh vegetation of the North American Pacific coast, in *Ecology of Halophytes*, Reimold, R. J. and Queen, W. H., Eds., Academic Press, New York, 1974, 175.
33. **Ungar, I. A.**, The vegetation of inland saline marshes of North America, north of Mexico, in *Grundfragen und Methoden in der Pflanzensoziologie*, Tuxen, R., Ed., Junk Publishers, The Hague, 1972, 397.
34. **Barbour, M. G., Craig, R. B., Drysdale, F. R., and Ghiselin, M. T.**, *Coastal Ecology, Bodega Head*, University of California, Berkeley, 1973.
35. **Eleuterius, L. N. and Eleuterius, C. K.**, Tide levels and salt marsh zonation, *Bull. Mar. Sci.*, 29, 394, 1979.
36. **Earle, J. C. and Kershaw, K. A.**, Vegetation patterns in James Bay coastal marshes. III. Salinity and elevation as factors influencing plant zonations, *Can. J. Bot.*, 67, 2967, 1989.
37. **Kolbek, J., Dostalek, J., Jarolimek, I., Ostry, I., and Sek-Ha, L.**, On salt marsh vegetation in North Korea, *Folia Geobotan. Phytotaxonom.*, 24, 113, 1989.
38. **Ellenberg, H.**, Sea marshes and inland saline habitats, in *Vegetation Ecology of Central Europe*, Cambridge University Press, Cambridge, 1988, 348.
39. **Binet, P.**, La Flore du Littoral, *Penn Ar Bed*, 25, 33, 1961.
40. **Beeftink, W. G.**, The coastal salt marshes of western and northern Europe: an ecological and phytosociological approach, in *Wet Coastal Ecosystems*, Chapman, V. J., Ed., Elsevier, Amsterdam, 1977, Chap. 6.
41. **Chapman, V. J.**, *Coastal Vegetation*, Pergamon Press, Oxford, 1964.
42. **Steiner, M.**, Zur Okologie der Salzmarschen der Nordostlichen Vereinigten Staaten von Nordamerika, *Jahrb. Wiss. Bot.*, 81, 94, 1934.
43. **Johnson, D. S. and York, H. H.**, The relation of plants to tide-levels, *Carnegie Inst. Wash. Pub.*, 206, 1, 1915.
44. **Kurz, H. and Wagner, K.**, Tidal marshes of the Gulf and Atlantic coasts of northern Florida and Charleston, South Carolina, *Florida State Univ. Studies*, 24, 1, 1957.
45. **Miller, W. R. and Egler, F. E.**, Vegetation of the Wequetequock-Pawcatuck tidal marshes, *Ecol. Monogr.*, 20, 143, 1950.
46. **Scrogonov, B. P.**, *Structure and Function of Plant Cells in Saline Habitats*, John Wiley & Sons, New York, 1973.
47. **Strogonov, B. P.**, *Physiological Basis of Salt Tolerance of Plants*, Israel Program for Scientific Translations, Jerusalem, 1964.
48. **Poljakoff-Mayber, A. and Gale, J.**, *Plants in Saline Environments*, Springer-Verlag, Berlin, 1975.
49. **Caldwell, M. M.**, Physiology of desert halophytes, in *Ecology of Halophytes*, Reimold, R. J. and Queen, W. H., Eds., Academic Press, New York, 1974, 355.
50. **Greenway, H. and Munns, R.**, Mechanisms of salt tolerance in nonhalophytes, *Annu. Rev. Plant Physiol.*, 31, 149, 1980.
51. **Albert, R.**, Halophyten, in *Pflanzenokologie und Mineralstoffwechsel*, Kinzel, E., Ed., Verlag Eugen Ulmer, Stuttgart, 1982, Chap. 3.
52. **Bernstein, L.**, Effects of salinity and sodicity on plant growth, *Annu. Rev. Plant. Phytopathology*, 13, 295, 1975.
53. **Brun, A.**, Mise au point bibliographique concernant l'etude des effects de la salinite sur les vegetaux, *Ann. Fac. Sci. Yaounde*, 28, 59, 1981.
54. **Binet, P.**, Aspects physiologiques de l'halophyte et de la resistance aux sels, *Soc. Bot. Fr. Actualites Bot.*, 3/4, 1, 1978.
55. **Flowers, T. J.**, Physiology of halophytes, *Plant Soil*, 89, 41, 1985.
56. **Flowers, T. J., Hajibagheri, M. A., and Clipson, J. W.**, Halophytes, *Q. Rev. Biol.*, 61, 313, 1986.
57. **Flowers, T. J.**, Halophytes, in *Ion Transport in Plant Cells and Tissues*, Baker, D. A. and Hall, J. L., Eds., North-Holland, New York, 1975, Chap. 10.
58. **Flowers, T. J., Troke, P. F., and Yeo, A. R.**, The mechanism of salt tolerance of halophytes, *Annu. Rev. Plant. Physiol.*, 28, 89, 1977.
59. **Munns, R., Greenway, H., and Kirst, G. O.**, Halotolerant eukaryotes, in *Encyclopedia of Plant Physiology*, Vol. 12C, Lange, O. L., Nobel, P. S., Osmond, C. B., and Ziegler, H., Eds., Springer-Verlag, Berlin, 1983, 59.
60. **Jefferies, R. L.**, Osmotic adjustment and the response of halophytic plants to salinity, *Bioscience*, 31, 42, 1981.

61. **Lerner, H. R.**, Adaptations to salinity at the plant cell level, *Plant Soil*, 89, 3, 1985.
62. **Wyn-Jones, R. G.**, Salt tolerance, in *Physiological Processes Limiting Plant Productivity*, Johnson C. B., Ed., Butterworths, London, 1981, Chap. 15.
63. **Jefferies, R. L.**, Aspects of salt-marsh ecology with particular reference to inorganic mineral nutrition, in *The Estuarine Environment*, Barnes, R. S. K. and Green, J., Eds., Applied Science Publishers, London, 1972, Chap. 5.
64. **Jennings, D. H.**, The effects of sodium chloride on higher plants, *Biol. Rev.*, 51, 453, 1976.
65. **Lauchli, A.**, Regulation des Salztransportes und Salzausschiessung in Glykophyten und Halophyten, *Ber. Deutsch. Bot. Ges.*, 92, 87, 1987.
66. **Jennings, D. H.**, Halophytes, succulence and sodium—a unified theory, *New Phytol.*, 67, 899, 1968.
67. **Smith, M. K. and McComb, J. A.**, Effect of NaCl on the growth of whole plants and their corresponding callus cultures, *Aust. J. Plant Physiol.*, 8, 267, 1981.
68. **Hedenstrom, H. and Breckle, S.-W.**, Obligate halophytes? A test with tissue culture methods, *Z. Pflanzenphysiol.*, 74, 183, 1974.
69. **Ellison, A. M.**, Effects of competition, disturbance, and herbivory on *Salicornia europaea*, *Ecology*, 68, 576, 1987.
70. **Bertness, M. D., Wise, S., and Ellison, A. M.**, Consumer pressure and seed set in a salt marsh perennial plant community, *Oecologia*, 71, 190, 1987.
71. **Bazely, D. R. and Jefferies, R. L.**, Changes in the composition and standing crop of salt-marsh communities in response to the removal of a grazer, *J. Ecol.*, 74, 693, 1986.
72. **Ungar, I. A., Benner, D. K., and McGraw, D. C.**, The distribution and growth of *Salicornia europaea* on an inland salt pan, *Ecology*, 60, 329, 1979.
73. **Ungar, I. A.**, Population ecology of halophyte seeds, *Bot. Rev.*, 53, 301, 1987.
74. **Ungar, I. A.**, Population dynamics of inland halophytic communities, *Bull. Soc. Bot. Fr.*, 121, 287, 1974.
75. **Ungar, I. A.**, Population characteristics, growth, and survival of the halophyte *Salicornia europaea*, *Ecology*, 68, 569, 1987.
76. **Abdel-Razik, M. S. and Ismail, A. M. A.**, Vegetation composition of a maritime salt marsh in Qatar in relation to edaphic features, *J. Vegetation Sci.*, 1, 85, 1990.

Chapter 2

SEED GERMINATION

I. INTRODUCTION

The success of halophyte populations, especially for annuals which have only one opportunity in their life history for reproducing, is greatly dependent on the germination responses of their seeds. Because the evaporative power of the air is greater during the summer months, surface salt marsh soils tend to have higher soil salinity and more negative water potentials in the summer than in the spring.[1] Seed germination usually occurs early in the growing season or during a period when soil salinity levels are reduced, allowing for the establishment of seedlings prior to the period of greatest salt stress. The soil salinity conditions at the germination stage of development are somewhat predictive of the edaphic conditions to which later developmental stages of the plant will be exposed. A number of studies have been done monitoring the periodicity of germination under field conditions, and these support the hypothesis that the survival of a seedling cohort to maturity is correlated with the time of seed germination.[2-5] Later seedling cohorts had much higher mortality and very few, if any, individuals that survived to the reproductive stage of development.

In general, researchers have concluded that salinity is inhibitory to the germination of halophyte seeds in two ways: (1) causing a complete inhibition of the germination process at salinities beyond the tolerance limits of a species, and (2) delaying the germination of seeds at salinities that cause some stress to seeds but do not prevent germination.[6,7] There is great variability among plants in terms of the level of salinity that will completely inhibit or delay germination, and even in the case of halophytes many seeds will not germinate at salinities of seawater strength (Table 1). It is, therefore, sometimes difficult to predict the salt tolerance of halophyte seeds based on the habitat location of mature plants in the field.[8,9]

Some aspects of the physiological responses of seeds to salinity have been summarized in recent reviews of the literature.[6,7] The population ecology of halophyte seeds has been described in some detail in a previous review, with emphasis placed on seed cycles and factors affecting seed storage in the soil.[1] In the following review of the current literature concerning the germination of halophytes, the emphasis will be on the relationship between physiological responses of seeds to the ecology of plant species growing in saline environments.

II. SALT TOLERANCE

The germination responses of seeds of both glycophytes and halophytes to salinity are highly variable and species specific.[6,7] Strogonov[10,11] has emphasized the significance of specific ion effects on the distribution and growth of halophytic species in the Soviet Union. The distribution of inland halophytes of North America indicates that these species are broadly tolerant of a number of ionic combinations.[12] Characteristic species of inland marshes such as *Puccinellia nuttalliana, Distichlis stricta, Hordeum jubatum, Suaeda depressa, Salicornia rubra, Atriplex triangularis, A. patula,* and *Triglochin maritima* are found to be broadly distributed in saline soils dominated by the anions chloride, sulfate, and carbonate and the cations sodium, potassium, and magnesium. Ungar[12] reported that these distributional data indicate that specific ion effects may be less significant in determining seed germination and growth responses of these inland halophytes of North American than have been reported for glycophytic species. Soil water potentials may be the most significant factor controlling seed germination, growth, and distribution of North American inland halophytes, while

10   *Ecophysiology of Vascular Halophytes*

TABLE 1
Percentage Germination at the Limits of Salt Tolerance for Selected Halophytes and
Recovery Percentage in Freshwater after Treatment in Saline Solutions

Species	Germination control (%)	Germination maximum salinity (%)	Maximum salinity (%)	Recovery germination (%)	Ref.
Armeria maritima	97.0	6.0	3.6	90.0	16
Aster tripolium	78.0	2.0	5.3	54.0	16
Atriplex nummularia	93.0	5.0	3.0	71.0	19
Crithmum maritimum	29.0	1.0	1.4	38.0	67
Hordeum jubatum	98.0	10.0	2.0	97.0	159
Juncus maritimus	84.0	1.0	3.6	47.0	16
Limonium vulgare	31.0	6.0	5.3	53.0	16
Melaleuca ericifolia	100.0	5.0	1.6	100.0	51
Mesembryanthemum australe	62.0	2.0	2.0	63.0	160
Plantago maritima	67.0	9.0	1.8	61.0	16
Puccinellia nuttalliana	86.0	5.2	2.0	87.6	66
Salicornia europaea	40.0	8.0	5.0	50.0	64
Scaveola taccada	57.5	1.0	1.7	47.5	161
Spartina alterniflora	42.0	2.0	6.0	38.0	162
Spergularia marina	48.0	2.0	2.0	46.0	64
S. media	79.0	1.0	2.0	75.0	120
Sporobolus virginicus	38.0	15.0	1.5	38.0	68
Suaeda depressa	47.0	2.0	4.0	54.0	163
S. linearis	30.0	1.0	5.0	34.0	64

coastal species are responding to both tidal inundation and the abrasive and depositional effects of tidal action, as well as to the salinity factor.

Seeds of halophytes vary greatly in their ability to germinate under hypersaline conditions. It is often difficult to predict from the germination responses of a species to salinity whether or not it will grow well under saline conditions. This is because seeds of many of the most salt-tolerant species, including *Suaeda depressa, Spartina alterniflora, Spartina patens, Salicornia europaea, Atriplex triangularis,* and *Juncus gerardii* do not germinate well at salinities approaching that of seawater.[13-16] However, some species such as *Suaeda maritima* are capable of germinating in seawater concentrations.[17] Ungar[7] concluded that an osmotically enforced dormancy is produced by hypersaline conditions, which prevents germination until salt stress is alleviated by an influx of freshwater that increases soil water potentials.

The effects of salinity and inundation on seed germination were studied for some halophytes and glycophytes by Rozema.[13] Three salt marsh species *J. gerardii, J. alpinoarticulatus, J. maritimus,* and the glycophyte *J. bufonius* were inhibited from germinating by increases in salinity, indicating that seeds of these species do not have a salt requirement but are tolerant of salinity. Inundation did not inhibit germination in any of the species, and for *J. gerardii* there appeared to be a stimulation of germination when seeds were flooded. The velocity of germination was reduced for all *Juncus* species when salinity was increased. Other halophytes studied, *Glaux maritima, Agrostis stolonifera,* and *Parnassia palustris,* were also inhibited by seawater concentrations of salts, demonstrating that the germination stage is not a good indicator of the position of species along a salt marsh gradient. It was concluded that seed germination responses were not a good indicator of plant zonation and that the tolerance of seedlings to salinity and flooding may be a better predictor of plant distribution on salt marshes.[13] *Atriplex rhagodioides* was found to be moderately salt tolerant at the germination stage, 70% germination at 3 mS/cm to 3% germination at 30 mS/cm, and seedling stages, but based on field distribution highly salt tolerant at later stages of development.[18]

Uchiyama[19] determined that seed germination of *Atriplex nummularia* is reduced by 75% at a salinity of 2.0% NaCl. The reduction in germination at higher salinities is accentuated by suboptimal temperature treatments, with 73% seed germination at 1.5% NaCl at 10°C and only 5% germination for the same salinity at 30°C. Although 16% of the seeds of *A. nummularia* could germinate at 3.0% NaCl, no seedling establishment occurred at this salinity. These data indicate that the tolerance of plants to salt stress may vary at different stages of development. The total number of seeds that germinated and the rate of germination of the seeds of *A. halimus* were inhibited by salinity increments from 1 to 5% NaCl, with seeds in the 4 and 5% NaCl treatment having less than 5% germination.[20] Ungerminated seeds from the 4 and 5% NaCl treatments were placed in distilled water and germinated to levels equivalent to that of the original distilled water controls >90%, indicating that the effect of NaCl was a reversible osmotic inhibition of germination.

Badger and Ungar[21] determined that seeds of *Hordeum jubatum* had optimal germination at all salinities tested, 0 to 2.0% NaCl, at a temperature regimen of 25°C day and 15°C night. At high temperatures 35°C day and 25°C night, no seeds germinated in 1.5% NaCl, while 60% of the seeds germinated at the more favorable lower temperature regimen (Figure 1). The interaction between salinity and temperature effects apparently had a significant effect on seed germination. The temperature that seeds were stored at and the length of storage also affected the germination responses of *H. jubatum* seeds, as was indicated by the variation in results obtained in a number of investigations.[12,22-24]

The effect of salinity on the seed germination of two species of *Salicornia* from the Bergen op Zoom marsh (Netherlands) was investigated: *Salicornia brachystachya* from the high marsh which varied significantly in salinity and *S. dolichostachya* from the low marsh which is exposed to less fluctuation in soil salinity.[25] Laboratory germination and field transplant studies indicated that both species could germinate equally well at chloride ion levels up to 1.4% (seawater concentration), indicating that seeds of these two species could germinate at the lower part of the marsh even though *S. brachystachya* is rarely found in the low marsh. Even though *S. brachystachya* is often subjected to higher soil salinities, laboratory studies demonstrated that it was somewhat more inhibited at the germination stage by high salt stress than was *S. dolichostachya*. Since seeds of both species could germinate in both the high and low marsh habitats, Huiskes et al.[25] concluded that the responses of seeds at the germination stage may not be the sole factor determining the distribution of populations of these two species in coastal salt marshes. Similar results were found by Khan and Weber[26] for the inland perennial halophyte *Salicornia pacifica* var. *utahensis*, with seed germination being gradually reduced from 55% in distilled water controls to 3% germination at 5% NaCl. They reported that seed germination occurred in early spring when salinity stress was reduced because of higher precipitation, whereas summer soil salinities would be inhibitory to seed germination. Khan et al.[27] reported that *Chrysothamnus nauseosus* was salt tolerant, but seed germination studies indicated that it behaved very much like a glycophyte and the germination percentages were greatly reduced at 87 mM NaCl. The rate of seed germination was also delayed with each salt increment. Only during periods of high precipitation, when soil salinity levels are reduced, could these seeds germinate under field conditions.

Germination of seeds of *Silene maritima* was strongly inhibited at 0.5 strength seawater, with germination percentages ranging from 0 to 20% in a number of different temperature regimens. Binet[28] reported that up to 88% of the seeds germinated in 0.25 strength seawater. These data indicate that *S. maritima* was tolerant of only moderate salinities during the germination stage of development.

The optimum germination percentages for *Crithmum maritimum* seeds collected in November from an Italian coastal area were in distilled water.[15] Salinity increments of up to 250 mM NaCl caused an 87% reduction in final germination percentages compared to the

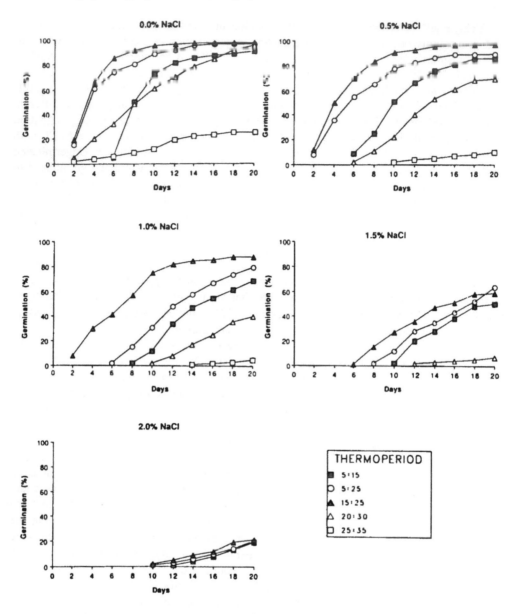

FIGURE 1.   Cumulative germination curves for fresh seeds of *Hordeum jubatum* exposed to 0.0, 0.5, 1.0, 1.5, and 2.0% NaCl for each thermoperiod tested. Temperature regimens represent dark to light photoperiods. (From Badger, K. S. and Ungar, I. A., *Can. J. Bot.*, 67, 1420, 1989. With permission.)

controls. A Timson[29] index indicated that the rate of germination was also delayed with increases in salinity. The level of dormancy varied in these seeds during storage, with seeds in January and February averaging 80 and 85% germination, respectively, while from June to August germination ranged from 5 to 7% in distilled water. It was concluded by Marchioni-Ortu and Bocchieri[15] that relative dormancy acts as a physiological control mechanism, which would prevent seeds from germinating during unfavorable edaphic conditions.

The germination response of seeds for the halophytes *Suaeda fruticosa*, *Cressa cretica*, *Trianthema triquetra*, and *Haloxylon recurvum* was studied at different osmotic potentials from soil extracts taken from the Pachpadra Salt Basin (India) by Rajpurohit and Sen.[30] All of the species investigated had a significant decrease in germination at −0.5 MPa, with very significant inhibition of seed germination at -0.7 MPa. High concentrations of NaCl in the soil were the chief inhibitory factor, and the authors observed that high germination percentages were only found in the field after heavy rains when the salt would be leached out of the surface soils.[30]

Studies with seeds from the Delta Marsh, Manitoba (Canada) indicated that germination of the seeds of *Atriplex patula* and *Phragmites communis* were reduced with increased seed burial.[24] Seeds of *A. patula* had 96% germination in surface soils, 4% at 3-cm and 0% at 4-cm depth, while *P. communis* seed germination ranged from 70% at the soil surface to 26% at 3-cm and 0% at 5-cm depth. A significant decrease in *A. patula* seed germination was found at 0.3% NaCl, while *P. communis* seeds were not inhibited by salinities up to 0.5% NaCl. Galinato and Van der Valk[24] concluded that germination of wetland species differed in response to depth of burial and soil salinity, which indicates that during a drawdown of a pond different species combinations could be established even though soils had identical seed banks.

Seeds of the halophyte *Suaeda nudiflora* collected from the New Port area of Bhavnagar (India) were found to germinate well (40%) at salinities up to 0.5% total salts by Joshi and Iyengar.[31] However, a sharp decrease in seed germination (6%) was found at salinities of 1.0% total salts, representing a 89.3% reduction in germination from the distilled water controls.

Cluff and Roundy[32] determined that reducing osmotic potentials, 0 to -2.0 MPa of NaCl, caused a decrease in the germination percentage of *Distichlis spicata* var. *stricta* seeds. Lowering the water potential to -2.0 MPa from the -0.1-MPa treatment reduced germination percentages to 20% of the control. Highest germination percentages were obtained at higher water potentials at $20 \times 30°C$ temperature regimens. Soil salinity increased and water potentials decreased in field sites from March to June, reaching levels beyond the range of salt tolerance of *D. stricta*, indicating that germination would have to take place early in the growing season or at less-saline sites during the summer months.

The effect of salinity on the germination of two *Puccinellia* species, *P. distans* and *P. lemmoni*, were examined by Harivandi et al.[33] over a salinity range from 0 to 100% seawater. Increases in seawater salinity caused a significant decrease in germination percentages for *P. lemmoni*, with only 2% of the seeds germinating in full strength seawater vs. 95% in distilled water. *Puccinellia distans* seeds were significantly inhibited at seawater concentrations of 50% seawater and above, with only 6% of the seeds germinating in 100% seawater vs. 97% in distilled water.[33] When germinated on soils, *Puccinellia distans* and *P. nuttalliana* had 48 and 47% germination, respectively, when watered with 75% seawater.[34] Seeds of *Puccinellia festucaeformis* were found to be highly salt tolerant, germinating at 22% in NaCl treatments of 750 mM.[35]

Seed germination of coastal species from the Frisian Islands (Netherlands) indicated that many of the species had at least some germination in up to 3.5% NaCl (seawater concentration), with the exception of the dune species *Gallium mollugo* and *Plantago lanceolata* and the marsh species *Spergularia salina* and *Aster tripolium*.[36] Some of the marsh and dune species had little inhibition of germination in up to 1.0% NaCl, including *Limonium vulgare*, *Plantago maritima*, *Rumex crispus*, and *Bromus hederaceus*, indicating that seed germination could take place in zones on the marsh in which species were not ordinarily established. Bakker et al.[36] concluded that the critical stage for many of these coastal species from the Netherlands was in most cases not germination but whether seedlings could become established in hypersaline conditions or in dunes where competition was a critical factor.

The seeds of *Halopeplis amplexicaulis* from a saline location in Algeria could germinate in salinities of up to 500 mM NaCl.[37] Germination percentages were reduced sharply at salinity concentrations above 300 mM, from about 85% at 300 mM to 45% at 400 mM and about 8% germination at 500 mM NaCl, with no seeds germinating at salinities >500 mM

NaCl. The rate of seed germination and the time for initial germination were delayed at higher salinities. Tremblin and Binet[37] reported that seed germination occurred under field conditions in May when soil salinity concentrations were reduced by precipitation to 0.9% total soluble salts, decreasing from 2.6% total soluble salts in March.

The seagrass *Najas marina* in Israel grows mainly in freshwater habitats.[38] These populations apparently maintain some salt tolerance in terms of germination responses, with germination in 74 mM NaCl being 66% of that in tap water. The germination of the seagrass *Cymodocea nodosa* in the French western Mediterranean is limited to April to October when salinity levels are reduced to 1.0 to 2.0% total salts. A prolonged period of up to 2 months is needed for seeds to obtain high germination percentages at 2.7% total salts: 7% germination after 18 d, 37% at 30 d, and 47% at 60 d. In freshwater treatments, 72% of the seeds germinated in 2 d, while at 1.0% total salts, 30% of the seeds germinated in 2 d, and no seeds germinated at 3.8% total salts. Salinity reductions to at least 2.7% would have to persist for at least 2 months to induce high germination percentages, limiting germination from April to October under natural conditions. This could happen near the mouths of large rivers, in very shallow waters, or near submarine freshwater springs.[39] Germination of *Ruppia maritima* seeds from a salt marsh pool in southern Brazil did not differ significantly over a salinity range from 0 to 4.0% total salts.[40] Estuarine seed sources had significantly higher germination percentages at lower salinities than when exposed to higher salinity values than did seed from the salt marsh pool population. Koch and Seeliger[40] concluded that because of large fluctuations in salinity in the salt marsh pool habitat, germination was less affected by changes in salinity level. *Ruppia maritima* var. *maritima* germinated best in freshwater, and some germination occurred at salinity levels up to 2.4% chlorinity.[41] Increased salinity was found to be inhibitory to seed germination of the western European *Ruppia* taxa R. *maritima* var. *maritima* and var. *longipes*. Submersed aquatic halophytes from Westernport Bay (Australia) *Ruppia polycarpa* and *Lepilaena cylindrocarpa* begin their seed germination in April in the field when salinities were about 4.0% total salts.[42] Both of these species had optimal germination at 20°C in freshwater under laboratory conditions. Soaking seeds in 2.25 and 4.50 × seawater concentration for 14 d did not inhibit germination when seeds were transferred to distilled water.

Scarification of *Scirpus robustus* seeds was effective in increasing seed germination from about 10% in unstratified to 85% after 1-month stratification.[43] Acid scarification for 7 min also alleviated seed dormancy and increased germination in nonstratified seeds from 10 to 80%. Optimal germination, 90%, was in tap water, with 40% germination in 0.5% NaCl, 5% germination in 1.0% total salts, and no seeds germinating at seawater (3.5% total salts) concentrations. Dietert and Shontz[43] concluded that germination requirements were best met during the spring in these Chesapeake Bay marshes when freshwater inflow and solar insolation were both high.

Zygophyllum dumosum seeds collected from Sede Boqer (Israel) had a decrease in germination percentages with increasing NaCl concentration.[44] Nonsaline treatments had 56% germination, 86 mM had 27% germination, 172 mM had 5% germination, and 516 mM had 0.5% germination.

III. POPULATION VARIATION

Several researchers have reported the occurrence of variation in salt tolerance between closely related species and among populations or ecotypes within a plant species.[8,45-49] Results of the research concerning population differentiation to salinity indicate that the source of seeds may be a critical factor in determining the level of salinity at which a seed germinates. It is most probable that one of two mechanisms has developed in populations to overcome salinity stress: either seeds become dormant under hypersaline conditions and do not germinate until the salt stress is removed, or genetic differentiation of seeds that can germinate at high salinities may occur. Evidence for increased salt tolerance in the germination of seeds has been reported for some taxa that grow in both saline and nonsaline habitats by Bazzaz,[46] Kingsbury et al.,[47] and Bulow-Olsen.[48]

Seed germination was investigated for populations of the submersed halophytic species *Zostera marina* in Chesapeake Bay (Maryland) by Orth and Moore.[50] No correlation was found between site salinity levels, ranging from 0.8 to 3.2% total salts, and the germination responses of seeds to salinity. Seed germination was monitored for seeds at the nine field sites, with germination ranging from 38.6% at the Clay Bank Site (1.2% total salts) to 23.8% at the Wachapreague Site (3.0% total salts). The majority of seeds, 66%, germinated from December to March when water temperatures ranged from 0 to 10°C. No seeds germinated when the water temperature was above 20°C.[50]

Germination responses of cliff (clay) and sand populations of *Crithmum maritimum* from Sardignia were different.[49] Seeds from the sand population germinated at salinities up to 120 mM NaCl, while the clay seed source had a maximum salt tolerance of 60 mM NaCl at constant temperatures of 10 and 20°C. At alternating temperatures of 20°C/10°C at 120 mM NaCl, seeds from the sand population had 36% germination, while the clay seed population had 6% germination.

Differences were found in the salt tolerance of *Festuca rubra* seeds from several populations on the salt marsh at Skalingen, Jutland (Denmark). Low dune and dry area populations had higher germination percentages (50%) in 25% seawater than seeds from wet area populations (25% germination).[48] At higher salinity concentrations, only 10% of the seeds germinated, and no trend was observed between the two populations. The germination of seeds in 25% seawater was reduced in dune populations when the husks were removed, indicating that a portion of the salt tolerance mechanism may be in this structure. Bulow-Olsen[48] determined that husk removal had no effect on germination in less salt-tolerant populations at 25% seawater concentrations, while husk removal in all populations led to a decrease in germination in distilled water treatments.

Germination of seeds of *Melaleuca ericifolia* collected from several sites in Australia were inhibited from germinating at seawater concentrations containing 1.8% total salts.[51] Seeds collected from two of the populations were completely inhibited from germinating at 1.6% total salts, but the Lake Wellington seed did germinate at this salinity level after 10 d. Seeds from the most saline Blind Bight population were more inhibited by the 0.8 and 1.2% total salt treatments than those from the less saline Koonwarra and Lake Wellington locations. The authors found that when all ungerminated seeds from the salt treatments were transferred to distilled water, all of the remaining seeds germinated. Ladiges et al.[51] concluded that the osmotic suppression of germination had ecological significance in that it assured that plant growth would take place under favorable conditions.

Germination was studied for seeds from different populations of *Carex lyngbyei*, an abundant species growing in intertidal salt marshes of the Pacific Northwest (Canada). Hutchinson and Smythe[52] found that prior exposure of seeds to 2.0% total salts produced different germination patterns in the four populations studied. At two sites, germination of seeds after pretreatment in 2.0% total salts did not differ from the distilled water pretreatments; at a third site, salt pretreatments inhibited final germination percentages, while seeds from a fourth site showed a significant augmentation of seed germination with the salt pretreatment. These data indicated that local genetic differentiation of populations could be occurring in *Carex lyngbyei* and that exposure to higher soil salinities prior to the germination season would stimulate germination of seeds in some sites in the spring when there was an order of magnitude reduction in the soil salinity level (from 1.8 to 0.3% total salts).[52]

Eurotia lanata seeds that were collected by Workman and West[8,9] from different locations in Utah varied in their salt tolerance, indicating that ecotypic differentiation occurred in relation to seed germination responses to salinity. Seeds from all populations had a decrease in seed germination with increased NaCl levels. However, the degree of salt tolerance of seeds from different populations at the germination stage of development did not correlate well with the soil salinity conditions in which plants were growing in the field. Seeds collected from the LaSal population were the most salt tolerant, germinating in salinities up to 4% NaCl, and yet the seed source was from plants growing in the least saline (3.0% total soluble salts) location. In contrast, seeds from the highly saline Cisco location (9.6% total soluble salts) had greatly reduced seed germination at 3% NaCl and no germination in 4% NaCl. Corroboration for these results was found in a study by Clark and West[45] of three populations of *E. lanata* from the Curlew Valley, UT in which seeds were collected from sites that were less than 1 mile apart. The populations differed significantly in their salt tolerance to NaCl and K_2SO_4, but not in response to Na_2SO_4 and $CaCl_2$. Clark and West[45] concluded that differences in germination responses were probably related to three factors: source of seeds, time of seed production, and climatic conditions. Ecotypic differentiation was evidently occurring in populations of *E. lanata* from saline sites that were less than 1 mile apart.

Germination percentages of seeds of *Kochia americana* varied with the time of seed collection in two populations from Utah.[53] The pure stand population had the highest germination in the September seed collection, while the mixed stand population had highest seed germination for the October seed collection. Salinity tolerance of seeds was up to 10% NaCl in September for the pure stand and in November for the mixed stand population. Clark and West[53] concluded that several variables, including seed source, salinity level, and time of seed collection, significantly affected the salt tolerance and germination percentages of seeds of *Kochia americana*.

Seeds were collected from southern saline and northern nonsaline soils of Iraq by Bazzaz[46] to determine if ecotypic differentiation was occurring in populations of *Prosopis farcta*. The northern Mosul population from a nonsaline habitat was more sensitive to salt stress than the two seed sources from saline locations. Although there was little variation in response for seeds from the three populations in up to 1% NaCl, germination percentages dropped to 27% in the Mosul population when seeds were treated with 2% NaCl, while seeds from the Basra population had 64% germination. These results indicated that salt ecotypes produced by the populations of *P. farcta* were correlated with field soil salinity conditions.[46] In contrast, seed germination responses to salinity for ten *P. farcta* populations from Israel indicated that seed germination responses to salinity were not significantly correlated to the field soil salinity conditions in which plants were growing.[54] Ecotypic differentiation was apparent in that four of the populations had a limit of salinity tolerance of 0.4 *M* NaCl (site salinities = 1.2 to 2.9 mS/cm) and the other six populations germinated in up to 0.6 *M* NaCl (site salinities = 4.0 to 37.3 mS/cm). The more salt-tolerant populations differed in their ability to germinate in 0.6 *M* NaCl, with only 6% germination for *P. farcta* seeds from the En Yahav (site salinity = 37.1 mS/cm) and 72% germination at Hamadya (site salinity = 4 mS/cm). Ecotypic differentiation in terms of seed germination responses to salinity was occurring in these populations of *P. farcta* from Israel, but Dafni and Negbi[54] concluded that the salt tolerance of seeds was not directly correlated with the soil salinity conditions in which plants were growing (Table 2). However, the three least salt-tolerant populations did originate from populations that were growing in soils with low salinity, which partially corroborates the results of Bazzaz.[46] The distribution of *P. farcta* indicates that mature plants are more tolerant of salinity than are seeds at the germination stage, but these differences may be related to different patterns of adaptation at the germination stage which are related to the development of either dormancy or higher germinability when seeds are exposed to hypersaline conditions.

Populations of *Lasthenia glabrata* ssp. *glabrata* (northern California alkali soils) and

TABLE 2
Germination Responses of *Prosopis farcta* Seeds from Different Populations

Population	Conductivity (mS/cm)	Germination % (M NaCl)			
		0	0.2	0.4	0.6
Jahula	1.5	65	58	30	0
Hefzi Bah	2.9	98	96	75	0
Hamadya	4.0	91	88	79	72
En Zin	20.2	90	79	56	33
Hazeva	33.3	98	84	69	10
En Yahaw	37.1	98	86	58	6

From Dafni, A. and Negbi, M., *Isr. J. Bot.*, 27, 147, 1978.

ssp. *coulteri* (southern California saline sites) were all inhibited at seawater concentrations of >20%.[47] No seeds germinated in the Yolo (northern California) population at 30% seawater, while both southern California populations germinated at about 25% at this salinity regime. Kingsbury et al.[47] determined that several adaptations were apparent in the southern populations of *L. glabrata* which allowed them to grow in habitats of higher salinity than did the northern populations: greater salt tolerance at the seed germination stage, reduced mortality of plants at high salt levels, and the ability to produce fertile seeds, even under 100% seawater conditions.

Populations of *Rumex crispus* were studied to determine the germinability of seeds from an inland and maritime location.[55] Seeds from the two populations had 87 and 82% germination in distilled water and 0% germination in seawater. Seeds were soaked in seawater for 10 d and then germinated in distilled water for 28 d, and it was found that seawater was not permanently inhibitory since the seeds from the inland source had 93% and the coastal source 86% germination. After a 70-d treatment with seawater, 61 and 65% of the seeds from the two sites, respectively, germinated in distilled water, indicating that seeds could survive long-term submergence in seawater by remaining dormant and then germinating when the salt stress was removed. Seeds of the maritime form germinated more rapidly than that of the inland form of *R. crispus*. All of the seeds of the maritime form of *R. crispus* remained floating in seawater for 34 d in the shaken test and 72 d in the unshaken test when the perianth segments were attached, while seeds of the inland population had only 3% floating after 34 d in the shaken test and 0% floating in the unshaken test after 72 d. Cavers and Harper[55] concluded that differences in germination and growth parameters between the two forms of *R. crispus* were never absolute, but variation in floating ability of fruits, germination time, root growth rate, and responses to salt spray all contributed to the success of coastal forms in the maritime sites.

Binet and Boucaud[56] demonstrated that taxa of *Suaeda* differed in their level of seed dormancy. *Suaeda maritima* var. *macrocarpa* and var. *flexilis*, *S. splendens*, and *S. fruticosa* were all reported to be highly salt tolerant. The level of seed dormancy differed among the species, with *S. maritima* var. *macrocarpa* having no apparent dormancy while the other taxa studied all had some level of seed dormancy. Boucaud[57] compared three varieties of *S. maritima* (var. *flexilis*, var. *macrocarpa*, and var. *vulgaris*) and found that the level of germination differed among the varieties even after a 60-d scarification treatment. Germination percentages ranged from 100% in var. *macrocarpa*, 50% in var. *flexilis*, and only 14% in var. *vulgaris*. Seed protein analysis for the three varieties of *S. maritima* indicated that they were closely related, with similarity coefficients between the three varieties ranging from 88.4 to 97.6%.[58] A mechanical dormancy produced by the hard testa was the primary factor influencing seed germination in *Suaeda* spp., and the inhibition of germination could be removed by scarification treatments.[57] Boucaud and Ungar[59,60] reported that both mechanically induced dormancy and dormancy caused by salt stress could be relieved by treatments with 0.001 *M* gibberellic acid.

Two varieties of *Arthrocnemum halocnemoides*, var. *pterygosperma* and var. *pergranulatum*, varied somewhat in their germination responses to salinity.[61] A 50% reduction in seed germination was reached for var. *pterygosperma* at 0.8% NaCl, while for var. *pergranulatum* a 50% decrease in germination did not occur until the 2.0% NaCl treatment. Malcolm[61] reported that no seed germination occurred in either variety at salinities >2.4% NaCl at any of the temperature regimes used.

Germination responses of two species of *Hordeum*, *H. marinum* and *H. murinum*, were found not to agree with their ecological distribution.[62] *H. marinum* grew in more saline habitats than *H. murinum*, but was less salt tolerant than the latter species at the germination stage of development. Final germination percentages in 2.6% NaCl solutions were 75% for the less salt-tolerant *H. murinum* and 22% for the more tolerant *H. marinum* after a 100-d after-ripening period. Onnis and Bellettato[62] concluded that both species had higher germination percentages when seeds were after-ripened for 100 d or more than for those either not after-ripened or those seeds held for shorter periods of time.

IV. OSMOTIC EFFECTS

A number of researchers have reported that seeds of halophytes remained ungerminated when exposed to high salinities. Transfer of ungerminated seeds to distilled water induced germination, indicating that high salinities caused an osmotically induced dormancy (Table 1).[7,16] Many species of halophytes have maximal germination under freshwater conditions, differing from the responses of less salt-tolerant glycophytes only in that they usually can germinate at higher levels of salinity.[12,63] Most of the British coastal marsh species examined by Woodell[16] had their highest seed germination in freshwater, and hypersaline conditions enforced a dormancy in seeds.

A significant characteristic of halophytes, which distinguishes them from the seeds of glycophytes, is their capacity to maintain seed viability for extended periods of time during exposure to hypersaline conditions and then to initiate germination when the salinity stress is reduced.[7,16] Investigations with seeds of *Crithmum maritimum*, *Halopeplis perfoliata*, *Limonium axillare*, *Centaurium littorale*, *Samolus valerandi*, *Parnassia palustris*, *Melaleuca ericifolia*, *Sporobolus virginicus*, *Juncus maritimus*, *Limonium vulgare*, *Armeria maritima*, *Plantago maritima*, *Rumex crispus*, *Atriplex triangularis*, *Salicornia europaea*, *Spergularia media*, *Puccinellia nuttalliana*, *Iva annua*, and *Hordeum jubatum* indicated that hypersaline conditions caused an enforced dormancy in the seeds of many halophytic species, but when the salinity stress was removed and seeds were placed in a freshwater medium, high germination percentages were obtained (Table 3).[7,16,51,64-73] The enforced dormancy response for halophyte seeds to saline conditions is of selective advantage to plants growing in highly saline habitats, because it guarantees that seeds will not germinate under extreme salt stress conditions that would expose seedlings and later growth stages to high risks of mortality.

Seeds of two perennial halophytes, *Limonium vulgare* and *Halopeplis perfoliata* from littoral salt marshes of the Red Sea (Saudi Arabia), were germinated at seawater concentrations ranging from 5 to 100%. Mahmoud et al.[70] reported that the salinity tolerance of seeds was much lower than the salinity levels found in the native habitats of plants; 8.9 to 11.2% total salts in *H. perfoliata* populations and 2.2 to 5.2% total salts in *L. axillare* populations. The maximum tolerance limits for germination of *Haplopeplis* were 1.8% total salts (40% seawater), and for *Limonium* the limit for germination was at 0.9% total salts (20% seawater). Recovery experiments were run for ungerminated seeds of both species from the 2.7% total salts (60% seawater treatment), and high germination percentages were

TABLE 3
Salinity Treatment Effects on the Germination
of *Spergularia media* and the Ability of Seeds
to Recover from Treatments at Various
Salinity Regimens

	Percentage germination		
NaCl treatment (%)	Salinity treatment (30 d)	Recovery in distilled water (21 d)	Total
0.0	79.0	10.0	89.0
1.0	41.0	55.0	96.0
2.0	1.0	75.0	76.0
3.0	0.0	94.0	94.0

Note: Germination was in a day-night regimen of 25°C for 12 h of light and 5°C for 12 h of dark.

From Ungar, I. A. and Binet, P.[120]

obtained, 97% for *Halopeplis* and 95% for *Limonium*. These data indicate that the low water potential of the medium inhibited germination but that this effect was not permanently toxic to seeds. Mahmoud et al.[70] concluded that leaching of the soil in the summer by heavy rains would be necessary for seeds of these two species to germinate. However, little seed germination occurred during the summer months because the rainfall is very low, and this is accompanied by high soil salinities that osmotically inhibit the germination of seeds.

Seeds of *Sporobolus virginicus* from False Bay (South Africa) were studied to determine the effect of inundation and salinity (0 to 3.0% NaCl) on the germination of seeds.[68] Germination percentages of *Sporobolus* seeds were reduced >50% at 1.5% NaCl, and no seeds germinated at 2.0% NaCl. Salinities ranging from 1.5 to 12.0% in the germination medium had no permanent inhibitory effect on seed germination. After a 7-week salt pretreatment, there was no reduction in seed viability when seeds were returned to freshwater conditions. Breen et al.[68] concluded that the principal effect of high salinity was osmotic and that there was no permanent toxic accumulation of specific ions. This was of ecological significance because seeds could withstand the highest salinity stress at the lake and provide a viable seed bank for recruitment of new individuals, but seed germination would be limited to periods when the soil salinity level was below 1.5% total salts.

A survey of the germination responses to salinity for 31 British coastal marsh species was done, and Woodell[16] determined that seeds of these species were not permanently inhibited by high salt treatments. Most of the coastal species studied either did not germinate or had low germination percentages at seawater concentrations or higher. Some recovery from salt-enforced dormancy was found in all species investigated, with 60% of the species reaching greater than 50% germination after a pretreatment with 950 mM NaCl (1 1/2 times seawater concentration). These results agree with the recovery test results of Ungar,[64] Ungar,[74] Ungar and Hogan,[65] and Macke and Ungar[66] for a number of inland halophytes. Other researchers have also reported high recovery percentages for seeds of halophytes after treatments with salinities that were beyond the species tolerance limits (Table 1).[6,7,16,51,68,70]

Seeds of *Atriplex hastata* ssp. *novae-zelandiae* required a minimum of 1-week stratification to attain maximal germination. Salinity treatments indicated that seed germination was completely inhibited in 2% NaCl, while 85% of the seeds germinated after 12 d in a 1% NaCl treatment. Germination was delayed in salinities up to 1% NaCl, but it was not inhibited significantly at this salt concentration.[75] Field conditions at the time of seed germination for *A. hastata* in New Zealand in June and July are characterized by low soil salinity (0.02 to 0.32% total salts) and relatively high soil moisture (5 to 13%), which correlates well with its laboratory germination responses for the species.[75]

The responses of seeds of nine halophytes and two glycophytes to salinity stress were investigated for plants growing on the Otago salt marshes (New Zealand).[76] Most halophyte seeds remained dormant at high salinities (3.5% NaCl) for several months but retained their viability. In contrast, seeds of *Triglochin striatum* and *Lolium perenne* had high mortality after being exposed to 1 to 4% NaCl and were significantly affected by the length of exposure to saline solutions. Partridge and Wilson[76] determined that the annual species *Atriplex prostrata* and *Polypogon monspeliensis* and short-lived perennials *Spergularia media* and *Cotula coronopifolia* were able to germinate at higher salinities than perennial species which had equivalent salt tolerance at the adult stage (*Selliera radicans, Sarcocornia quinquefolia, Juncus maritimus, Plagianthus divaricatus*). The short-lived plants had similar salt tolerance at the germination and growth stages of development, while long-lived perennials were reported to be more sensitive to salinity at the germination stage than were annuals. The authors concluded that perennial species were able to delay germination until soil salinity levels were greatly reduced because seedling establishment was only occasional, while the colonizing annual species were able to germinate more rapidly in open habitats and under more saline conditions.[76]

Species occurring in southern California salt marshes had a wide range of germination responses to treatments from 0 to 2% sea salt salinity.[77] Of the 14 species of salt marsh plants examined, only 50% could germinate at 2% salinity, 36% were restricted to 1% or less salinity, and 14% were limited to 0.5% or lower salinity. At 3.5% total salts, only seeds of *Salicornia bigelovii* and *S. virginica* germinated. Based on these germination results, it is clear that a substantial gap in hypersaline conditions was necessary, where soil salinity conditions were below 1.0% total salts, for the recruitment of species in the San Diego Bay salt marshes. Reduced soil salinity below 1.0% total salts explained the invasion of species onto the marsh in 1980, and the lack of establishment of species in other years was due to high soil salinity. Zedler and Beare[77] concluded, based on the inhibition or delay of seed germination by salinity stress, that 2 to 3 months of brackish conditions (1 to 2% total salts) or a 2- to 3-week freshwater influx from winter rains was necessary for most species to become established on the San Diego Bay salt marshes.

V. SPECIFIC ION TOXICITY

Three types of experiments have been commonly used to determine if a specific ion toxicity is the cause of reduction in germination percentages for seeds of halophytes when they are exposed to hypersaline conditions: (1) seeds have been pretreated in hypersaline solutions, and the recovery of seed germination in distilled water has been determined;[16] (2) comparative studies have been carried out of the differences in germination responses of seeds in NaCl solutions vs. their responses to isotonic solutions of organic solvents such as mannitol, polyethylene glycol, and sugars;[14,78,79] (3) comparisons have been made of the germination responses of halophyte seeds to isotonic solutions of inorganic compounds such as NaCl, KCl, $MgCl_2$, Na_2SO_4, $MgSO_4$, and $NaCO_3$.[66,80-82] These three types of experiments have yielded significant results in clarifying the relationship between the reduction in seed germination percentages caused by osmotic inhibition vs. the effects of specific ion toxicities.

Onnis et al.[80] reported that seeds of the halophyte *Puccinellia festucaeformis*, a coastal halophyte from Tuscany (Italy), had a high tolerance to eight different inorganic salts in up to 250 mM concentrations. Germination percentages ranged from a low of 67% in 250 mM $CaCl_2$ to a high of 91% in 250 mM $NaNO_3$, with 84% of the seeds germinating in a 250 mM NaCl treatment. Greater inhibition of seed germination was observed when seeds were

TABLE 4
**Percentage Germination (± Standard Error) of *Puccinellia nuttalliana*
Seeds after 25 d at Different Osmotic Concentrations**

	Osmotic potential (MPa)					
	0	−4	−8	−1.2	−1.6	−2.5
Na$_2$SO$_4$	86.0 ± 2.0	83.6 ± 2.0	79.2 ± 2.0	40.0 ± 2.8	1.6 ± 0.8	0.0
NaHCO$_3$	86.0 ± 2.0	88.0 ± 1.6	87.2 ± 1.6	31.0 ± 5.1	2.8 ± 1.2	0.0
NaCl	86.0 ± 2.0	90.0 ± 3.2	78.8 ± 2.8	34.0 ± 4.2	5.2 ± 0.8	0.0

From Macke, A. J. and Ungar, I. A., *Can. J. Bot.*, 49, 515, 1971.

exposed to 500-mM concentrations of salts: less than 10% germination was found in CaCl$_2$,
MgCl$_2$, and Na$_2$SO$_4$; seeds in NaCl, KCl, KNO$_3$, and NaNO$_3$ had from 19 to 28% germi-
nation; and seeds in MgSO$_4$ had 61% germination. These data indicated that seeds of
Puccinellia were responding to a specific ion toxicity at salinity concentrations of 500 mM.[80]
It is interesting that although seeds of *P. festucaeformis* would normally be exposed to NaCl
salinity, it is best adapted at the seed germination and growth stages of development to
MgSO$_4$ salinity. It would be difficult to predict the behavior of seeds in inland saline habitats
where soils are commonly characterized by high sulfate concentrations, from plants currently
distributed in coastal habitats whose soils are characterized by NaCl salinity. *Puccinellia
nuttalliana* responded to three inorganic ions in the same manner (Table 4), while *Iva annua*
was inhibited more by NaHCO$_3$ and ethylene glycol than by NaCl or Na$_2$SO$_4$ (Figure 2).[65,66]
Specific ion toxicity effects were found by Hyder and Yasmin[78] to be more important
than osmotic effects in determining seed germination responses to salinity of the inland
North American halophyte *Sporobolus airoides*. The limits of salt tolerance for this species
was 375 mM NaCl, with each salinity increment from 25 to 325 mM causing a reduction
in the germination percentage of seeds of *S. airoides*. A −0.3 MPa isotonic treatment of
mannitol, NaCl, CaCl$_2$, KCl, and MgCl$_2$ indicated that *S. airoides* was responding to a
specific ion toxicity produced by inorganic ions rather than an osmotic inhibition of seed
germination. In −0.3 MPa mannitol 50% of the seeds germinated, while in NaCl only 17%
germinated and in other salts less than 15% of the seeds germinated, with MgCl$_2$ being most
inhibitory. Experiments to determine if the inhibitory effects of MgCl$_2$ could be counteracted
by NaCl or CaCl$_2$ were carried out, and it was determined that the recovery of germination
was better in the Ca than Na treatments. The success of seed germination of *S. airoides* and
other halophytic species may therefore depend on the ionic composition of the soil, with
Ca acting to ameliorate the toxic effects of other cations present in the soil substrate.[78]
Germination responses of seeds from three Oregon populations of the salt desert shrub
Sarcobatus vermiculatus to treatments with polyethylene glycol (PEG), PEG-NaCl, and
PEG-KCl were determined by Romo and Haferkamp.[79] Osmotic potentials of solutions ranged
from −0.3 to −2.2 MPa, and it was determined that the osmotic potential of the medium
was the controlling factor in determining the germination response of seeds. In the Great
Basin region, soil water potentials are highest and soil salinity levels are lowest in the spring.
The limits for seed germination of *S. vermiculatus* in these populations were determined to
be −1.6 MPa. Both NaCl and KCl treatments did not differ significantly in their inhibitory
effects on seed germination than that of isotonic treatments with PEG, indicating that the
primary effect of inorganic ions was osmotic rather than a specific ion toxicity. NaCl and
Na$_2$SO$_4$ were found to be less inhibitory than PEG to seed germination in a study using
seeds from a Montana population of *S. vermiculatus*. Studies of seed germination with
populations of *S. vermiculatus* from New Mexico indicated higher germination percentages
at −1.6 MPa than for Montana, and germination occurred at osmotic potentials as low as

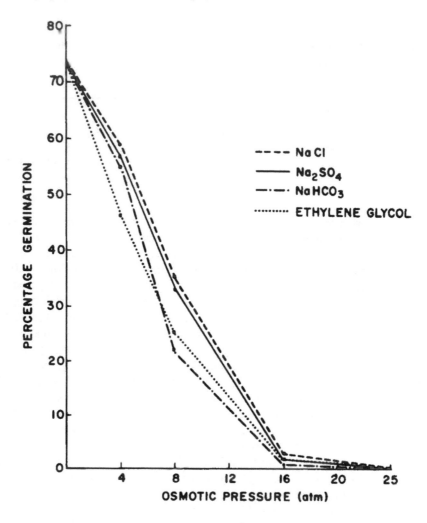

FIGURE 2.   Percentage seed germination of *Iva annua* at varied water potentials after a 20-
d treatment period. (From Ungar, I. A. and Hogan, W., *Ecology*, 51, 150, 1970.

−3.6 MPa.[83,84] These data demonstrate that ecotypic differentiation may be occurring in
different parts of the range of *S. vermiculatus*.[85] Final germination percentages of *Kochia
prostrata* seeds were also affected by the osmotic potential of NaCl-PEG and KCl-PEG
solutions.[79] Seeds and fruits germinated equally well at osmotic potentials as low as −2.0
MPa, with intact fruits having slower rates of germination at 20°C but equivalent rates at
10 and 30°C. Fruit bracts could play a significant role in early mineral nutrition during the
germination phase of halophytes.[79] Seed appendages were found to be a source of Na for
seedlings of *Sarcobatus vermiculatus*.[85] The Na content of fruits was measured, and it was
determined that embryos had the lowest Na content (1843 μg/g) and bracts the highest
(53,100 μg/g), while mean values for the entire utricle were 38,900 μg/g Na. Eddleman
and Romo[85] hypothesized that Na from the bracts may function to promote germination at
low soil water potentials by promoting a favorable water balance at the germination and
seedling growth stages of development. High salt levels in the bracts may be a positive
adapation that chenopodiaceous halophytes have developed for successful recruitment of
individuals in saline habitats.

Both *Sporobolus arabicus* and *Suaeda fruticosa* grow in saline soils of Pakistan, with
soil salinity levels ranging from 5.9 to 35 mS/cm in the *Suaeda* habitats and 6.5 to 17.7

mS/cm in the *Sporobolus* habitat. Sheikh and Mahmood[86] found that the highest germination percentages for both species were obtained in the lowest salinity treatment (2 mS/cm with NaCl and $CaCl_2$). Calcium chloride was less inhibitory than NaCl to *Sporobolus* under salinity treatments from 6.5 to 24.0 mS/cm. *Sporobolus* had only 11% of the control germination in 24 mS/cm NaCl, while *Suaeda* was less inhibited, having 51% of the control seed germination under this treatment. The results for $CaCl_2$ treatments followed the same pattern, with *Sporobolus* having 16% of the control and *Suaeda* having 56% of the control seed germination at 23 mS/cm $CaCl_2$. *Suaeda fruticosa* was found to be more tolerant of salinity and sodicity than *Sporobolus arabicus*, based on these germination experiments, and these data agree well with the species distribution in saline and sodic soils of Pakistan.[86]

Salicornia brachiata, a tropical salt marsh species (India) had 63% germination in seawater salinities (3.6% total salts), which was equivalent to a 36% reduction in germinability from the distilled water treatments.[81] Germination in this species was delayed at higher salinities. Isotonic solutions of NaCl, Na_2SO_4, and dextran at -0.45 MPa were not inhibitory to germination and yielded similar germination responses, 99, 84, and 87% for the three osmotica, respectively. Joshi and Iyengar[81] determined that high salinity tolerance in *S. brachiata* reflects the high soil salinities it must adjust to, ranging from 43.8 to 216.6 mS/cm in saturation extracts from soil samples in its native habitat.

Cluff et al.[87] reported that the germination of seeds of *Distichlis spicata* var. *stricta* was inhibited and delayed at osmotic potentials lower than -0.1 MPa, and no seeds germinated at -2.0 MPa. The effect of PEG 6000 and NaCl solutions as osmotic agents was tested, but no significant differences ($p < 0.05$) in germination percentages were found between isotonic solutions of the two osmotica. These results indicate that sodium chloride was acting as an osmotic agent in inhibiting *D. spicata* var. *stricta* seed germination rather than as a toxic specific ion. Soil salinities in the Nevada seed collection sites ranged from -4.0 to -7.5 MPa in the surface soils and -2.0 to -3.0 MPa at 5 cm below the soil surface. These soil salinity values are beyond the range of seed germination potential for *Distichlis*, even though adult plants grow in these habitats, indicating that germination must take place after salts are leached from the soil during periods of precipitation early in the spring.[32,87] In the glycophyte *Oryza sativa* treated with isotonic solutions of NaCl and PEG 6000 (0.2 to 0.8 MPa), NaCl was more inhibitory to seed germination than PEG. Germination in PEG 6000 (-0.8 MPa) was 70 to 80%, while at the same osmotic potential only 5 to 20% of the seeds germinated in NaCl treatments, indicating that there was a specific ion effect induced by NaCl.[88] These results with *O. sativa* corroborate other studies with glycophytes, indicating that specific ion toxicity may be a more significant factor in the seed germination of salt-intolerant species.[7]

Securigera securidaca seeds were inhibited from germinating at NaCl concentrations below -0.6 MPa and at -1.0 MPa $MgSO_4$.[89] Combinations of salts, Na_2SO_4 + $MgSO_4$ and NaCl + $MgSO_4$, increased the limits of germination tolerance to -1.4 and -1.5 MPa, respectively, for the two salt mixtures. Similar amelioration of single salt effects were obtained with mixtures of $MgCl_2$ + NaCl and $CaCl_2$ + $MgCl_2$, indicating that the inhibitory effect of single salts can be considerably alleviated in natural soil systems by synergistic interactions between salts.[90]

VI. IONIC CONTENT OF SEEDS

The ionic content of seeds has been determined for a number of species of halophytes (Table 5). Elemental distribution in seeds of *Atriplex canescens* and *Salicornia pacifica* var. *utahensis* was studied with energy dispersive X-ray microanalysis by Khan et al.[91] Seeds of *A. canescens*, a cold desert shrub, had low levels of elements in all tissues when compared to *S. pacifica* var. *utahensis*. Compartmentation of elements occurred in seeds of *Salicornia*,

TABLE 5
Ion Content of Halophyte Seeds

Species	Ion (mg/g dry weight)					Ref.
	Ca	Mg	K	Na	Cl	
Eurotia lanata	0.11	0.21	1.30	0.04	—	93
Cakile maritima	1.26	3.20	7.27	0.21	0.65	96
Arctotheca populifolia	1.21	4.78	10.20	0.47	1.19	96
Atriplex glabriuscula	—	—	6.30	0.69	0.39	97
Salsola kali	—	—	6.30	0.34	0.57	97
Atriplex triangularis	—	—	5.00	6.00	5.00	74
Salicornia europaea	1.24	3.23	8.40	0.89	1.59	164
S. europaea	0.10	3.70	2.10	7.20	—	95
Atriplex laciniata	—	—	5.47	0.34	0.57	97
Cakile maritima	—	—	5.63	0.76	0.50	97

with the seed coats containing high levels of Na and Cl when compared with the embryo and endosperm tissues of the same seeds. The differences in seed ion accumulation between the two species were related to the differences in the habitats the two species occupied, with *S. pacifica* occurring in moist-saline sites and *A. canescens* in drier, less-saline habitats.[91]

The ionic content of air-dried seeds of *Atriplex triangularis* was somewhat different for the two seed morphs, with an Na content of 0.7 and 0.4%, a K content of 0.4 and 0.3%, and a Cl content of 1.6 and 1.0% for the small and large seed morphs, respectively.[73] Seeds of *A. triangularis* collected from field sites with a high-salt content had a mean total inorganic ion content of 2.1%, while seeds collected from a low-salt habitat had 1.7% total salts. Ungar[74] determined that the ionic content of *A. triangularis* seeds varied somewhat in laboratory salinity gradients. Seeds from plants grown in nutrient control treatments had 0.2% Na, 0.6% K, 0.1% Cl, and an estimated 1.0% total salts. When the medium salinity was raised to 2.0%, the seeds contained 0.5% Na, 0.6% K, 0.4% Cl, and an estimated 2.3% total salts.[74] The K content of seeds remained relatively constant in all salinity treatments, while the Na and Cl content of seeds increased with salt increments in the laboratory investigations. Similar increases in Na and Cl content were found for *A. triangularis* seeds collected from high-salt field habitats when compared with plants from low-salt habitats.

Several investigations have been carried out to attempt to partition the ionic content in different tissues of the seed. Using energy dispersive X-ray microanalysis techniques, Khan et al.[92] found higher concentrations of Na, Cl, and Ca in the seed coat than in the endosperm or embryo of *A. triangularis*. Similar results were found in the analysis of seeds of the shrub *A. confertifolia*, but the concentration of Na, Cl, and Ca was lower in the embryo and endosperm of this species than in seeds of *A. triangularis*. Khan et al.[92] concluded that *Atriplex* species were able to partition Na and Cl in seeds, with higher content of Na and Cl in the seed coats and reduced levels of these ions in the embryo. The elemental content of seeds of *Salicornia pacifica* var. *uthensis* was higher than that for seeds of *Atriplex canescens*.[91] Seed coats of *S. pacifica* had higher amounts of Na, Cl, K, and Ca than did the endosperm of embryo tissue. Seeds of *A. canescens* had a relatively low element content and did not have any significant difference in Na and Cl content in the three tissue types investigated. The maintenance of reduced ionic content in the embryo prevents the development of osmotic or specific ion antagonisms during the germination of seeds. The ionic content in seeds of *S. pacifica* and *A. canescens* is apparently correlated to the environments in which they grow. *Atriplex canescens* is found in dry-saline sites, which have lower Na and Ca content in their soils than do the moist-saline habitats occupied by *S. pacifica*.[91]

The ionic content of seeds and bracts of *Eurotia lanata* collected from nine locations in Wyoming, Colorado, and Utah was determined by Booth.[93] The Na content of bracts ranged from 390 to 760 µg/g and in seeds from 290 to 540 µg/g, with a mean value for bracts of 590 µg/g and 420 µg/g for seeds. The mean values for K were 10,630 µg/g in bracts and 12,570 µg/g in seeds. Seeds were imbibed in 50, 500, and 5000 µg/g Na in the laboratory, and the resulting Na content for seeds was 660, 2530, and 15,500 µg/g for each treatment, respectively. Bracts contained significantly higher Ca concentrations (9580 µg/g) than did seeds (1070 µg/g), and it was concluded that the bracts could serve as a significant nutrient source for germinating seeds.[93]

The cation content of a relatively salt-tolerant variety of *Ricinus communis* seeds was measured for plants that were grown in the presence of 160 mM NaCl.[94] The Na content in the endosperm and embryo of seeds was 1.43 µM/g dry weight compared to 165 µM/g of K, yielding a K to Na ratio of 115. Seed coats contained 31 µM/g of Na and 250 µM/g of K, with a K to Na ratio of 8.1. Plants were able to transport K selectively to young parts of the shoot, while retaining Na and Cl in older tissues. The plants also effectively controlled the movement of ions into seeds, since the K to Na ratio and low Na content of seeds was similar to that found for *R. communis* plants cultivated in nonsaline conditions.[94]

Poulin et al.[95] found that *Salicornia europaea* seeds collected from saline environments contained 0.77% Na, while vegetative organs had from 8 to 15% Na. Seeds of *Cakile maritima* also had a relatively low Na content 0.02% and a Cl content of 0.07% when compared with a leaf Na content of 6.8% and Cl content of 14.1%.[96]

The effect of salinity on the germination of seeds of four annual British strand-line species *Atriplex glabriuscula*, *A. laciniata*, *Cakile maritima*, and *Salsola kali* were investigated by Ignaciuk and Lee[97] to determine whether these species had different responses to salt stress. Germination under field conditions occurred in June when soil salinity levels were reduced to 100 mM total salts. All four of the strand-line species were inhibited by salinities of 240 mM NaCl, with germination of 2% for *C. maritima*, 57% for *A. glabriuscula*, 61% for *A. laciniata*, and 52% for *S. kali*. The highest percentage germination at 600 mM NaCl was 0, 1, 22, and 21% for the four species, respectively. All species showed some delayed germination when the salinity concentration was increased. Recovery experiments in distilled water after 30 d in a 600 mM NaCl treatment indicated that the two *Atriplex* spp. were not permanently inhibited by hypersaline conditions, but germination for *C. maritima* and *S. kali* seeds was reduced to 50 and 73% of the control values in distilled water. The loss of viability for seeds of the latter two species was correlated with high increases in Na and Cl content when the seeds were soaked in 600 mM NaCl. The Na and Cl content increased for all species from that in the distilled water treatment when they were exposed to 600 mM NaCl, but the increases for *C. maritima* and *S. kali* were much greater than the values for the *Atriplex* spp: *A. glabriuscula*, 11 to 30 µM Cl/g, 30 to 59 µM Na/g; *A. laciniata*, 16 to 46 µM Cl/g, 15 to 91 µM Na/g; *C. maritima*, 14 to 110 µM Cl/g, 33 to 243 µM Na/g; *S. kali*, 16 to 154 µM Cl/g, 15 to 177 µM Na/g dry weight. The K content of seeds for all four species did not differ greatly in the two treatments, averaging 151 µM/g in distilled water and 132 µM/g in the 600 mM NaCl treatment. Ignaciuk and Lee[97] concluded from the results of the recovery experiments and the relatively low accumulation of ions in seeds exposed to 600 mM NaCl that the inhibition of germination in the *Atriplex* spp. was osmotic in nature, while the high accumulation of ions and reduced recovery percentages for seeds of *C. maritima* and *S. kali* indicated that a specific ion inhibition could be causing the inhibition of germination of seeds in these species. Beadle[98] considered high concentrations of Na in the bracteoles surrounding the seeds of *Atriplex* spp. from Australia to be inhibitory to germination in *A. nummularia*, *A. vesicaria*, *A. semibaccata*, *A. inflata*, and *A. spongiosa*. The NaCl content of bracteoles ranged from 0.44 to 0.90 M in *A. inflata* and from 0.48 to 0.72 M for the other four species. Seed germination was low when fruits of all five *Atriplex* spp. were waterlogged, ranging from 1 to 16% compared with high germination percentages 76 to 100% for fruits soaked for 24 h in which the excess water was removed. Molar concentrations required to prevent seed germination in the five species investigated ranged from 0.4 to 0.7 *M* NaCl, indicating that the bracteoles contained sufficient NaCl to inhibit the germination of these species of *Atriplex*.[98] The NaCl content of bracteoles of the French coastal species *A. tornabeni* was determined by Binet[99] to be 12.7% on an air dry mass basis. When hydrated, the bracts absorbed 2.8 times their initial dry mass in water, yielding a NaCl content of 4.5% or 0.77 *M*. The high salt content could initially impede germination of *A. tornabeni* seeds in nature, but with precipitation the NaCl would be readily leached from the bracteoles.[99] The fruits of *A. polycarpa* collected in California were found to produce a water-soluble inhibitory substance that reduced seed germination. Seed germination in bracteole leachates was about 25% of that in the distilled water controls.[100] In contrast to these results, it was determined that utricles collected from Kern County, CA (a location with a soil electrical conductivity of 2.4 mS/cm) did not exhibit a dormancy and had 85% germination of seeds with intact bracteoles 1 week after collection. Sankary and Barbour[101] concluded that the bracteoles probably did not contain any inhibitor substances. The seeds of *A. polycarpa* were sensitive to salt stress, and only 17% of the seeds germinated in 1% NaCl, which was only 21% of the germination determined for the distilled water controls (80% germination).[101] Differences in the results from the two investigations with *A. polycarpa* might be related to the amount of leaching in the germination experiments or the time of seed collection. Seeds of *A. dimorphostegia* from Israel had 8% germination when bracteoles were intact and 52% germination when the bracteoles were removed from the seeds.[102] Bracteoles surrounding seeds of *A. rosea* were reported to contain 3.6% Cl on an air-dried mass basis.[103] Bracteole extracts were found to inhibit seed germination, but when the extracts were filtered through charcoal the inhibitory effect was lost, even though the Cl content of the extracts remained constant. Kadman-Zahavi[103] concluded that some other inhibitory factor was present in the bracteoles, since the NaCl concentration in bract extracts was not sufficiently high to inhibit the germination of *A. rosea* seeds. Air-dried bracteoles of *A. dimorphostegia* had an ash content of 27.7%, with 5.4% Na and 12.2% Cl. Koller[102] reported that the osmotic pressure of the bract extracts (-0.6 MPa) was considered to be sufficient to inhibit seed germination in this species, but speculated that an organic inhibitor could also be present in the bracts. Since the inhibitory substance in the bracts was found to be water soluble, the germination of seeds would be determined by the quantity of rainfall necessary to leach the inhibitor from seeds.[102]

Fernandez et al.[104] determined that the bracteoles of *A. repanda* contained 0.93 to 1.23% Na, 0.70 to 0.95% Cl, and 8.42 to 11.27% saponins on a dry mass basis. These concentrations of NaCl and saponins were deemed sufficient to inhibit the seed germination of *A. repanda*. The inhibitor substances present in the bracteoles surrounding the seeds of *A. repanda* were reported to inhibit seed germination by as much as 60%.[105] Both the final total percentage germination and the rate of germination of seeds were inhibited by saponin and 1% NaCl, two substances found in the bracteoles surrounding the seeds of *A. repanda*. Germination in distilled water was 82%, while in 1% NaCl it was 48%, and in 10% saponin it was 34%. A synergistic inhibitory effect was found when seeds were subjected to a combination of 1% NaCl and 10% saponin, yielding ony 5% germination. Fernandez et al.[105] concluded that the low natural reseeding percentages for *A. repanda* in Chile may be due to the presence of inhibitory substances in the bracts surrounding the seeds. The ecological significance of inhibitors in bracteoles may be great in determining the periodicity of seed germination, because seeds could only germinate when there was sufficient moisture to leach the inhibitory substances out of the bracteoles.

VII. DORMANCY

One of the characteristics of many halophytes is that their seeds are subjected to some degree of seed dormancy. Binet[106,107] summarized the various forms of dormancy that have evolved in plants growing in saline environments. The two basic mechanisms by which seeds develop dormancy are either because of some morphological or biochemical characteristic of the diaspore producing a primary dormancy (fruit or seed) or due to some environmental trigger that puts seeds into a secondary dormancy.[108] Dormancy in halophyte seeds can be classified into three categories: (1) innate dormancy because of immature embryos or pericarp and testa impermeability to oxygen or water, (2) induced dormancy by low or high temperature and lack of illumination, and (3) enforced dormancy due to high soil salinity, low water potentials, low winter temperatures, and reduced illumination because of seed burial in the soil.[109] Some species of halophytes do not produce dormant seeds, and others such as *Rhizophora mangle* exhibit vivipary.[63]

In the genus *Suaeda* the testa was found to be inhibitory to the germination of seeds and that acid or mechanical scarification was necessary to obtain germination in a number of species, including *S. depressa*, *S. maritima* var. *flexilis*, and *S. maritima* var. *vulgaris*.[57,110,111] The pericarp of *Triglochin maritima* was determined to inhibit seed germination.[112] Embryonic dormancies have been reported for *Halimione portulacoides*, *Plantago maritima*, *Juncus maritimus*, *J. gerardi*, and *Statice limonium*.[106,113]

Binet[114,115] and Ungar[72] reported that other species of halophytes such as *Juncus maritimus*, *Triglochin maritimum*, and *Spergularia marina* have dormancy induced by darkness, which in nature is probably triggered by the burial of seeds in the soil. The induced dormancy for *J. maritimus* seeds in the dark can be alleviated by stratification of seeds for 30 d, yielding 56 to 88% germination, but seeds returned to light without stratification had only 0 to 8% germination.[115] A number of environmental treatments could break the dormancy in *Spergularia marina* seeds, including exposure of seeds to light, stratification, and the exposure to alternating temperature regimens.[72] It was also determined that treatments with 1000 mg/l gibberellic acid broke the dormancy of seeds in the dark, with germination increasing from 0% in dark controls to 91% for seeds treated with the growth regulator.

The fruits of *Triglochin maritimum* were more dormant than those of *T. palustre*, probably because of the more resistant sclerenchyma tissue in the pericarp of the former species. Binet[114] concluded that the weak parenchyma tissue in the pericarp of *T. palustre* could be more readily penetrated by the embryo than the hard pericarp of *T. maritimum*. In *Crambe maritima* both the pericarp and testa contributed to the dormancy of seeds, but stratification removed the inhibitory effect of both tissues.[107]

Seeds of *Cochlearia anglica* were able to germinate at low temperatures, 5 to 15°C, and in diluted seawater concentrations. Binet[106] found that high temperatures, 20 to 25°C, impeded seed germination and induced a dormancy that could be relieved by stratification. Dark-induced seed dormancy in *Juncus maritimus* was less pronounced when seeds were exposed to a seawater pretreatment.[115] Seeds in the dark in freshwater had 18% germination, while those exposed to seawater pretreatments had 74% germination after being returned to the light and placed in freshwater. The recovery of seeds or stimulation of germination after exposure to saline conditions beyond their tolerance limits has been reported for a number of species of halophytes.[6,7,64]

Dormancy in seeds of halophytes is a significant factor in the ecophysiology of salt marsh species. It permits seeds to remain viable in the soil during periods when the environment is not suitable for germination. This is especially significant in the salt marsh habitat because flooding, hypersaline conditions, or burial of seeds by tidal deposition may make environments unsuitable for seed germination or the normal development of plants. Some seeds remain dormant because of pericarp, testa, or embryonic restrictions to germination (innate dormancy), while in other cases the seeds are responding to some environmental factor when they become dormant (induced dormancy and enforced dormancy). In unpredictable salt marsh and salt desert habitats, seed dormancy provides a mechanism for plant species to overcome periods when germination is not favorable, and then seeds can germinate

TABLE 6
The Interaction between
Hormonal and Salinity Treatments
and Their Effects on Germination
of *Spergularia media*[a]

Hormonal	NaCl treatment (%)			
treatments	0.0	0.5	1.0	1.5
H_2O	76.0	64.0	40.0	36.0
GA_3	77.0	72.0	85.0	52.0
Kinetin	63.0	80.0	72.0	24.0
GA + K	91.0	90.0	80.0	54.0

[a] Germination was at a day-night regimen of
light at 25°C for 12 h and dark at 5°C for
12 h.

From Ungar, I. A. and Binet, P., *Aquatic Bot.*,
1, 45, 1975.

TABLE 7
The Modification of Salinity Stress Effects on Seed
Germination of *Hordeum jubatum* by Gibberellic
Acid

Treatment	Days				
	5	15	25	35	45
1.5% NaCl	0.0	20.0	52.0	76.0	87.0
1.5% NaCl (800 mg/l GA_3)	0.0	18.0	89.0	98.0	99.0
2.0% NaCl	0.0	5.0	36.0	67.0	76.0
2.0% NaCl (800 mg/l GA_3)	0.0	2.0	48.0	89.0	94.0
3.0% NaCl	0.0	0.0	1.0	2.0	3.0
3.0% NaCl (800 mg/l GA_3)	0.0	0.0	7.0	40.0	73.0

at a later time when conditions are more suitable for the growth and development of a species. Ungar[1] has concluded that local extinctions of populations will occur in those instances where plant species, especially in the case of annuals, have not developed some mechanism to avoid germination under hypersaline conditions.

VIII. GROWTH REGULATOR-SALINITY INTERACTIONS

Researchers have indicated that in some glycophytes cytokinins could alleviate the inhibitory effects of high salinity on seed germination, but seeds of halophytes have not generally been reported to be stimulated to germinate under conditions of salt stress when they were treated with cytokinins (Table 6).[116-119] However, a number of investigations indicate that the halophytes studied were stimulated to germinate under salt stress conditions by treatments with gibberellic acid (GA_3) (Table 7).[7,14,26,72,120,121] In a recent study with *Salicornia pacifica* var. *utahensis*, Khan and Weber[26] indicated that GA_3 partially alleviated the inhibitory effect of NaCl but that kinetin was ineffective in overcoming the osmotically induced dormancy. GA_3 treatments increased germination percentages from 3%, in saline controls containing 2% NaCl, to 25% in GA_3-treated seeds of *Salicornia*. In the less salt tolerant cold desert shrub, *Chrysothamnus nauseosus* var. *viridulus* both kinetin and GA_3

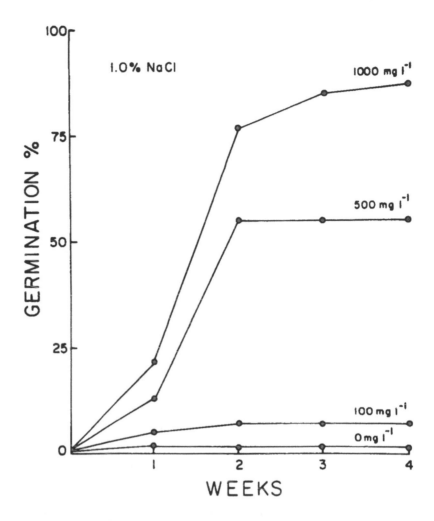

FIGURE 3. The effects of various GA₃ concentrations on the germination of *Spergularia marina* seeds in a medium with 1.0% NaCl.

were found to stimulate seed germination at 51 mM NaCl.[27] Seeds from saline controls containing 51 mM NaCl had 16% germination, while kinetin treatments had 48% and GA₃ treatments, 36% germination.

Ungar[72] determined that GA₃ overcame the inhibitory effect of darkness on the germination of *Spergularia marina* seeds, with germination increasing from 0% in the dark to 91% in the 1000 mg/l GA₃ treatment. Germination in salt treatments of 1.0% NaCl increased from 4% in the controls to 87% in the 1000 mg/l GA₃ treatments (Figure 3). However, at 2 and 3% NaCl, GA₃ was not capable of alleviating the stressful affect of NaCl, and no seeds germinated. Boucaud and Ungar[60] demonstrated that GA₃ relieved salt-induced suppression of seed germination in *Suaeda* species.

The role of growth regulators in alleviating salt stress-induced inhibition of seed germination for the halophyte *Atriplex triangularis* was investigated by Khan and Ungar.[121] Untreated small seeds had only 8% germination in 345 mM NaCl, while seeds treated with 2.9 mM GA₃ had 27% germination. Treatments with kinetin were also found to alleviate the osmotically induced dormancy for small seeds of *A. triangularis*. Small seeds germinating in 345 mM NaCl had an increase from 7% germination in the control to 69% germination in the 4.7 μM kinetin treatment.[121]

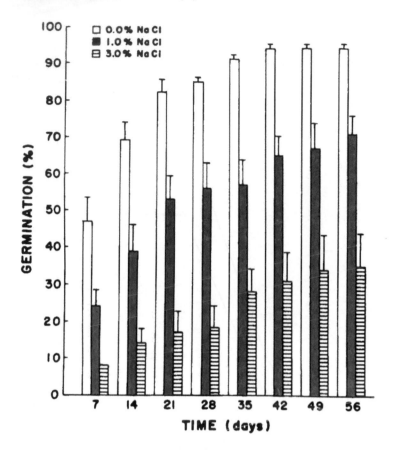

FIGURE 4.   The effect of salinity on seed germination of the large seed morph of *Salicornia europaea*.

Large seeds of *Salicornia europaea* could germinate at salinities of 5% NaCl, but seed germination was somewhat inhibited with increases in salinity (Figure 4).[69] Germination of small seeds occurred in up to 3% NaCl, but germination percentages were reduced markedly with salinity increments of 1% NaCl to about 25% of distilled water control values (Figure 5). Seeds were not permanently inhibited from germinating when exposed to high salinities, since seed germination was similar to control treatments in distilled water after exposure to 100 d in 5 and 10% NaCl. Ungar[69] determined that GA_3 (500 mg/l) stimulated germination of small seeds of *S. europaea* in the controls and 1% NaCl treatments, but the cytokinin zeatin riboside (100 mg/l) was not found to stimulate the germination of *S. europaea* seeds. Similar results were found in an earlier investigation of *S. europaea* when the effects of kinetin were compared with GA_3.[122]

Seeds of the salt marsh species *Distichlis spicata* were determined to be dormant at maturity.[123] The growth regulators GA_3 and kinetin were found to be ineffective in breaking the dormancy in seeds of *D. spicata*. Leaching seeds in estuarine water for 16 weeks did not promote seed germination. Treatment with 0.01 *M* NaCl yielded 22% germination compared with 64% germination in the $NaNO_3$ treatment, while seeds stratified for 4 weeks had 88 and 94% germination in the two salt treatments, respectively. Amen et al.[123] concluded that dormancy is caused by a restrictive seed coat and the accumulation of a biochemical inhibitor in *D. spicata* seeds. Application of nitrate salts overcame the biochemical block, possibly by stimulating nitrate reductase activity in seeds.

Okusanya and Ungar[124] reported that the month in which seeds were produced on the

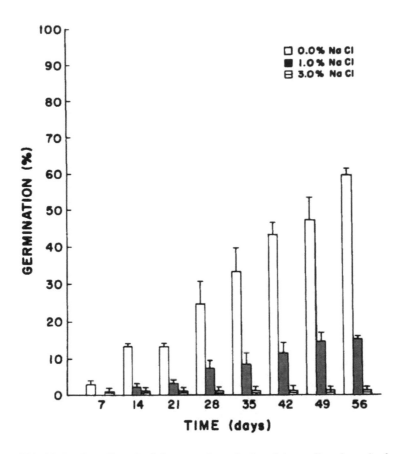

FIGURE 5. The effect of salinity on seed germination of the small seed morph of *Salicornia europaea*.

parent plant influenced the germination requirement for the seeds of *Spergularia marina*. Seed germination was inhibited at salinities above 1% NaCl, indicating that *S. marina* behaves like a facultative halophyte at the germination stage of development. Seeds of *S. marina* have a light requirement for germination, and treatments with GA_3 were found to overcome the inhibitory effect of darkness on the germination of *S. marina* seeds.[72] Treatments with 1000 mg/l GA_3 increased the seed germination of *S. marina* from 0 to 91% in the dark and also stimulated germination of seeds in the light.[72] At 1.5% NaCl, germination increased from 3% in controls to 32% in the GA_3-treated seeds. However, at higher salinities of 2 and 3% NaCl, GA_3 could not alleviate the inhibitory affect of NaCl, and no seeds germinated. Salt-induced suppression of seed germination was relieved by GA_3 in *Suaeda* spp.[60]

IX. SEED BANK

The literature concerning the occurrence and the extent of seed banks in saline environments has been recently reviewed.[1,125,126] Data collected in investigations from inland and coastal saline sites indicate that there are seeds producing a seed bank in many of the sites studied (Table 8), including California coastal salt marshes,[127,128] Baltic seashore meadows,[129] British estuarine marshes,[130] New England coastal marshes,[131] and inland saline marshes.[1,132] The extent of the seed banks in coastal marshes varied considerably, ranging from a low of 47 seeds/m² in a Massachusetts salt marsh to 140,000 seeds/m² in a Baltic salt marsh.[129,131] The majority of investigations have not looked at the seasonal change in

TABLE 8
Seed Banks Reported for Coastal and Inland Salt
Marshes

Plant community	Location	Seeds/m²	Ref.
Spartina alterniflora	Massachusetts	47	131
Juncus gerardii	Sweden	140,000	129
Salicornia europaea	England	23,500	135
Spergularia marina	California	83,050	127
Spartina patens	New Jersey	10,470	140
Salicornia virginica	California	380	128
Atriplex triangularis	Ohio	21,496	136
Salicornia europaea	Ohio	123,006	125
Spergularia marina	Ohio	471,135	144
Distichlis spicata	Utah	850	133
Scirpus maritimus	Utah	2,194	133

seed banks, which is essential for understanding the dynamics of seeds in the salt marsh ecosystem. Most studies have involved only single samples of salt marsh seed banks, which provide some interesting data regarding the relationship between the above-ground vegetation and the flora of the seed bank.

The species composition of seed banks and the above-ground dominants of plant communities were not highly correlated in six salt marsh communities investigated at Ogden Bay, UT by Smith and Kadlec.[133] The dominant species of the salt marsh communities were usually not the dominant species in their seed banks. Two species, *Hordeum jubatum* and *Tamarix pentandra*, were present in the seed bank but not in the above-ground vegetation, and while *Typha* spp. was not present in half of the community types, it was well represented in the seed banks of all communities. High soil salinities probably prevented the germination of seeds of less tolerant species in some of the community types in the field. However, laboratory germination studies at moderate soil salinities, 9 mS/cm, allowed some of the species that could not withstand hypersaline conditions to germinate.[133]

Dovey estuarine marshes were reported to have an underrepresentation of some of the dominant salt marsh species in the seed bank, while other species were well represented in the salt marsh communities in which they occurred.[130] The seed bank did not reflect the current plant communities at the site. The plant communities closest to the river on the Dovey salt marsh were characterized by *Puccinellia maritima*, *Armeria maritima*, *Aster tripolium*, and *Salicornia europaea*. Seed banks contained large numbers of *P. maritima* (71% of total) and small numbers of the other species, with very few *A. maritima* which was abundant in the above-ground vegetation. Milton[130] determined that several species were overrepresented, including *Glaux maritima*, *Agrostis* spp., and *P. maritima*, while some of the other abundant species such as *Aster tripolium*, *Armeria maritima*, and *Festuca rubra* were underrepresented in several of the salt marsh zones.[130] Preliminary results of soil surveys of salt marsh sediments at La Perouse Bay, Canada indicate that a seed bank of *S. europaea* was present.[134] However, a persistent seed bank was not found by Jefferies et al.[135] in the *Salicornia* community after the normal germination period on the estuarine marshes at Stiffkey, England. In March the upper marsh contained a seed bank of 23,500 *Salicornia europaea* seeds/m², whereas by June after seedling emergence no seeds were found in the soil samples.

It is critical to know how many seeds remain in the soil after the normal germination period is over, since these seeds could form a persistent seed bank. Laboratory investigations with seeds of halophytes indicate that many of the salt marsh species can withstand exposure to hypersaline conditions for relatively long periods of time and then germinate when the

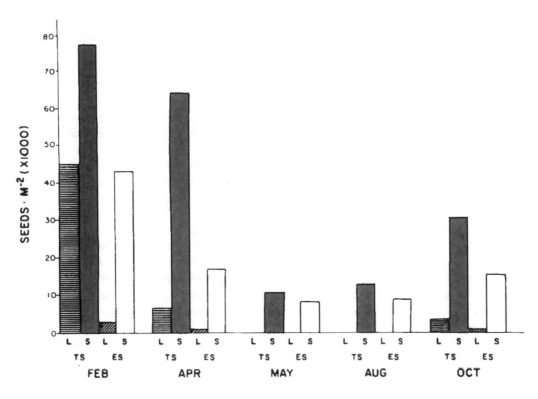

FIGURE 6. Seasonal distribution of large (L) and small (S) *Salicornia europaea* seeds in the seed bank of two populations; TS = Tall, ES = Edge. (From Ungar, I. A., *Ecology*, 68, 569, 1987.)

salinity stress is removed.[7,16] Seed burial experiments for wetland species from the Delta marsh, Manitoba (Canada) indicated that some of the larger seeded emergent annuals *Scolochloa festucacea* and *Hordeum jubatum* could germinate and emerge from burial at a depth of 5 cm.[24] However, smaller seeds of *Typha glauca*, *Atriplex patula*, and *Chenopodium rubrum* only germinated within the upper few centimeters of soil. These results indicate that only the species producing smaller seeds would be capable of forming a persistent seed bank when buried to a depth of 5 cm or greater. In studies with dimorphic seeds of *Salicornia europaea* and *Atriplex triangularis*, Khan and Ungar[121] and Ungar[1] determined that the smaller seeds were relatively dormant when buried in the soil, while the larger seeds for both species germinated readily with burial. Wertis and Ungar[136] and Ungar[1,125] reported that the seasonal pattern of germination indicates that all of the larger seeds of *A. triangularis* and *S. europaea* germinate early in the spring, while the small seeds that do not germinate remain in the soil as a persistent seed bank (Figure 6). Small seeds in both of these species have a light requirement for inducing seed germination, and burial of seeds in the soil evidently prevents seed germination because light cannot penetrate to the depth that seeds are buried.

Preliminary results of a study of seed banks in a community dominated by *S. europaea* in a Rittman, OH salt marsh indicated that seeds of *S. europaea* were chiefly confined to the upper 2 cm of soil.[137] The seeds stored in the soil had relatively high germinability, with small seeds averaging 60% germination in both June and November and large seeds averaging 93% germination. Ungar[125] found a seed bank of 123,006 seeds/m² in February in a community dominated by *S. europaea* on a Rittman, OH inland salt marsh. The large seeds made up 37% of the total seed bank in February, and by May there were no large seeds remaining in the seed bank, indicating that they all germinated by this time. The small seed morph comprised the seed bank in May and June and had a mean of about 10,000 seeds/m². Experimental data indicated that the smaller seed morph of *S. europaea* remained dormant and ungerminated when buried in the marsh soil. Only 4% of the small seeds germinated, while 60% of the large seeds germinated when they were buried at a depth of 4 cm. These results add additional support to the seed bank data, indicating that large seeds of *S. europaea* are more readily germinable, while small seeds are more dormant and have a light and stratification requirement in order to reach their optimal germination percentages (Table 9).[125,137] Small seeds may ensure survival of populations of the annual *S. europaea* from year to year in these highly stressful salt marsh habitats.

An example of the significance of seed banks for inland salt marsh habitats was described for a saline site at Stink Lake, SD by Ungar.[138] In 1967, the annual *Salicornia rubra* made up 90% of the relative cover and had a frequency of 89% in quadrats in a community it dominated. Soil salinity in this zonal community ranged from 2.3 to 8.0% total salts during the growing season. In 1968, both *S. rubra* and another salt pan annual *Suaeda depressa* were completely absent from this area because of high soil salinity in the spring when germination normally occurs. Hypersaline conditions were caused by below normal rainfall from October to March in 1967 to 1968, averaging 49% of normal precipitation. Prior to the growing season in 1969, precipitation was 142% of normal, and a large population of *S. rubra* established itself at this site from the seed bank produced in 1966.[12]

Rapid changes in the species composition of zones on the Rittman, OH salt marsh can also be explained by the nature of the seed bank.[7,139] At this location increases in soil salinity led to a community dominated by *Salicornia europaea*, and as the salinity level dropped *Atriplex triangularis* and then *Hordeum jubatum* became the established dominants. Changes can occur rapidly on this marsh from one year to the next, indicating that the seed bank plays a significant role in the recruitment of species in this system with continuously changing edaphic conditions.[7,139]

Wertis and Ungar[136] reported that a community dominated by *Atriplex triangularis* had a seed bank of 21,496 seeds/m² in January on an Athens, OH salt pan (Figure 7). Of the seeds present in the soil, 17,196 were of the small seed morph, and 4300 were of the large seed morph. After the normal spring germination period, in July the number of seeds in the soil cores dropped to a mean of 923 seeds/m², and no large seeds were found in the summer samples. As for *S. europaea* seed banks, it is apparently only the small seed morph that is capable of producing a seed bank in *A. triangularis* populations.[136]

A study of the composition of the seed bank of a coastal marsh at Absecon, NJ indicated that there was a seed bank of 6225 seeds/m² in the short *Spartina alterniflora* zone and 10,470 seeds/m² in the *Spartina patens* zone of the salt marsh.[140] The higher seed bank values were obtained using direct seed counts from the soil cores, while greenhouse seed germination estimates produced much lower values, averaging about 5% of the totals found from direct seed counts (259 seeds/m² for *S. alterniflora* zone and 601 seeds/m² for the *S. patens* zone). Not only were the number of seeds present underestimated by the germination assay, but twice as many species were also found when seeds were extracted from soils.[140] *S. alterniflora* and *S. patens* accounted for 96.5 and 85% of the cover in the zones they dominated and were underrepresented in the seed bank (14 and 22% for each species, respectively), while species such as *Pluchea purpurascens* and *Salicornia europaea* were overrepresented in the New Jersey salt marsh seed bank. Engel[140] determined that field seed germination did not reflect the composition of the zonal communities, since only *P. purpurascens* (55%) and *S. europaea* (45%) germinated in the *S. alterniflora* zone, and these species also comprised the majority of seedlings (97.4%) in the *S. patens* zone of the salt marsh.

Most recolonization of bare ground in the Great Sippewissett (Massachusetts) salt marsh dominated by *Spartina alterniflora* was by vegetative expansion of this dominant species.[131]

TABLE 9
Percentage Germination (± SE) of Seeds of Edge *Salicornia* Plants at the End of 5 Weeks With (+) or Without (−) Cold Pretreatment at Different Temperatures and Salinities

Cold	Temp (°C)	Seed size[a]	Salinity (% NaCl)			
			0	1	3	5
−	5—15	L	48 ± 4.0	17 ± 5.3	4 ± 2.3	0 ± 0.0
		S	0 ± 0.0	0 ± 0.0	0 ± 0.0	0 ± 0.0
−	5—25	L	35 ± 4.4	19 ± 10.4	1 ± 1.0	1 ± 1.0
		S	3 ± 1.0	1 ± 1.0	0 ± 0.0	0 ± 0.0
−	15—25	L	40 ± 3.7	23 ± 1.9	6 ± 1.2	0 ± 0.0
		S	3 ± 1.9	1 ± 1.0	0 ± 0.0	0 ± 0.0
+	5—15	L	89 ± 1.9	78 ± 6.0	82 ± 4.2	43 ± 11.1
		S	41 ± 9.1	32 ± 4.9	10 ± 3.5	0 ± 0.0
+	5—25	L	75 ± 9.1	78 ± 3.8	40 ± 7.1	2 ± 1.2
		S	30 ± 4.8	12 ± 1.6	1 ± 1.0	0 ± 0.0
+	15—25	L	54 ± 3.5	39 ± 3.0	38 ± 4.2	2 ± 1.2
		S	5 ± 1.9	3 ± 1.9	0 ± 0.0	0 ± 0.0

[a] S = small seeds; L = large seeds.

From Philipupillai, J. and Ungar, I. A., *Am. J. Bot.*, 71, 542, 1984.

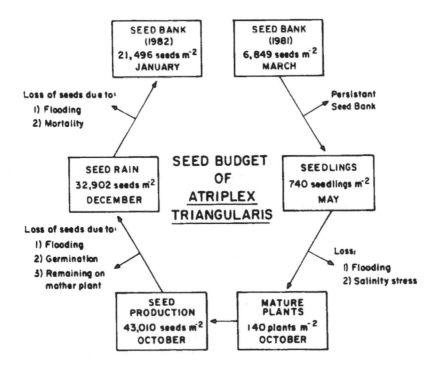

FIGURE 7. The seed budget for a population of *Atriplex triangularis*. (From Wertis, B. and Ungar, I. A., *Am. Midl. Nat.*, 116, 152, 1986.)

A low density seed bank of 27.8 to 47 seeds/m², with a mean density of 34.8 seeds/m², was found in the three zonal communities dominated by *S. alterniflora*. The seed bank did not contain any seed of the dominant perennial grasses and was dominated by the succulent annual *S. europaea*. At field sites, 95% of the seedlings found were of *Salicornia europaea*, and a single seedling each was found of *S. bigelovii*, *Gerardia maritima*, and *Aster tenuifolius*. Hartman[131] found no seeds of *S. alterniflora* in the seed bank or in the field germination observations, indicating that the recolonization of bare patches on the marsh by the dominant *S. alterniflora* was by vegetative growth of its rhizomes into denuded areas. Disturbed areas on the Great Sippewissett salt marsh in Massachusetts caused by wrack coverage were found to have up to 25,000 *S. europaea* seeds/m² in the seed bank after 3 years.[141]

Seeds of many of the dominant perennial grasses growing in salt marshes either have a small number of buried seeds or do not usually maintain a persistent seed bank.[128,131] The composition of the seed bank and its relationship to above-ground vegetation was determined on the China Camp salt marsh in San Francisco Bay by sampling the soils in October and February.[128] In the two zonal communities investigated, the dominants maintained above-ground cover of 99.3% for the zone dominated by *Spartina foliosa* and 96.6% cover for the zone dominated by *Salicornia virginica*. The predominant grass species was not found in the seed bank. However, in the October soil sample there was a mean of 380 seeds/m², with *S. virginica* comprising 96.7% of the total seed bank. In the February soil samples *S. virginica* made up 92.3% of the total seed bank, *Rumex crispus* 3.5%, and *Distichlis spicata* 1.7%. The other 14 species found on the China Camp salt marsh had a relative density of less than 0.3% for each of the species, comprising 2.5% of the total. The October soil collections contained 700 seeds/m², with the maximum density at one location being 3175 seeds/m². The seed bank of the *S. virginica* zonal community corresponded well with the above-ground composition of the community, containing 88% of the site dominants in the seed bank. Hopkins and Parker[128] concluded that the seed bank is apparently a low density and low diversity source for the occasional establishment of individuals on these salt marshes that are dominated by perennial species which spread vegetatively. Other research in California indicates that saline tidal marsh restoration areas contained large numbers of annuals, maintaining seed densities in the soil as high as 83,050 seeds/m² (*Spergularia marina*, 76,300; *Cotula coronopifolia*, 6600; and *Polypogon monspeliensis*, 150 seeds/m²).[127]

Seed banks from salt marshes in The Netherlands contained most of the species in the above-ground vegetation.[36] A number of species present in the seed bank of salt marshes such as the dune species were not found represented as adult plants. Many salt marsh species were found higher up on the marsh in the seed bank than they occurred as adults in the above-ground plant communities, including *Plantago maritima*, *Atriplex prostrata*, *Juncus gerardii*, *Spergularia salina*, *Salicornia europaea*, *Limonium vulgare*, and *Glaux maritima*. These data indicated that the coastal species had a greater potential to occupy zones on the marsh than their actual distribution in the above-ground vegetation indicated. Bakker et al.[36] concluded that the limiting factors on the salt marshes were hypersaline conditions at different stages of plant development and competition in less saline habitats.

An investigation at the Delta marsh (Canada) indicated that on drawdown in parts of the experimental marsh, the recruited vegetation could be predicted from the seed bank.[142] Seedlings of some species such as *Typha glauca*, *Chenopodium rubrum*, and *Scirpus lacustris* were predicted from seed banks to occur at higher densities than they were found in the emergent seedlings, while other species were present in higher densities in the emergent seedlings than was predicted from the seed bank, including *Hordeum jubatum*, *Atriplex patula*, and *Scolochloa festucacea*. Van der Valk and Pederson[142] concluded that because species have different environmental requirements for germination, a number of environmental conditions may affect the emergence of seeds from the seed bank in wetland habitats. These include reduction of illumination due to the plant canopy or litter deposition, lack of soil moisture, temperature, nutrients, flooding, soil salinity, and photoperiod.

A brackish marsh wetland in Manitoba, Canada was reported to have a relatively small seed bank, with a total of 56 seedlings emerging from 25 soil samples.[142] Forty-six of the total seedling number were for *Scirpus maritimus* seedlings, comprising 82% of the total. *Atriplex patula* made up 9% of the total seedling population, while other species such as *Hordeum jubatum*, *Typha* spp., *Zanichellia palustris*, and *Potamogeton pectinatus* each individually formed less than 4% of the total species association in the seed bank.[142]

Hutchings and Russell[143] determined the production of seeds at three sites on the Hayling Island salt marsh. Seed rain averaged 19,967, 16,221, and 11,060 seeds/m² on the high, middle, and low sites, respectively. The seed banks in the three sites had a decrease of 92, 72, and 36% in density of seeds from the seed rain, respectively. The seed density from seed banks in the high, middle and low sites was 1614, 4563, and 7108 seeds/m² for all species, respectively. For most of the species peak values were obtained in fall and winter. Summer seed bank values for *Armeria maritima* from the middle site declined to 12% of the maximum, and for *Puccinellia maritima* at the low site it dropped to 4% of the peak value. Seedling establishment at the three sites was 294, 118, and 220 seedlings/m² or only 1.5, 0.7, and 2.0% of the total seed rain in the high, middle, and low sites, respectively. Similarity coefficients between the above-ground vegetation and seed banks at the three sites ranged from 44 to 73%, indicating a reasonably close relationship between stored seed and the current flora. Although seeds of the halophytic species could remain viable in saline soils, the accumulation of a large persistent seed bank might be limited because of the abrasive action of tides. The fact that the surface soils remain wet and cracks do not develop might also prevent the downward migration of seed in the soil. Hutchings and Russell[143] concluded that the physical conditions on tidal salt marshes could be the limiting factor which prevents the accumulation of buried seed banks in coastal marshes.

Juncus gerardii was reported to be the dominant species in the seed bank of a Baltic seashore meadow.[129] Of the 13 species that germinated in June, *J. gerardii* accounted for 86.8% of the seeds (130,000 seeds/m²). Two other species of significance were *Triglochin maritimum* with 5.6% and *Glaux maritima* with 5.9% of the total seed bank in areas dominated by *J. gerardii*. A comparison of the above-ground vegetation with the seed bank indicated that there was considerable variation in the abundance of some of the species. In this Baltic seashore meadow, *J. gerardii* was overrepresented in the seed bank, while other species such as *Festuca rubra*, *Agrostis stolonifera*, and *Plantago maritima* were underrepresented. Jerling[129] noted that the plants that dominated this seed bank, including *J. gerardii*, *T. maritimum*, and *G. maritima*, all had small seeds with innate or induced dormancies, and the plants reproduced vegetatively.

Storage of seeds in the soil is a significant factor in the recruitment of plant populations in some saline habitats, because their establishment is often difficult in these highly stressful environments. If because of hypersaline conditions an entire cohort of plants dies, the only way populations can recover quickly is by the production of a persistent seed bank. Seed germination and establishment occurs when salinity levels are reduced and environmental conditions become more favorable. Annuals, such as *Spergularia marina*, have only one chance to reproduce before they become senescent or the population becomes locally extinct due to the development of hypersaline conditions on a salt marsh. The development of a persistent seed bank provides *S. marina* with the opportunity for recruitment of a new population later in the same growing season or in subsequent years. Ungar[144] reported the seasonal change in the number of seeds in the seed bank for *S. marina* in an inland Rittman, OH salt marsh, with 379,268 seeds/m² in February and a depletion by seed germination or other factors to 38,506 seeds/m² in June. Field data for 1989 indicated that bare areas in the *S. marina* dominated zone contained 541,347 seeds/m² after the normal germination period in the spring. These areas were invaded by *Salicornia europaea*, indicating that the edaphic conditions were not suitable for *Spergularia*, but that a persistent seed bank would be available for the development of a large *S. marina* population when the soil salinity levels are reduced and the edaphic conditions become more favorable for seed germination and establishment.[144]

Seed banks for the two submerged aquatic halophytes *Zostera marina* and *Zostera noltii* from Yerseke (Netherlands) appeared to be annual. Hootsmans et al.[117] determined that the magnitude of the seed bank varied during the growing season for *Z. noltii*, ranging from <10 to 150 seeds/m², while *Z. marina* had <10 seeds/m² in its seed bank from May to August. These seed bank values are much lower than the potential size of the seed bank based on the number of inflorescences per plant, 200 seeds/m² for *Z. marina* and 9000 seeds/m² for *Z. noltii*. Differences from the predicted values could be due to movement of seeds in the soil or the fact that seeds may be deposited in September and October.[145]

One of the factors affecting the size of the seed bank is the number of seeds that are produced by the above-ground vegetation. Recent investigations by Ellison[146] and Bertness et al.[147] indicated that herbivory could play a significant role in limiting the number of viable seeds produced by plants in salt marshes and thus reduce the potential for the production of large seed banks. Herbivore pressure was not constant but varied with the location of *Salicornia europaea* plants on a salt marsh at Barrington, RI.[146] In the short *Spartina alterniflora* zone no insect damage was observed on *S. europaea*, while in the *S. patens* zone up to 20% of the internodes were damaged by beetles (*Erynephala maritima*). Moth larvae (Coleophorids) ate significantly more *Salicornia* seeds in a *J. gerardi* zone in which the grass was removed than beneath the *Juncus* canopy. Herbivore pressure on *S. europaea* was greater in open patches and when the perennial graminoid canopy was removed, because it may be easier for herbivores to find plants when they are in patches than beneath a plant canopy.[146] However, the survival percentages of plants under the perennial canopies and in removal areas were the same, indicating that insect herbivory may have been the cause of mortality in the removal areas.

The absence of seed banks or reductions in their population size in salt marshes could be due to consumption of seeds by predators.[146,147] Predation of plants on a salt marsh at Barrington, RI reduced the reproductive capacity of a number of dominant graminoids. Most of the insect damage was caused by generalist feeding grasshoppers (*Conocephalus spartinae*). Ovule loss ranged from 51 to 80% in the four perennial graminoids; losses were 54% for *Juncus gerardi*, 80% for *Spartina patens*, 51% for *Distichlis spicata*, and 50% for *S. alterniflora*. The predispersal consumption of seeds by predators has a significant effect on the availability of seeds for the seed bank. Bertness et al.[147] hypothesized that the limitation of seed production by consumers could minimize the importance of seedlings in the establishment of species in coastal marshes, increasing the importance of clonal growth for the colonization of bare areas.

Germination cues in wetland habitats may include light, temperature, oxygen, and salinity. Seed dormancy may be of primary or secondary origin, but in either case serves to store seeds in the soil when environmental conditions are not suitable for germination. Leck et al.[148] described the relationship between seed banks and plant communities as involving (1) seed longevity and germination characteristics, (2) surface vegetation, (3) seed rain, (4) age of wetland, and (5) herbivory.

X. PARENTAL EFFECTS

Germination of halophyte seeds may be affected by the parental environments to which seeds are exposed.[1] It is somewhat difficult to sort out the direct effect of the parent plant on seed responses vs. that of the physical environment to which seeds are exposed when they are still attached to the parent plant. *Spergularia* is a useful organism to study the effects of environment on seed germination, because it produces flowers and seeds from

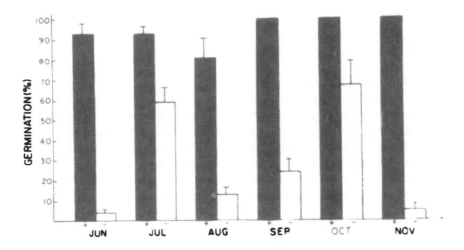

FIGURE 8. The effect of a 30-d 5°C treatment on germination of *Spergularia marina* seeds at 15°C/12 h-dark and 25°C/12 h-light. + = stratified (black bar), − = unstratified (white bar). Mean values ± SE. (From Ungar, I. A., *Bot. Gaz.*, 149, 432, 1988.)

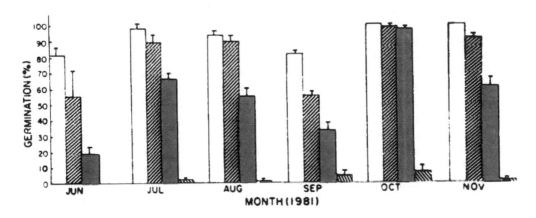

FIGURE 9. The effect of salinity on the seed germination of monthly collections of *Spergularia marina* seeds. White bar = 0% NaCl, left diagonal bar = 1.0% NaCl, black bar = 1.5% NaCl, right diagonal bar = 2.0% NaCl. (From Ungar, I. A., *Bot. Gaz.*, 149, 432, 1988.)

June through November on its indeterminate inflorescence. Okusanya and Ungar[124] and Ungar[72] determined that seeds collected from plants at different times during the growing season had different levels of dormancy depending on the month of seed collection.[149] Seeds collected from plants during the months of June and November were most dormant, followed by August and summer collections in their degree of dormancy. The dormancy for all monthly seed collections could be alleviated with a 30-d stratification period pretreatment, followed by a 5°C-night/15°C-day temperature regimen (Figure 8). Seed germination was inhibited at salinities above 1% NaCl, indicating that *S. marina* behaves like a facultative halophyte in terms of its germination response to salinity. The germination response of seeds to salinity varied both monthly and with collections from one year to the next, indicating that the salt tolerance of seeds was also affected by parental environments (Figure 9). Highest germination in NaCl treatments was from the October 1981 *S. marina* seeds and lowest germination was in the June 1981 seeds. Seeds of *S. marina* harvested from laboratory-grown plants initiated from the original monthly seed collections were uniformly dormant at the previously stimulatory temperature regimen of 5°C-night/15°C-day and would only germinate when seeds were exposed to a stratification pretreatment.[149] These data for field collections of *S. marina* seeds indicated that there was a parental environment control of seed germination when the seeds were still attached to parent plants. The parental control of the germination requirements for seeds has ecological significance, because it determined whether seeds would be able to germinate in the same growing season in which they were produced and under what specific environmental conditions germination could take place. Variation in dormancy levels within monthly cohorts of seeds could provide numerous opportunities for the establishment of plants within a population during the growing season. Because of high seedling mortality in stressful salt marsh habitats, Ungar[144] concluded that annual halophytes have developed a number of mechanisms to spread out their germination period over time.

Uchiyama[19] determined that the culture medium that *Atriplex nummularia* plants were grown in, profoundly affected the germination responses of seeds produced on these plants when they were exposed to high salinity levels. Only 4% of the seeds germinated in 2% NaCl from parental plants grown in the 0% NaCl controls, while 41% of the seeds germinated at 2% NaCl from plants cultivated in the 2% NaCl treatment.

Although the Ohio population of *Spergularia marina* did not demonstrate any morphological seed dimorphism,[149] both winged and unwinged seeds have been found on individual plants in populations from The Netherlands, England, and Sweden by Salisbury,[150] Sterk,[151] and Telenius and Torstensson.[152] Winged seeds had an averge length of 1.49 mm and weight of 103 μg, while wingless seeds were 0.59 mm long and weighed 68 μg.[152] Experiments were carried out to determine the effect of wings on seeds on wind dispersal (1 and 4 m/s), and it was ascertained that in air currents winged seeds dispersed further (19.8 cm) than unwinged seeds (16.6 cm) and seeds with wings removed (15.9 cm). Both seed forms were found to float in water for 2 weeks, and no differences were found in disperal distances when seeds were exposed to a current in the water. Since 95% of the seeds were lost from the seed bank because of flooding of the salt marsh in autumn at Tullgan, Sweden, it could be that the winged seeds had a selective advantage, because more of them were retained in the plant debris on the salt marsh than were wingless seeds. The evolution of the seed wing in *S. marina* might be coupled with an increase in seed size, and Telenius and Torstensson[152] concluded that this characteristic could have evolved to reduce the negative effect of increased seed size on the wind dispersal of seeds.

Salicornia europaea exhibited a seed dimorphism that affects the germination responses of seeds. Philipupillai and Ungar[137] determined that the small seeds from lateral flowers were more dormant and apparently less salt tolerant than the large seeds from the medial flowers. Under optimal temperature conditions 5°C/15°C with a stratification pretreatment, the large seeds had 43% germination and the small seeds 0% germination in a 5% NaCl treatment. The maximum salt tolerance for small seeds was 10% seed germination in a 3% NaCl treatment, while without stratification small seeds had 0% and large seeds 6% germination at 3% NaCl (Table 9). All of the large seeds of *S. europaea* germinated in the field by March, while small seeds were present in the soil throughout the year, providing a persistent seed bank and accounting for seedlings that appear from May to July or later in the growing season.[137]

A number of plant species found growing in saline habitats exhibit some form of seed dimorphism or polymorphism that serves to spread out the germination of a species both spatially and temporally.[7] Differences in seed size and time of germination among the seeds produced by a single parent have been shown to influence the population biology of progeny, by resulting in a differential in competitive ability, survivorship, and fecundity.[5,109,153-157] Harper[109] hypothesized that seed polymorphism is the optimal strategy for plant species growing in fluctuating environments since it allows an organism to allocate its resources toward different ends to satisfy the different environmental conditions. By producing seeds with varying germination requirements, *S. europaea* allocates part of its resources to the

TABLE 10
Index of Germination Velocity of *Atriplex triangularis* Seeds from Various Salinity and Temperature Treatments

Temperature night-day	Salinity (%)	Seed size		
		Small	Medium	Large
5—15°C	0.0	18.95	20.90	15.70
	0.5	12.22	15.50	11.60
	1.0	9.70	9.85	4.40
	1.5	1.50	8.00	1.15
5—25°C	0.0	37.15	33.90	43.95
	0.5	24.50	26.25	41.75
	1.0	9.30	21.10	41.30
	1.5	1.90	12.35	31.75
20—30°C	0.0	33.00	40.10	44.05
	0.5	18.65	28.15	42.00
	1.0	7.90	17.11	15.60
	1.5	4.05	12.75	16.10
10—20°C	0.0	18.80	24.25	32.95
	0.5	12.05	16.65	23.85
	1.0	2.60	9.85	13.95
	1.5	1.40	8.90	8.20

From Khan, M. A. and Ungar, I. A., *Bot. Gaz.*, 145, 487, 1984.

large seeds that display a K-strategy by preempting environmental resources. If low rainfall and high soil salinities prevent the establishment of the early large seed cohort or do not permit the establishment of any cohort in a year, a persistent seed bank of r-strategy small seeds of *S. europaea* ensures the potential for the reestablishment of the population at a later date.[137]

One mechanism for regulating spatial and temporal germination patterns is to develop some form of seed dimorphism or polymorphism. Polymorphic seeds of the halophyte *Atriplex triangularis* were investigated to determine the differences in germination responses of the different seed morphs.[73] Both the rate of and percentage germination of seeds decreased for all seed morphs when salinity stress was increased (Table 10). Large seeds had higher germination percentages in the salt treatments than did the small seeds, and the optimal temperature for germination was an alternating temperature of 5°C-night/25°C-day. The fact that the salt tolerance of the different seed morphs varies from one another is of ecological significance. Germination of seeds of halophytes can occur over a broad salinity range, 200 to 1700 mM NaCl. However, for a large number of halophytes germination does not occur at seawater salinity concentrations (600 mM).[6] *Atriplex triangularis* falls into this latter group of species, with salinities above 1% NaCl being very inhibitory to seed germination. Seeds responded to a gradual increase in medium salinity with a decrease in percentage of the seeds germinating, a delay in the rate of germination, and a decrease in mean seedling length.[73] The germination of seeds of halophytes during periods of reduced salinity in early spring is of ecological significance, because establishment of seedlings during this period of low soil salinity and high soil moisture assures that some of the early cohort will survive to the end of the growing season and reach reproductive maturity. Cohorts of seedlings that initiate later in the summer are exposed to extremely high levels of salt stress on inland salt pans, and even if seeds germinated during periods of rainfall there would be almost no chance of these seedlings surviving to reach reproductive maturity.[73,136]

The degree of salt tolerance for seeds of *A. triangularis* was related to seed size.[73] Small seeds, mean dry mass 0.6 mg, and large seeds with a mean dry mass of 2.4 mg had tolerance limits to salinity of 2.0 and 5.0% NaCl, respectively. The sharpest distinction between the two seed morphs was at 2.0% NaCl, where small seeds had a mean germination of 8% and large seeds had a mean of 60%. The velocity of germination decreased with increasing salinity for all seed morphs, but the rate of germination was lower for the small seed morph. Seeds of *A. triangularis* that did not germinate at high salinity concentrations after 20 d were placed in distilled water to determine the recovery percentages. Germination in these recovery tests ranged from 87 to 100% and did not differ significantly from the original distilled water controls. Both the rate of germination and the final germination percentages were reduced by salt stress, but the effect of *A. triangularis* seed morphs was chiefly a transitory osmotic effect, and there was no permanent specific ion toxicity observed.[73]

A field germination experiment with *A. triangularis* indicated that seeds buried in the soil vary in their capacity to germinate, depending on several biotic and abiotic characteristics, including seed size, availability of light, and level of soil salinity stress.[5] Seed germination experiments in field plots that were initiated in April generally had lower germination percentages than those started in November when the seeds of *A. triangularis* received natural stratification, indicating the normal seasonal pattern of seed germination for the species. Burial of seeds, small seed size, and high salinity all negatively affected germination, with the lowest germination for all treatments usually occurring in the high salt habitats, when seeds were buried in the soil and for small seeds. After the normal stratification period over winter, in March in the low salt habitat 71.6% of the surface small seeds and only 17.2% of the buried small seeds germinated, while for the large seeds there was no difference in the germination percentages for surface or buried seeds. The same pattern was followed for the high salt habitat with fewer seeds germinating in both surface and buried cohorts; 34.2 and 8.4% for small seeds and 64.4 and 59.6% of the large seeds germinating at the two soil depths, respectively.[5] Larger seeds of *A. triangularis* germinated earlier than the smaller seeds, which is ecologically significant in these highly saline habitats, since seedling mortality was significantly higher for seedling cohorts that developed later in the growing season.

XI. CONCLUSIONS

The germination of seeds for most halophytes occurs during periods of the year when soil salinity levels are reduced.[6] Laboratory investigations with halophytes indicate that optimal germination percentages are usually found in nonsaline conditions. At some point along a salinity continuum the final germination percentages of halophyte seeds declined or the rate of seed germination was reduced (Table 1).[7] However, the seeds of halophytes can generally tolerate higher salinity concentrations than those of glycophytes. The primary effect of salt stress on seeds is apparently osmotic,[6,16,158,165,166] but some cases of specific ion toxicity have been reported.[78,79,84] A significant characteristic of seeds of halophytes is their ability to withstand long periods of immersion in hypersaline conditions and then germinate when environmental conditions become more favorable.[7,16,51,64-73] Seeds of halophytes tend to accumulate less than 10% of the ionic content reported for their shoots, indicating that they have developed some mechanism to prevent excess ion accumulation in the environment of their developing embryos.[72,92-97] Many species of halophytes have dormancy mechanisms, innate, induced, or enforced, which permits seeds to remain dormant in the seed bank until favorable environmental conditions prevail.[7] There is some indication that ecotypic differentiation for salt tolerance of seeds occurs in highly saline habitats,[46,47,55] but other studies indicate that the parental habitat may not be directly correlated with salinity tolerance at the germination stage.[45,53,54] More research is needed to determine the population dynamics of seeds in saline environments and how seed mortality is affected by environmental variables. We need more information concerning the effect of predation on seed production and loss during the seed cycle. The influence of parental environments on the germination responses of seeds needs a greater research effort so that we can predict how plastic species are in their responses to environmental conditions.

REFERENCES

1. **Ungar, I. A.,** Population ecology of halophyte seeds, *Bot. Rev.,* 53, 301, 1987.
2. **Riehl, T. E. and Ungar, I. A.,** Growth and ion accumulation in *Salicornia europaea* under saline field conditions, *Oecologia,* 54, 193, 1982.
3. **Riehl, T. E. and Ungar, I. A.,** Growth, water potential, and ion accumulation in the inland halophyte *Atriplex triangularis* under saline field conditions, *Acta Oecol. Oecol.Plant.,* 4, 27, 1983.
4. **Wertis, B. and Ungar, I. A.,** Seed demography and seedling survival in a population of *Atriplex triangularis* Willd., *Am. Midl. Nat.,* 116, 152, 1986.
5. **Khan, M. A. and Ungar, I. A.,** Life history and population dynamics of *Atriplex triangularis, Vegetatio,* 66, 17, 1986.
6. **Ungar, I. A.,** Halophyte seed germination, *Bot. Rev.,* 44, 233, 1978.
7. **Ungar, I. A.,** Germination ecology of halophytes, in *Contributions to the Ecology of Halophytes,* Sen, D. N. and Rajpurohit, K., Eds., Junk, The Hague, 1982, 143.
8. **Workman, J. P. and West, N. E.,** Germination of *Eurotia lanata* in relation to temperature and salinity, *Ecology,* 48, 659, 1967.
9. **Workman, J. P. and West, N. E.,** Ecotypic variation of *Eurotia lanata* populations in Utah, *Bot. Gaz.,* 130, 26, 1969.
10. **Strogonov, B. P.,** *Physiological Basis of Salt Tolerance of Plants,* Israel Program for Scientific Translation, Jerusalem, 1964.
11. **Strogonov, B. P.,** *Structure and Function of Plant Cells in Saline Habitats,* John Wiley & Sons, New York, 1973.
12. **Ungar, I. A.,** Inland halophytes of the United States, in *Ecology of Halophytes,* Reimold, R. J. and Queen, W. H., Eds., Academic Press, New York, 1974, 235.
13. **Rozema, J.,** The influence of salinity, inundation and temperature on the germination of some halophytes and non-halophytes, *Oecol. Plant.,* 10, 341, 1975.
14. **Ungar, I. A.,** Salinity, temperature, and growth regulator effects on seed germination of *Salicornia europaea* L. *Aquatic Bot.,* 3, 329, 1977.
15. **Marchioni-Ortu, A. and Bocchieri, E.,** A study of the germination responses of a Sardinian population of sea fennel *(Crithmum maritimum), Can. J. Bot.,* 62, 1832, 1984.
16. **Woodell, S. R. J.,** Salinity and seed germination patterns in coastal plants, *Vegetatio,* 61, 223, 1985.
17. **Okusanya, O. T.,** An experimental investigation into the ecology of some maritime cliff species. II. Germination studies, *J. Ecol.,* 67, 293, 1979.
18. **Mahmood, K. and Malik, K. A.,** Salt tolerance studies on *Atriplex rhagodioides* F. Muell., *Environ. Exp. Bot.,* 27, 119, 1987.
19. **Uchiyama, Y.,** Salt tolerance of *Atriplex nummularia, Tech. Bull. Trop. Agric. Res. Ctr. Jpn.,* 22, 1, 1987.
20. **Zid, E. and Boukhris, M.,** Quelques aspects de la tolerance de l'*Atriplex halimus* L. au chlorure de sodium; multiplication, croissance, composition minerals, *Oecol. Plant.,* 12, 351, 1977.
21. **Badger, K. S. and Ungar, I. A.,** The effects of salinity and temperature on the germination of the inland halophyte *Hordeum jubatum, Can. J. Bot.,* 67, 1420, 1989.
22. **Cords, H. P.,** Factors affecting the competitive ability of foxtail barley *(Hordeum jubatum), Weeds,* 8, 636, 1960.
23. **Banting, J. D.,** Germination, emergence and persistence of foxtail barley, *Can. J. Plant Sci.,* 59, 35, 1979.
24. **Galinato, M. I. and Van der Valk, A. G.,** Seed germination traits of annuals and emergents recruited during drawdowns in the Delta marsh, Manitoba, Canada, *Aquatic Bot.,* 26, 89, 1986.
25. **Huiskes, A. H. L., Stienstra, A. W., Koustaal, B. P., Markusse, M., and Van Soelen, J.,** Germination ecology of *Salicornia dolichostachya* and *Salicornia brachystachya, Acta Bot. Neerl.,* 34, 369, 1985.
26. **Khan, M. A. and Weber, D. J.,** Factors influencing seed germination in *Salicornia pacifica* var. *utahensis. Am. J. Bot.,* 73, 1163, 1986.
27. **Khan, M. A., Sakhla, N., Weber, D. J., and McArthur, E. D.,** Seed germination characteristics of *Chrysothamnus nauseosus* ssp. *viridulus* (Astereae, Asteraceae), *Great Basin Nat.,* 47, 220, 1987.

28. Binet, P., Observations sur l'aptitude a germer des graines de *Silene maritima* With., *Bull. Soc. Linn. Normandie*, 7, 142, 1966.
29. Timson, J., New method of recording germination data, *Nature (London)*, 207, 216, 1965.
30. Rajpurohit, K. S. and Sen, D. N., Soil salinity and seed germination under water stress, *Trans. Indian Soc. Desert Techn. Univ. Centre Desert Studies*, 2, 106, 1977.
31. Joddi, A. J. and Iyengar, D. N. N., Germination of *Oenothera rosea* Moq., *Oecolog.*, 1, 367, 1977.
32. Cluff, G. J. and Roundy, B. A., Germination responses of desert saltgrass to temperature and osmotic potential, *J. Range Manage.*, 41, 150, 1988.
33. Harivandi, M. A., Butler, J. D., and Soltanpour, P. N., Effects of sea water concentrations on germination and ion accumulation in alkaligrass, *Commun. Soil Sci. Plant Anal.*, 13, 507, 1982.
34. Harivandi, M. A., Butler, J. D., and Soltanpour, P. M., Salt influence on germination and seedling survival of six cool season turfgrass species, *Commun. Soil Sci. Plant Anal.*, 13, 519, 1982.
35. Onnis, A. and Miceli, P., *Puccinellia festucaeformis* (Host) Parl.: dormienza e influenza della salinita sulla germinazione, *Giorn. Bot. Ital.*, 109, 27, 1975.
36. Bakker, J. P., Dijkstra, M., and Russchen, P. T., Dispersal, germination and early establishment of halophytes on a grazed and abandoned salt-marsh gradient, *New Phytol.*, 101, 291, 1985.
37. Tremblin, G. and Binet, P., Installation d'*Halopeplis amplexicaulis* (Vahl) Ung. dans une sebkha algerienne, *Acta Oecol. Oecol. Plant.*, 3, 373, 1982.
38. Agami, M. and Waisel, Y., Germination of *Najas marina L.*, *Aquatic Bot.*, 19, 327, 1984.
39. Caye, G. and Meinesz, A., Experimental study of seed germination in the seagrass *Cymodocea nodosa*, *Aquatic Bot.*, 26, 79, 1986.
40. Koch, E. W. and Seeliger, U., Germination ecology of two *Ruppia maritima* L. populations in southern Brazil, *Aquatic Bot.*, 31, 321, 1988.
41. Van Vierssen, W., Van Kessel, C. M., and Van Der See, J. R., On the germination of *Ruppia* taxa in western Europe, *Aquatic Bot.*, 19, 381, 1984.
42. Vollebergh, P. J. and Congdon, R. A., Germination and growth of *Ruppia polycarpa* and *Lepilaena cylindrocarpa* in ephemeral saltmarsh pools, Westernport Bay, Victoria, *Aquatic Bot.*, 26, 165, 1986.
43. Dietert, M. F. and Shontz, J. P., Germination ecology of a Maryland population of saltmarsh bulrush (*Scirpus robustus*), *Estuaries*, 1, 164, 1978.
44. Agami, M., The effects of different soil water potentials, temperature and salinity on germination of seeds of the desert shrub *Zygophyllum dumosum*, *Physiol. Plant.*, 67, 305, 1986.
45. Clark, L. D. and West, N. E., Further studies on *Eurotia lanata* germination in relation to salinity, *Southwestern Nat.*, 15, 371, 1971.
46. Bazzaz, F. A., Seed germination in relation to salt concentration in three populations of *Prosopis farcta*, *Oecologia*, 13, 73, 1973.
47. Kingsbury, R. W., Radlow, A., Mudie, P. J., Rutherford, J., and Radlow, R., Salt stress responses in *Lasthenia glabrata*, a winter annual composite endemic to saline soils, *Can. J. Bot.*, 54, 1377, 1976.
48. Bulow-Olsen, A., Germination response to salt in *Festuca rubra* in a population from a salt marsh, *Holarctic Ecol.*, 6, 194, 1983.
49. Bocchieri, E. and Marchioni-Ortu, A., *Crithmum maritimum* L.: comportamento alla germinazione di popolazioni sarde di sabbia e di rupe, *Rendiconti Semin. Fac. Sci. Univ. Cagliari*, 54, 42, 1984.
50. Orth, R. J. and Moore, K. A., Seed germination and seedling growth of *Zostera marina* L. (eelgrass) in the Chesapeake Bay, *Aquatic Bot.*, 15, 117, 1983.
51. Ladiges, P. Y., Foord, P. C., and Willis, R. J., Salinity and waterlogging tolerance of some populations of *Melaleuca ericifolia* Smith, *Aust. J. Ecol.*, 6, 203, 1981.
52. Hutchinson, I. and Smythe, S. R., The effect of antecedent and ambient salinity levels on seed germination in populations of *Carex lyngbyei* Hornem, *Northwest Sci.*, 60, 36, 1986.
53. Clark, L. D. and West, N. E., Germination of *Kochia americana* in relation to salinity, *J. Range Manage.*, 22, 286, 1969.
54. Dafni, A. and Negbi, M., Variability in *Prosopis farcta* in Israel: seed germination as affected by temperature and salinity, *Isr. J. Bot.*, 27, 147, 1978.
55. Cavers, P. B. and Harper, J. L., Comparative biology of closely related species living in the same area. IX. *Rumex*: the nature of adaptation to a sea-shore habitat, *J. Ecol.*, 55, 73, 1967.
56. Binet, P. and Boucaud, J., Dormance, levee de dormance et aptitude a germer en milieu sale dans le genre *Suaeda* Forsk, *Bull. Soc. Fr. Physiol. Veget.*, 14, 125, 1968.
57. Boucaud, J., Etude morphologique et ecophysiologique de la germination de trois varietes de *Suaeda maritima* Dum., *Bull. Soc. Linn. Normandie*, 3, 63, 1962.
58. Ungar, I. A. and Boucaud, J., Action des fortes teneurs en NaCl sur l'evolution des cytokinines au cours de la germination d'un halophyte: le *Suaeda maritima* (L.) Dum. var. *macrocarpa* Moq., *C. R. Acad. Sci. Paris*, 281, 1239, 1975.
59. Boucaud, J. and Ungar, I. A., The role of hormones in controlling the mechanically induced dormancy of *Suaeda* spp., *Physiol. Plant.*, 29, 97, 1973.

60. Boucaud, J. and Ungar, I. A., Hormonal control of germination under saline conditions of three halophytic taxa in the genus *Suaeda, Physiol. Plant.*, 37, 143, 1976.
61. Malcolm, C. V., Effect of salt, temperature and seed scarification on germination of two varieties of *Arthrocnemum halocnemoides, J. R. Soc. West. Aust.*, 47, 72, 1964.
62. Onnis, A. and Bellettato, R., Dormienza e alotolleranza in due specie spontanee di *Hordeum, Giorn. Bot. Ital.*, 106, 101, 1972.
63. Waisel, Y., *Biology of Halophytes*, Academic Press, New York, 1972.
64. Ungar, I. A., Influence of salinity on seed germination in succulent halophytes, *Ecology*, 43, 763, 1962.
65. Ungar, I. A. and Hogan, W., Seed germination in *Iva annua* L., *Ecology*, 51, 150, 1970.
66. Macke, A. J. and Ungar, I. A., The effects of salinity on germination and early growth of *Puccinellia nuttalliana, Can. J. Bot.*, 49, 515, 1971.
67. Okusanya, O. T., The effect of sea water and temperature on the germination behaviour of *Crithmum maritimum, Physiol. Plant.*, 41, 265, 1977.
68. Breen, C. M., Everson, C., and Rogers, K., Ecological studies on *Sporobolus virginicus* (L.) Kunth with particular reference to salinity and inundation, *Hydrobiologia*, 54, 135, 1977.
69. Ungar, I. A., Seed dimorphism in *Salicornia europaea* L., *Bot. Gaz.*, 140, 102, 1979.
70. Mahmoud, A., El Sheikh, A. M., and Baset, S. A., Germination of two halophytes: *Halopeplis perfoliata* and *Limonium axillare* from Saudi Arabia, *J. Arid Environ.*, 6, 87, 1983.
71. Schat, H., Germination ecology of some dune slack pioneers, *Acta Bot. Neerl.*, 32, 203, 1983.
72. Ungar, I. A., Alleviation of seed dormancy in *Spergularia marina, Bot. Gaz.*, 145, 33, 1984.
73. Khan, M. A. and Ungar, I. A., Seed polymorphism and germination responses to salinity stress in *Atriplex triangularis* Willd., *Bot. Gaz.*, 145, 487, 1984.
74. Ungar, I. A., Autecological studies with *Atriplex triangularis* Willdenow, in *Symposium on the Biology of Atriplex and Related Chenopods*, Tiedemann, A. R., McArthur, E. D., Stutz, H. C., Stevens, R., and Johnson, K. L., Eds., General Tech. Rep. INT-172, Forest Service, U.S. Department of Agriculture, Ogden, UT, 1984, 40.
75. Ward, J. M., Studies in ecology on a shell barrier beach. III. Chemical factors of the environment, *Vegetatio*, 15, 77, 1967.
76. Partridge, T. R. and Wilson, J. B., Germination in relation to salinity in some plants of salt marshes in Otago, New Zealand, *N. Z. J. Bot.*, 25, 255, 1987.
77. Zedler, J. B. and Beare, P. A., Temporal variability of salt marsh vegetation: the role of low-salinity gaps and environmental stress, in *Estuarine Variability*, Wolfe, D. A., Ed., Academic Press, New York, 1986, 295.
78. Hyder, S. Z. and Yasmin, S., Salt tolerance and cation interaction in alkali sacaton at germination, *J. Range Manage.*, 25, 390, 1972.
79. Romo, J. T. and Haferkamp, M. R., Forage *Kochia* germination response to temperature, water stress, and specific ions, *Agronomy J.*, 79, 27, 1987.
80. Onnis, A., Pelosini, F., and Stefani, A., *Puccinellia festucaeformis* (Host) Parl.: germinazione e crescita iniziale in funzione della salinita del substrato, *Giorn. Bot. Ital.*, 115, 103, 1981.
81. Joshi, A. J. and Iyengar, E. R. R., Effect of salinity on the germination of *Salicornia brachiata* Roxb., *Indian. J. Plant Physiol.*, 25, 65, 1982.
82. Romo, J. T. and Haferkamp, M. R., Effects of osmotic potential, potassium chloride, and sodium chloride on germination of greasewood, *(Sarcobatus vermiculatus), Great Basin Nat.*, 47, 110, 1987.
83. Sabo, D. G., Johnson, G. V., Martin, W. C., and Aldon, E. F., Germination requirements of 19 species of arid land plants, Rocky Mt. Exp. Sta. Pap., RM-210, Forest Service, U.S. Department of Agriculture, Rocky Mt., NC, 1979.
84. Romo, J. T. and Eddleman, L. E., Germination response of greasewood *(Sarcobatus vermiculatus)* to temperature, water potential and specific ions, *J. Range Manage.*, 38, 117, 1985.
85. Eddleman, L. E. and Romo, J. T., Sodium relations in seeds and seedlings of *Sarcobatus vermiculatus, Soil Sci.*, 143, 120, 1987.
86. Sheikh, K. H. and Mahmood, K., Some studies on field distribution and seed germination of *Suaeda fruticosa* and *Sporobolus arabicus* with reference to salinity and sodicity of the medium, *Plant Soil*, 94, 333, 1986.
87. Cluff, G. J., Evans, R. A., and Young, J. A., Desert saltgrass seed germination and seedbed ecology, *J. Range Manage.*, 36, 419, 1983.
88. Bal, A. R. and Chattopadhyay, N. C., Effect of sodium chloride and PEG 6000 on germination and seedling growth of rice, *Biol. Plant.*, 27, 65, 1985.
89. Al-Jibury, L. K., Clor, M. A., and Talabany, D., Effects of certain salts and their combinations on germination and seedling development of *Securigera securidaca* Linn, *Arab Gulf J. Sci. Res.*, 4, 5, 1986.
90. Al-Jibury, L. K. and Clor, M. A., Interaction between sodium, calcium and magnesium chlorides affecting germination and seedling growth of *Securigera securidaca* Linn, *Ann. Arid Zone*, 25, 105, 1986.

91. **Khan, M. A., Weber, D. J., and Hess, W. M.**, Elemental distribution in seeds of the halophytes *Salicornia pacifica* var. *utahensis* and *Atriplex canescens*, *Am. J. Bot.*, 72, 1672, 1985.
92. **Khan, M. A., Weber, D. J., and Hess, W. M.**, Elemental comparmentalization in seeds of *Atriplex lentigula* and *Atriplex confertifolia*, *Great Basin Nat.*, 47, 91, 1987.
93. **Booth, D. T.**, Seedbed ecology of winterfat: cations in diaspore bracts and their effect on germination and early plant growth, *J. Range Manage.*, 10, 178, 1989.
94. **Jeschke, W. D. and Wolf, O.**, Effect of NaCl salinity on growth, development, ion distribution, and ion translocation in castor bean *(Ricinus communis)*, *J. Plant Physiol.*, 132, 45, 1988.
95. **Poulin, G., Bourque, D., Eid, S., and Jankowski, K.**, Composition chimique de *Salicornia europaea* L., *Nat. Can.*, 105, 473, 1978.
96. **Hocking, P. J.**, Salt and mineral nutrient levels in fruits of two strand species, *Cakile maritima* and *Arctotheca populifolia*, with special reference to the effect of salt on the germination of *Cakile, Ann. Bot.*, 50, 335, 1982.
97. **Ignaciuk, R. and Lee, J. A.**, The germination of four annual strand-line species, *New Phytol.*, 84, 581, 1980.
98. **Beadle, N. C. W.**, Studies in halophytes. I. The germination of the seed and establishment of the seedlings of five species of *Atriplex* in Australia, *Ecology*, 33, 49, 1952.
99. **Binet, P.**, Etude de quelques aspects physiologiques de la germination chez *Atriplex tornabeni* Tin., *Bull. Soc. Bot. Nord Fr.*, 18, 40, 1965.
100. **Cornelius, D. R. and Hylton, L. O.**, Influence of temperature and leachate on germination of *Atriplex polycarpa*, *Agronomy J.*, 61, 209, 1969.
101. **Sankary, M. N. and Barbour, M. G.**, Autecology of *Atriplex polycarpa* from California, *Ecology*, 53, 1155, 1972.
102. **Koller, D.**, Germination-regulating mechanisms in some desert seeds. IV. *Atriplex dimorphostegia* Kar. et Kir., *Ecology*, 38, 1, 1957.
103. **Kadman-Zahavi, A.**, Notes on the germination of *Atriplex rosea*, *Bull. Res. Counc. Isr.*, 4, 375, 1955.
104. **Fernandez, G., Johnston, M., and Olivares, A.**, Rol del pericarpio de *Atriplex repanda* en la germinacion. III. Estudio histologico y quimico del pericarpio, *Phyton*, 45, 165, 1985.
105. **Fernandez, G., Olivares, A., Johnston, M., and Contreras, P.**, Rol del pericarpio de *Atriplex repanda* en la germination, *Phyton*, 46, 19, 1986.
106. **Binet, P.**, La germination des semences des halophytes, *Bull. Soc. Fr. Physiol. Vegetale*, 10, 253, 1965.
107. **Binet, P.**, Dormances et aptitude a germer en milieu sale chez les halophytes, *Bull. Soc. Fr. Physiol. Veget.*, 14, 115, 1968.
108. **Bewley, J. D. and Black, M.**, *Physiology and Biochemistry of Seeds in Relation to Germination*, Springer-Verlag, Berlin, 1982, 375.
109. **Harper, J. L.**, *Population Biology of Plants*, Academic Press, New York, 1977.
110. **Binet, P.**, La dormance des semences de *Suaeda vulgaris* Moq. et de *Suaeda macrocarpa* Moq., *Bull. Soc. Bot. Fr.*, 107, 159, 1960.
111. **Ungar, I. A.**, An ecological study of the vegetation of the Big Salt Marsh, Stafford County, Kansas, *Univ. Kansas Sci. Bull.*, 46, 1, 1965.
112. **Binet, P.**, Dormance primaire et secondaire des semences de *Triglochin maritimum* L.: Action du froid et de lumiere, *Bull. Soc. Linn. Normandie*, 10, 131, 1959.
113. **Malcoste, R.**, Germination des semences d'*Obione portulacoides* Moq., *Bull. Soc. Linn. Normandie*, 3, 9, 1962.
114. **Binet, P.**, Les semences de *Triglochin palustre* et de *Triglochin maritimum* L.: Etude comparee de leur germination, *Bull. Soc. Linn. Normandie*, 2, 148, 1961.
115. **Binet, P.**, Induction d'une dormance chez les graines de *Juncus maritimus* Lamk. Role protecteur de l'eau de mer, *Rev. Gen. Bot.*, 69, 54, 1962.
116. **Kaufmann, M. R. and Ross, K. J.**, Effects of water potential on germination of lettuce, sunflower, and citrus seeds, *Can. J. Bot.*, 47, 1761, 1970.
117. **Hegarty, T. W. and Ross, H. A.**, Use of growth regulators to remove the differential sensitivity to moisture stress of seed germination and seedling growth in red clover *(Trifolium pratense* L.), *Ann. Bot.*, 43, 657, 1979.
118. **Ross, H. A. and Hegarty, T. W.**, Action of growth regulators on lucerne germination and growth under water stress, *New Phytol.*, 85, 495, 1980.
119. **Bozcuk, S.**, Effects of kinetin and salinity on germination of tomato, barley and cotton seeds, *Ann. Bot.*, 48, 81, 1981.
120. **Ungar, I. A. and Binet, P.**, Factors influencing seed dormancy in *Spergularia media, Aquatic Bot.*, 1, 45, 1975.
121. **Khan, M. A. and Ungar, I. A.**, The role of hormones in regulating the germination of polymorphic seeds and early seedling growth of *Atriplex triangularis* under saline conditions, *Physiol. Plant.*, 63, 109, 1985.

122. **Ungar, I. A.,** Salinity, temperature, and growth regulator effects on seed germination of *Salicornia europaea* L., *Aquatic Bot.,* 3, 329, 1977.

123. **Amen, R. D., Carter, G. E., and Kelly, R. J.,** The nature of seed dormancy and germination in the salt marsh grass *Distichlis spicata, New Phytol.,* 69, 1005, 1970.

124. **Okusanya, O. T. and Ungar, I. A.,** The effects of time of seed production on the germination response of *Spergularia marina, Physiol. Plant.,* 59, 335, 1983.

125. **Ungar, I. A.,** Population characteristics, growth, and survival of the halophyte *Salicornia europaea, Ecology,* 68, 569, 1987.

126. **Leck, M. A.,** Wetland seed banks, in *Ecology of Soil Seed Banks,* Leck, M. A., Parker, V. T., and Simpson, R. L., Eds., Academic Press, New York, 1989, Chap. 13.

127. **Josselyn, M. N. and Perez, R. J.,** Sediment characteristics and vegetation colonization, in *The Hayward Regional Shoreline Marsh Restoration,* Niesen, T. and Josselyn, M. N., Eds., Tech. Rep. 1, Tiburon Center for Environmental Studies, Tiburon, CA, 1981, 7.

128. **Hopkins, D. R. and Parker, V. T.,** A study of the seed bank of a salt marsh in northern San Francisco Bay, *Am. J. Bot.,* 71, 348, 1984.

129. **Jerling, L.,** Composition and viability of the seed bank along a successional gradient on a baltic sea shore meadow, *Holarctic Ecol.,* 6, 150, 1983.

130. **Milton, W. E. J.,** The occurrence of buried viable seeds in soils at different elevations and on a salt marsh, *J. Ecol.,* 27, 149, 1939.

131. **Hartman, J. M.,** Recolonization of small disturbance patches in a New England salt marsh, *Am. J. Bot.,* 75, 1625, 1988.

132. **Ungar, I. A. and Riehl, T. E.,** The effect of seed reserves on species composition in zonal halophyte communities, *Bot. Gaz.,* 141, 447, 1980.

133. **Smith, L. M. and Kadlec, J. A.,** Seed banks and their role during drawdown of a North American marsh, *J. Appl. Ecol.,* 20, 673, 1983.

134. **Jefferies, R. L., Jensen, A., and Bazely, D.,** The biology of the annual *Salicornia europaea* agg. at the limits of its range in Hudson Bay, *Can. J. Bot.,* 61, 762, 1983.

135. **Jefferies, R. L., Davy, A. J., and Rudmik, T.,** Population biology of the salt marsh annual *Salicornia europaea* agg., *J. Ecol.,* 69, 17, 1981.

136. **Wertis, B. and Ungar, I. A.,** Seed demography and seedling survival in a population of *Atriplex triangularis* Willd., *Am. Midl. Nat.,* 116, 152, 1986.

137. **Philipupillai, J. and Ungar, I. A.,** The effect of seed dimorphism on the germination and survival of *Salicornia europaea* L. populations, *Am. J. Bot.,* 71, 542, 1984.

138. **Ungar, I. A.,** Species-soil relationships on sulfate dominated soils in South Dakota, *Am. Midl. Nat.,* 83, 343, 1970.

139. **Ungar, I. A.,** Seed bank investigations with soils from an inland salt pan, *Bull. Ecol. Soc. Am.,* 60, 134, 1979.

140. **Engel, J.,** The ecology of seed banks of the upland-salt marsh transition zone, M.S. thesis, Rutgers University, New Brunswick, NJ, 1983.

141. **Hartman, J. M. and Engler, M.,** Density effects on growth and survival of *Salicornia bigelovii* and *S. europaea, Biol. Bull.,* 163, 365, 1982.

142. **Van der Valk, A. G. and Pederson, R. L.,** Seed banks and the management and restoration of natural vegetation, in *Ecology of Soil Seed Banks,* Leck, M. A., Parker, V. T., and Simpson, R. L., Eds., Academic Press, New York, 1989, Chap. 15.

143. **Hutchings, M. J. and Russel, P. J.,** The seed regeneration dynamics of an emergent salt marsh, *J. Ecol.,* 77, 615, 1989.

144. **Ungar, I. A.,** A significant seed bank for *Spergularia marina, Ohio J. Sci.,* 88, 200, 1988.

145. **Hootsmans, M. J. M., Vermaat, J. E., and Van Vierssen, W.,** Seed bank development, germination and early seedling survival of two seagrass species from the Netherlands: *Zostera marina* L. and *Zostera noltii* Hornem, *Aquatic Bot.,* 28, 275, 1987.

146. **Ellison, A. M.,** Effects of competition, disturbance, and herbivory on *Salicornia europaea, Ecology,* 68, 576, 1987.

147. **Bertness, M. D., Wise, C., and Ellison, A. M.,** Consumer pressure and seed set in a salt marsh perennial plant community, *Oecologia,* 71, 190, 1987.

148. **Leck, M. A., Parker, V. T., and Simpson, R. L.,** *Ecology of Soil Seed Banks,* Academic Press, New York, 1989.

149. **Ungar, I. A.,** Effects of the parental environment on the temperature requirements and salinity tolerance of *Spergularia marina* seeds, *Bot. Gaz.,* 149, 432, 1988.

150. **Salisbury, E. J.,** *Spergularia salina* and *Spergularia marginata* and their heteromorphic seeds, *Kew Bull.,* 1, 41, 1958.

151. **Sterk, A.,** Biosystematic studies on *Spergularia media* and *S. marina* in the Netherlands. II. The morphological variation of *S. marina, Acta Bot. Neerl.,* 18, 325, 1969.

152. **Telenius, A. and Torstensson, P.**, The seed dimorphism of *Spergularia marina* in relation to dispersal by wind and water, *Oecologia*, 80, 206, 1989.

153. **Black, J. N.**, Competition between plants of different initial seed size in swards of subterranean clover (*Trifolium subterraneum*) with particular reference to leaf area and the light microclimate, *Aust. J. Agric. Res.*, 9, 299, 1958.

154. **Baskin, J. M. and Baskin, C. C.**, Influence of germination date on survival and seed production in a natural population of *Leavenworthia stylosa*, *Am. Midl. Nat.*, 88, 318, 1972.

155. **Baker, H. G.**, The evolution of weeds, *Annu. Rev. Ecol. Syst.*, 5, 1, 1974.

156. **Cook, R. E.**, Pattern of juvenile mortality and recruitment in plants, in *Topics in Plant Population Biology*, Solbrig, O. T., Jain, S., Johnson, G. B., and Raven, P. H., Eds., Columbia University Press, New York, 1979, 207.

157. **Cideciyan, G. P. and Malloch, A. J.**, Effects of seed size on the germination, growth, and competitive ability of *Rumex crispus* and *Rumex obtusifolius*, *J. Ecol.*, 70, 227, 1982.

158. **Myers, B. A. and Couper, D. I.**, Effects of temperature and salinity on the germination of *Puccinellia ciliata* (Bor) cv. Menemen, *Aust. J. Agric. Res.*, 40, 561, 1989.

159. **Ungar, I. A.**, The effect of salinity and temperature on seed germination and growth of *Hordeum jubatum*, *Can. J. Bot.*, 52, 1357, 1974.

160. **MacKay, J. B. and Chapman, V. J.**, Some notes on *Suaeda australis* Moq. var. *nova zelandica* va. nov. and *Mesembryanthemum australe* So. ex Forst F., *Trans. R. Soc. N.Z.*, 82, 41, 1954.

161. **Lesko, G. L. and Walker, R. B.**, Effect of seawater on seed germination in two Pacific atoll beach species, *Ecology*, 50, 730, 1969.

162. **Mooring, M. T., Cooper, A. W., and Seneca, E. D.**, Seed germination response and evidence for height ecophenes in *Spartina alterniflora* from North Carolina, *Am. J. Bot.*, 58, 48, 1971.

163. **Williams, M. D. and Ungar, I. A.**, The effect of environmental parameters on the germination, growth, and development of *Suaeda depressa* (Pursh) Wats., *Am. J. Bot.*, 59, 912, 1972.

164. **Austenfeld, F.-A.**, Nutrient reserves of *Salicornia europaea* seeds, *Physiol. Plant.*, 68, 446, 1986.

165. **Khan, M. A. and Ungar, I. A.**, The effect of salinity and temperature on the germination of polymorphic seeds and growth of *Atriplex triangularis*, *Am. J. Bot.*, 71, 481, 1984.

166. **Myers, B. A. and Morgan, W. C.**, Germination of the salt tolerant grass *Diplachne fusca*. II. Salinity responses, *Aust. J. Bot.*, 37, 239, 1989.

Chapter 3

GROWTH

I. INTRODUCTION

Growth of halophytes in several genera, including *Salicornia, Suaeda,* and *Atriplex,* is stimulated by some level of salinity (Figure 1).[1-6] However, most species of halophytic flowering plants are inhibited by salt increments, with none making maximal growth at seawater concentrations (Table 1). The characteristic that distinguishes halophytes from glycophytes is that species in the latter category have relatively low salt-tolerance limits, usually less than 1.0% total soluble salts, while halophytes can tolerate hypersaline conditions ranging up to >5% total soluble salts for at least short periods of time. Salinity tolerance varies at different stages of plant development. Seeds of halophytes may remain dormant at salinity levels two or three times that of seawater and then germinate later when soil salinity concentrations are reduced. Seedlings are often reported to be the most sensitive stage of development, because young roots are exposed to the higher salinity levels in the surface few centimeters of soil. Therefore, when describing the salt tolerance of an organism one must be careful to define the stage of development being tested, since the level of salt tolerance and growth responses may differ significantly at the various developmental stages.

The inhibition of plant growth under saline conditions may be due to osmotic effects or the effects of a specific ion toxicity. Lagerwerff[7] and Billard et al.[8] concluded from reviews of the literature that the level of salt tolerance of a plant should be defined in terms of the major ecological conditions influencing growth during the period that plants are being exposed to conditions of salt stress, because the environment in which a plant is found growing may strongly influence its optimal growth parameters and degree of salt tolerance.

II. EFFECT OF SALINITY ON PLANT GROWTH

Partridge and Wilson[9] determined that the salt tolerance of 29 halophyte species decreased from the lower to the upper marsh, which correlated well with changes in field salinity conditions. Some species such as *Suaeda novae zelandiae* and *Salicornia quinquefolia* maintained 50% of maximal yield at salinities of >3.5% NaCl, whereas for 22 species death occurred at a salinity below 3% NaCl. The 50% growth limit was reached at below 2.0% NaCl for most species. Best dry mass yields were obtained for all species, except *Suaeda novae zelandiae* (optimum at 1.5% NaCl) at below 1% NaCl, and for half of the species investigated it was reached in the nutrient controls.[9]

Clough[10] reported that both *Avicennia marina* and *Rhizophora stylosa* from Cape Ferguson (Australia) were found to have their optimal biomass production at 25% seawater concentrations. Dry mass yields for both of the latter species were higher at full-strength seawater concentrations than that of controls in freshwater. Burchett et al.[11] also reported that the dry mass production in the two mangrove species *Avicennia marina* and *Aegiceras corniculatum* was maximal in 25% seawater. Growth of *Avicennia marina* was slightly reduced (11%) in 100% seawater from the tap water control, with most of the decrease being due to a 38% decrease in root dry mass. *Aegiceras corniculatum* had a 31% decrease in dry mass when grown in seawater compared with the tap water controls, with the majority of this decrease being accounted for by a drop in leaf dry mass. Root respiration did not vary significantly in the seawater treatments. Reduced root dry mass for both mangrove species might be the result of a reduced supply of substrate from the leaves to the roots.[11]

A tropical leguminous halophyte *Canavalia obtusifolia* from the coast of Cameroon behaved like a facultative halophyte.[12] After 7 weeks of growth, the total biomass was found

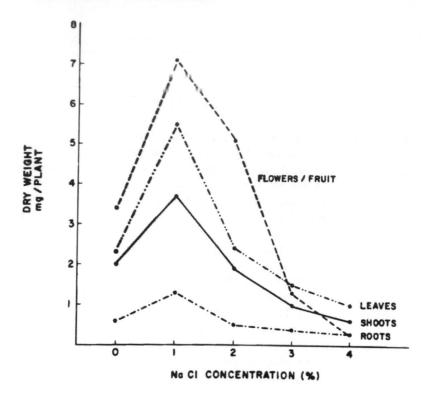

FIGURE 1.   The effect of salinity on the mean dry mass (mg/plant) of *Suaeda depressa* grown at different salinities. (From Williams, M. D. and Ungar, I. A., *Am. J. Bot.*, 59, 912, 1972. With permission.)

to decrease with each salt increment from 25 to 250 mM NaCl. Dry mass production at 250 mM NaCl was 50% of that for the controls treated with nutrient solutions. Sodium content of leaves was higher than that of roots, while K remained relatively constant in leaves at all NaCl concentrations. Deficiencies in K were therefore not responsible for the reduction in growth of plants at higher salinities.[12] Hocking[13] reported that *Cyperus involucratus*, an emergent aquatic plant from Australia, was inhibited by increases in medium salinity, with dry mass production in 150 mM NaCl being 27% of that for treatments with 1 mM NaCl. In NaCl treatments above 25 mM, both roots and culms had higher Na and Cl concentrations than leaves, indicating that *C. involucratus* was capable of sequestering these elements in its roots. Relatively high K concentrations were maintained under all NaCl treatments, also indicating that K deficiencies were not limiting growth.

Barbour[14] reported that five salt marsh species, *Mesembryanthemum chilense*, *Salicornia virginica*, *Frankenia grandifolia*, *Jaumea carnosa*, and *Distichlis spicata*, were inhibited by salinity increments ranging from 1/9 strength seawater (0.5% total salts) to 8/9 strength seawater (2.2% total salts). All of the species studied had optimal growth at 0.1% total salts, with none surviving in the 2.2% total salt treatment after 4 weeks of subirrigation. It was concluded that none of the species studied was an obligate halophyte, because there was no salt stimulation of growth and none of the species required seawater for normal growth.[14,15] The New Zealand coastal species *Tetragonia trigyna* also had its highest dry mass production in nutrient solution controls, but was able to tolerate salinities of up to 300 mM NaCl (Table 1).[16] Dry mass production was reduced from 2.0 g/plant in controls to 0.64 g/plant in the 300 mM NaCl treatment, decreasing to only 43.5% of the control in leaves, 10.2% in stems, and 33.3% in roots (Table 2).

TABLE 1
The Effect of Salinity on the Dry Mass Production of Halophytes
(g/plant)

Species	Salinity (%)						Ref.
	0.0	1.0	2.0	3.0	4.0	5.0	
Aster tripolium	21.1	80.4	53.6	—	—	—	31
Atriplex gmelini	4.0	2.5	—	1.0	—	—	19[a]
A. hastata	20.0	7.0	8.0	0.5	—	—	20[a]
A. inflata	9.4	9.3	8.7	6.6	—	—	17
A. nummularia	22.8	18.5	12.7	8.8	4.2	1.9	23
A. triangularis	12.4	10.3	5.6	4.3	—	—	106
A. vesicaria	23.0	20.0	11.0	7.0	—	—	21[a]
Distichlis spicata	0.2	0.2	0.2	0.1	—	—	66
Hordeum jubatum	0.9	0.2	0.0	—	—	—	107
Phragmites communis	0.5	—	0.5	0.4	—	—	69
Salicornia brachystachya	0.4	1.3	1.1	0.5	0.3	—	52
S. europaea	0.3	0.9	0.9	0.5	—	—	45
S. herbacea	0.8	7.3	10.6	—	—	—	31
S. patula	0.4	2.2	1.5	0.7	0.5	—	52
Selliera radicans	13.5	11.5	6.0	2.5	1.7	1.7	9[a]
Spartina alterniflora	1.0	0.8	0.5	0.2	—	—	66
S. anglica	18.0	30.0	23.0	12.0	7.0	4.0	9[a]
S. cynousoroides	1.9	1.4	0.9	0.4	—	—	66
Suaeda maritima	0.1	0.3	0.3	—	0.2	—	54[a]
S. monoica	4.5	—	—	7.0	—	2.0	27[a]
S. salsa	0.9	2.1	1.6	—	—	—	32[a]
Tetragonia trigyna	2.0	1.1	0.6	—	—	—	16

[a] Values estimated from graphs.

TABLE 2
Dry Weight of Plant Organs of Tetragonia trigyna
Grown for 6 Weeks on Hoagland Nutrient Solution
with Various NaCl Concentrations

NaCl (mM)	Leaves (g)	Stems (g)	Roots (g)	Total (g)
0	1.08a	0.59a	0.33a	2.00
60	0.99a	0.22b	0.24b	1.45
150	0.79b	0.13c	0.21b	1.13
300	0.47c	0.06d	0.11c	0.64

Note: Values indicated by the same letter in a column are not significantly different at $p < 0.05$.

From Watkins, C. B., Brown, J. M. A., and Dromgoole, F. I., N.Z. J. Bot., 26, 153, 1988.

III. GROWTH RESPONSES OF SPECIES OF ATRIPLEX TO SALINITY

The addition of 50 mM NaCl to the nutrient medium stimulated the growth of two Australian saltbushes, Atriplex inflata and A. nummularia.[17] Dry mass production in the 50 mM NaCl was greater than 100% of controls. Ashby and Beadle[17] determined that higher salinities up to 400 mM gave similar growth stimulation in plants, and at 600 mM growth of both *A. inflata* and *A. nummularia* was still significantly greater than in nutrient controls. *A. halimus* reached optimal growth at salinities between 1.0 and 2.0% NaCl,[18] with the maximum limits of tolerance being 3.0% NaCl. Growth of *A. gmelini* was enhanced by relatively low salt concentrations up to 50 mM NaCl, but higher concentrations of NaCl up to 500 mM were inhibitory to growth. Other salts including Na_2SO_4, KCl, or K_2SO_4 also stimulated the growth of *A. gmelini* in 50 mM treatments.[19] Biomass yields of both roots and shoots of *A. hastata* were stimulated by 100 mM NaCl, but salinities of 200 to 300 mM were inhibitory (Table 1).[20] At still higher salinities from 400 to 600 mM, growth was very strongly inhibited. In contrast, the desert xerophyte, *A. vesicaria* apparently reached its maximum yields in treatments between 20 to 50 mM NaCl, with salt increments ranging from 100 to 1000 mM NaCl showing an inhibition of growth that was linearly related to the salt level of the growth medium.[21]

Greenway[22] reported that optimal growth for *A. nummularia* occurred at between 100 and 200 mM/l NaCl, KCl, or $MgCl_2$. Dry mass at 1 and 10 mM NaCl was equivalent to that accumulated at 300 mM NaCl. Calcium chloride was determined to be more inhibitory to shoot and root growth than chloride salts of other cations. Greenway[22] concluded that the reported high calcium accumulation in the shoots of *Atriplex* was the cause of the specific toxic effect of $CaCl_2$ to *A. nummularia*. It was hypothesized that halophytes such as *A. nummularia* are electrolyte accumulators and are therefore able to maintain turgor and high growth rates at relatively high soil salinities. In contrast, salt-excluding halophytes and glycophytes have reduced growth rates at higher salinities because of their inability to accumulate solutes rapidly.

The Australian perennial shrub *A. nummularia* was found to vary in its salt tolerance at different stages of development, with maximal limits of tolerance being 3.0% NaCl for germination, 2.0% NaCl for seedlings, and 5% NaCl for adult plants (Table 3).[23] Dry mass production of shoots was reduced from 22.8 g/plant in controls to 8.8 g/plant at 3.0% NaCl, with plants at 5.0% NaCl producing only 1.9 g/plant dry mass. Adult plants were all able to survive salinities up to 4.0% NaCl, and some individuals survived at 5.0% NaCl after cultivation for 1 year.[23] Optimal dry mass production for *A. nummularia* at salinities ranging from 0 to 1.0% NaCl was at a temperature of 25°C.

A. amnicola (= *A. rhagodioides*) had maximal biomass production at salinities from 25 to 50 mM NaCl, with a reduction in biomass yield of from 85 to 90% in the 750 mM NaCl treatment (Table 4).[24] Na leaf content was as high as 300 to 400 μM/g fresh weight for plants growing at 25 mM NaCl, while for plants grown at 400 mM NaCl the oldest leaves had Na values reaching 900 to 1000 μM/g fresh weight. Concentrations of starch and sugars were similar for plants grown at 25 and 400 mM NaCl, indicating that the availability of photosynthate was not limiting growth when plants were exposed to high salinity concentrations. Aslam et al.[24] concluded that reduced growth at high salinity levels was due to a diversion of carbohydrates from cell growth into the synthesis of solutes for osmotic regulation, an inadequate rate of accumulation of osmotic solutes, or excess ion accumulation.[24]

Growth of *A. halimus* was stimulated by the addition of NaCl or KCl to the medium, but a mixture of NaCl and KCl produced a significant >100% increase in the shoot dry weight and a >90% increase in root dry weight.[25] Treatments with NaCl and KCl (−0.1 MPa) individually yielded a 50% increase in shoot dry weight when compared to controls (−0.05 MPa). Leaf sap osmotic potentials increased from −0.5 MPa in the controls to −0.8 MPa in the 206 meq/l treatment of NaCl or KCl. Under all but the highest salinity treatment, leaf sap osmotic potential was significantly lower than that of the medium osmotic concentration. However, plant leaves exposed to mixtures of NaCl and KCl did not reach levels as negative as that in single salt treatments.[25]

Several species of *Atriplex*, *A. nitens*, *A. calotheca*, and *A. halimus*, were grown at

TABLE 3
Effect of NaCl on Seed Germination and Growth of *Atriplex nummularia*

	NaCl (%)					
	0	1.0	2.0	3.0	4.0	5.0
Seed germination (%)	96 ± 3	80 ± 11	28 ± 10	16 ± 6	0	0
Seedling establishment (%)	96 ± 3	30 ± 10	10 ± 4	0	0	0
Dry mass (g/shoot)	22.8	18.5	12.7	8.8	4.2	1.9

From Uchiyama, Y., *Tech. Bull. Trop. Agric. Res. Ctr. Japan*, 22, 1, 1987.

TABLE 4
Ethanol-Insoluble Dry Weight of Different Plant Parts of *Atriplex amnicola* Grown for 32 d at 25 or 400 mol m^{-3} NaCl

	Ethanol-insoluble dry weight (mg plant^{-1})		Weight as % of whole plant	
Plant part	25 mol m^{-3} NaCl	400 mol m^{-3} NaCl	25 mol m^{-3} NaCl	400 mol m^{-3} NaCl
Fully expanded leaves (main stem)	114	71	19.3	20.8
Recently expanded leaves (main stem)	7.4	6.2	1.26	1.82
Rapidly expanding leaves (main stem)	1.3	0.72	0.22	0.21
Buds (main stem)	0.19	0.25	0.03	0.07
Stems (main stem)	85	40	14.3	11.8
Branches	255	112	43.1	33.0
Expanded root tissue	129	110	21.8	32.7
Root tips	0.37	0.15	0.06	0.04
Total plant	592	341	—	—

From Aslam, Z., Jeschke, W. D., Barrett-Lennard, E. G., Setter, T. L., Walkin, E., and Greenway, H., *Plant Cell Environ.*, 9, 571, 1986.

salinities ranging from 0 to 750 m*M* NaCl.[26] The three species varied from one another in their degree of salt tolerance. *A. halimus* had the least decrease in dry mass production (40%), *A. calotheca* (67%) and *A. nitens* (80%) at 750 m*M* NaCl, but all three species were able to survive in the highest salt treatment. The Na and Cl contents of tissues were relatively high in the three species, ranging from 52.9 mg/g dry weight for *A. calotheca* to 127 mg/g dry weight for *A. halimus* at 500 m*M* NaCl.[26] Both *A. spongiosa* and *Suaeda monoica* were able to grow in medium containing greater than 600 m*M* NaCl, with dry mass production decreasing by 50% at 800 m*M* NaCl for *A. spongiosa* and at 1350 m*M* NaCl for *S. monoica*.[27] Growth of *A. hortensis* var. *cupreata* was inhibited by increments of NaCl from 100 to 500 m*M*.[28] Dry mass production of *A. hortensis* at 500 m*M* NaCl was about 25% of that for nonsaline controls.

IV. SALT STIMULATION OF GROWTH IN *SALICORNIA* and *SUAEDA*

A high concentration of NaCl is probably not essential for the optimal growth of most halophytes, but some species such as *Suaeda maritima* and *Salicornia herbacea* grew larger and benefited from NaCl concentrations above the minimal amounts required as a micronutrient in plants.[29] Nutrient deficiencies could be a factor controlling the growth of annuals on the salt marsh. Fertilization experiments with *Suaeda maritima* indicated upper marsh untreated controls had yields five times lower (23 mg/plant) than lower marsh plants, and that a nitrogen + phosphorus fertilization produced yields in the upper marsh (108 mg/plant) that were nearly equivalent to biomass production in the lower marsh (122 mg/plant).[11] Similar increases in yields, up to five times that of untreated upper marsh controls, were obtained for N + P fertilization treatments with *Salicornia europaea*.

Dry mass production for *Suaeda maritima* increased to >150% of controls when salinity was increased from 10 to 170 mM NaCl.[30] Further increases in salinity from 170 to 680 mM caused a gradual decrease in biomass production to less than 50% of the control value. The greatest increase in biomass was found at 30 mM NaCl, with further increments in yield up to the maximal at 170 mM NaCl. Although maximal biomass yields were produced at 170 mM NaCl, 90% of this increase was found at NaCl concentrations that were only 10% of this concentration.[30] Dry mass production of *Salicornia herbacea* was stimulated at NaCl concentrations up to 1.5%.[31] Net photosynthesis on a dry mass basis reached its peak at 2.0% NaCl for *S. herbacea*, while for *Aster tripolium* maximum photosynthetic levels on a dry mass basis were reached in the nonsaline controls. Highest dry mass yields for *A. tripolium* were obtained in the 1.0% NaCl treatments (Table 1), where there was a 30% decrease in photosynthesis from controls on a dry weight basis, but equal levels of net photosynthesis to controls on a leaf area basis.[31]

A comparison of growth responses was made between a Normandy (France) coastal salt marsh species *Suaeda maritima* var. *flexilis* and an inland saline area species from the Azerbagian steppes (Iran) *S. maritima* var. *salsa*.[32] Both species of *Suaeda* had optimal shoot elongation at salinities above that of the nutrient solution controls, with an optimum at 0.1% NaCl for var. *salsa* and equivalent stimulation of shoot elongation in 0.1 and 0.75% NaCl for var. *flexilis*. Dry mass production reached an optimal level at 0.75% NaCl for var. *flexilis* 3960 mg/plant compared with 830 mg/plant in the controls, while biomass production for var. *salsa* reached its optimum at 0.1% NaCl 5236 mg/plant compared with 779 mg/plant in the nutrient controls.[32]

Biomass production of *S. aegyptiaca* and *S. monoica* from salt flats on the Dead Sea (Israel) was stimulated 10 and 5 times, respectively, by 150 mM NaCl, whereas plants treated with isotonic solutions of KCl had growth responses similar to freshwater controls.[33] Both NaCl and KCl stimulated a 100% increase in fresh weight production of *Salsola kali*. Growth enhancement in the *Suaeda* spp. was apparently due to a specific effect of Na, but Na and K did not play equivalent roles in stimulating growth as these ions did in *Salsola*.[33] The biomass yields of *S. aegyptiaca* increased gradually with NaCl treatments up to a 125-meq/l optimum.[34] Potassium chloride was not stimulatory, and Na_2SO_4 stimulated growth but not to the same level as NaCl. It was concluded that Na had a specific promotive effect on *S. aegyptiaca* at the macronutrient level, which was not due to a simple replacement of K by Na, since increasing K concentrations did not further stimulate growth.[33] These data indicate that *S. aegyptiaca* responded in a similar manner to other Na-requiring annual species of *Suaeda*, including *S. depressa* (Figure 1),[35] *S. maritima*,[36] and *Salicornia fruticosa*.[37]

Little research has been done to determine why some species are inhibited by the lack of NaCl in the medium. However, in a recent investigation by Hagege et al.,[38] it was reported that *Suaeda maritima* growth was limited in nonsaline and low-salt (17 mM NaCl) treatments because growth was limited for new internodes. Plants exposed to no NaCl in the medium produced more ethylene than 130 mM NaCl-treated plants, which could inhibit shoot elongation. Biomass production in nonsaline controls was 20 times less than that in the 130 mM NaCl treatments. Lignin content of shoots was higher in the nonsaline controls and decreased with increased salinity. Higher peroxidase activity may be associated with the catabolism of auxin in *S. maritima*, and lower auxin levels as well as the increase in ethylene production could account for the reduced growth in nonsaline controls.[38]

TABLE 5
The Effect of Salinity on Growth Parameters
of *Salicornia europaea*[a]

NaCl (mM)	Dry mass (mg)	Height (cm)	Nodes (no.)	Lateral branches (no.)
0.0	338	4	5	3
170.0	922	11	9	9
340.0	881	9	8	7
510.0	484	5	6	2

[a] Each value represents a mean of ten plants.

From Ungar, I. A., *Bull. Soc. Bot. Fr. Actualites Bot.*, 124, 95, 1978.

A number of researchers have reported that species of *Salicornia* have increased biomass production with salt increments in the growth medium ranging from 170 to 340 mM NaCl (Tables 1 and 5).[39-45] The nature of the NaCl-stimulated growth is not clearly understood, but Caldwell[46] has suggested that the function of high salinity may be to provide the rapidly permeating electrolytes that are necessary for osmotic adjustment and maintenance of turgor during the growth of halophytes. The distribution of *S. europaea* is often limited to the most saline zones of coastal marshes[47] and inland salt marshes.[48] Hypersaline conditions in the field sites, ranging from 500 to 1000 mM total salts, were found to inhibit the biomass production of *S. europaea*.[49] Dry mass production of *S. europaea* on an Ohio salt pan was found to vary considerably with the salt content of soils.[50] Highest dry mass yields were obtained in the zone with the lowest soil salinity levels (Figure 2). Growth of *S. herbacea* plants from a Japanese salt marsh population was stimulated by some seawater concentrations when compared with nonsaline controls.[51] Maximum biomass yields were obtained in the treatment with a 50% seawater concentration. Growth responses for two species, *S. brachystachya* and *S. patula*, collected from salt marshes in the Mediterranean region of France were investigated using salinity levels that ranged from 0.005 to 3.5% NaCl.[52] Optimal biomass yield for *S. patula* was at 0.8% NaCl (2197 mg/plant) compared with 525 mg/plant at 3.5% NaCl (Table 6). *S. brachystachya* had its highest biomass production at 0.3% NaCl (1652 mg/plant). Controls for the two species had low biomass yields, averaging about 384 mg/plant in the 0.005% treatment. Grouzis et al.[52] concluded that both species of *Salicornia* were obligate halophytes and that they required a certain level of salt in the environment to attain full development.

The optimal conditions for hypocotyl elongation in *S. herbacea* seedlings were at 100 to 200 mM NaCl.[53] In the light, NaCl and GA_3 synergistically promoted hypocotyl elongation when given to seeds simultaneously. The promotive action of NaCl could be replaced by the inorganic salt KCl but not by mannitol. Kawasaki et al.[53] concluded that the promotive effect of NaCl for *S. herbacea* hypocotyl elongation in the dark resulted from a lowering in water potential due to an uptake of NaCl. For the action of GA_3 to be effective, NaCl is required to loosen cell walls.

Flowers[54] determined that the dry mass production of the halophyte *Suaeda maritima* reached optimal levels in treatments receiving between 170 and 340 mM NaCl (2.5 times greater biomass than that of nutrient controls). Biomass yields in 680 mM NaCl were 1.5 times greater than the nutrient controls. Stimulation of growth in *S. maritima* is not primarily due to ion exclusion, because these plants can tolerate Na tissue levels reaching 600 meq/g dry mass in their shoots. Flowers[54] hypothesized that the increased growth at relatively high salinities may be related to the reversal of ATPase activity by Na. The salt tolerance

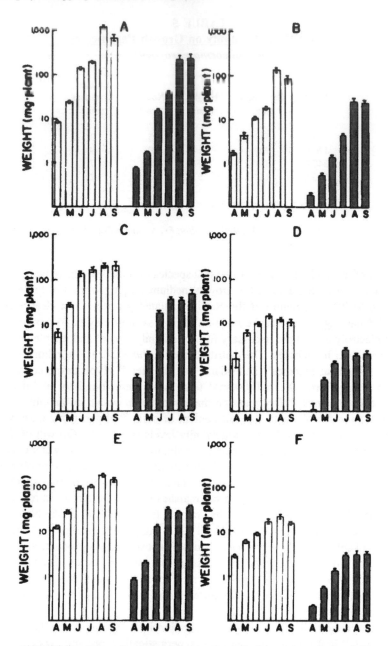

FIGURE 2.   Seasonal patterns of shoot and root fresh mass (open bar) and dry mass (closed bar) for *Salicornia europaea* plants. (A) Tall *Salicornia* shoots; (B) tall *Salicornia* roots; (C) edge *Salicornia* shoots; (D) edge *Salicornia* roots; (E) submerged *Salicornia* shoots; (F) submerged *Salicornia* roots. (From Riehl, T. E. and Ungar, I. A., *Oecologia*, 54, 193, 1982. With permission.)

of *S. maritima* varied with the habitat in which the species was found growing on the Orne estuary, and Boucaud[55] determined that two varieties behaved somewhat differently. Ecologically these varieties were found in different habitats, with *S. maritima* var. *flexilis* being found in sandy habitats at the upper limits of the high marsh and *S. maritima* var. *macrocarpa* being found on the low marsh.[55,56] The mean annual Na soil content was 0.06% and Cl 0.08% in the var. *flexilis* zone (maximum levels for Na were 0.13% and Cl 0.20%) compared with 0.73% for Na and 0.82% for Cl in the soils of the var. *macrocarpa* zone (maximum

TABLE 6
The Effects of NaCl on the Dry Mass Production (mg/plant) of
Salicornia patula and *S. brachystachya* After a 66 d Growth Period

	NaCl (%)			
	0.005	0.8	1.8	3.5
Salicornia patula				
Shoot (mg)	311 ± 7.6	2058 ± 174.3	1408 ± 105.9	488 ± 41.9
Root (mg)	90 ± 2.9	139 ± 11.9	104 ± 7.6	37 ± 3.9
Total (mg)	401	2197	1512	525
S. brachystachya				
Shoot (mg)	271 ± 6.4	1170 ± 60.7	1012 ± 66.9	283 ± 16.0
Root (mg)	95 ± 2.6	137 ± 7.1	135 ± 11.0	40 ± 2.9
Total (mg)	366	1307	1147	323

From Grouzis, M., Heim, G., and Berger, A., *Oecol. Plant.*, 12, 307, 1977.

levels for Na were 1.15% and Cl 1.1%). These soil data agree well with the types of growth responses obtained for the two varieties in the laboratory.[55-59] *S. maritima* var. *flexilis* was reported to have equally high shoot length growth at salinities ranging from 0.0 to 0.75% NaCl, with reductions in shoot growth from 1.5 to 3.0% NaCl.[55] Dry mass production for var. *flexilis* followed a similar pattern to that of shoot elongation, while var. *macrocarpa* dry mass production was stimulated at salinities between 0.75 and 1.5% NaCl when compared with nutrient controls and the 2.3 to 3.0% NaCl treatments. Maximal salt tolerance limits for growth for var. *flexilis* were 2.3% NaCl and for var. *macrocarpa* 3.0% NaCl.[8] Absence of NaCl from the medium greatly restricted the growth of *S. maritima* var. *marcocarpa*, and a Ca-induced toxicity was apparent in treatments with >1.0 meq/l Ca.[59] Calcium was not found to alleviate the NaCl requirement for optimal growth of *S. maritima*, which was estimated to be 129 mM NaCl. Significantly higher levels of Ca were accumulated by *S. maritima* plants in the nonsaline controls than at higher salinities, with sodium apparently strongly interfering with Ca uptake at salinities of 129 and 390 mM NaCl.[59] In contrast, Billard et al.[8] reported that the addition of calcium chloride to the medium ameliorated the growth of *Cochlearia anglica* in 1.0% NaCl. As the salinity of the medium increased, the optimal calcium chloride content for growth increased.

V. IONIC EFFECTS AND INTERACTIONS WITH OTHER FACTORS

Growth of halophytes may be stimulated by a specific ion such as sodium or may represent a general response to the increased ionic content of the medium. Other environmntal factors may also influence the response of halophytes to salt stress. A concentration of 20 mM NaCl, KCl, and Na$_2$SO$_4$ stimulated the growth of *Sarcobatus vermiculatus*.[60] Salt increments from 63 to 1004 mM NaCl (−0.45 to −4.37 MPa) inhibited dry mass production. The ash weight of leaf samples ranged from 40 to 55% of the total dry mass of leaves for *S. vermiculatus*.

Spartina foliosa, a California coastal salt marsh species, had about 50% less dry mass production in seawater (3.3% total salts) than in the freshwater treatment, with only 39% of the plants surviving in the seawater treatment.[61] Lipid content of plants was higher in the freshwater treatment (17.9%) than in the seawater treatment (3%), while protein content of plants averaged 17.9% in freshwater and 12.4% in seawater treatments. The K to Na ratio ranged from 14:1 in freshwater- to 0.37:1 in seawater-treated *S. foliosa* plants. There was a sharp decrease in K with increases in Na to 50% seawater, while at higher seawater concentrations, K concentrations remained relatively stable.[61] Field measurements of above-ground production of *S. foliosa* and *Salicornia virginica* were made for a population from San Francisco Bay.[62] Maximal biomass yields of *Spartina foliosa* ranged from 270 to 690 g/m^2, while minimal yields (<50 g/m^2) occurred when *S. foliosa* was growing with *Salicornia virginica*. Above ground yields of *S. virginica* ranged from 550 to 960 g/m^2. The zonation of these species was probably affected by physical factors such as salinity rather than an interspecific competition, because both species decreased to minimal biomass yields at the middle of the ecotone.

The highest dry mass for *Spartina alterniflora* plants was in treatments with 0.5% NaCl and 14 mg/l NH$_4$-N.[63] Salinity concentrations of 4.0% were very inhibitory to biomass production, as were reductions in nitrogen levels to 0.14 mg/l NH$_4$-N. Haines and Dunn[63] concluded that the interaction between salinity and nitrogen may be the most significant factor in determining biomass yields of *S. alterniflora*.

Dry mass production of *Halopeplis amplexicaulis* reached its maximum 197 mg/plant at 100 mM NaCl (at 28°C and 55% relative humidity), while at 200 mM NaCl dry mass yield was only 40 mg/plant, and in nutrient controls biomass was 25 mg/plant.[64] These data indicate some salt stimulation of growth in *H. amplexicaulis* at moderate salinities, but an inhibition of growth at higher salinities. At higher relative humidity (80%), biomass production reached its maximum at 150 mM NaCl, indicating that the relative humidity of the air may interact with soil conditions to determine yields of plants.

Neales and Sharkey[65] reported that the growth of the halophyte *Disphyma australe* was stimulated by salt increments up to 200 mM NaCl. Total plant dry mass decreased about 30% at 500 mM NaCl from the optimal values, with most of the decrease being because of about a 50% reduction in leaf number at the higher salinity. Inorganic ion content of leaves ranged from 14.1% in 0 mM NaCl to 55.9% in the 500-mM NaCl treatment. The Na and Cl content of leaves increased with salinity increments up to 200 mM NaCl, while K showed a decrease in up to 200 mM NaCl.

The effect of salinity and drainage on the growth of salt marsh species *Spartina alterniflora, Distichlis spicata,* and *S. cynosuroides* was investigated by Parrondo et al.[66] The species varied in their response to salinity stress, and the relative decrease in growth from control treatments at seawater salinities (3.2% NaCl) was 80.9% in *S. cynosuroides,* 79.6% in *S. alterniflora,* and 34.2% in *D. spicata* (Table 1). Dry mass production of *S. alterniflora* was stimulated in flooded sediments, mainly due to an increase in root dry mass. However, biomass of *S. cynosuroides* decreased 45% in flooded sediments, with root biomass decreasing 41% and shoot biomass decreasing 45% in flooded treatments compared with drained sediment controls. Parrondo et al.[66] concluded that tolerance to salinity was probably not the most limiting factor to species distribution and that tolerance to flooding was a stronger factor determining the relative plant distribution in Louisiana salt marshes. Growth of four salt-excreting halophytic grasses, *S. alterniflora, S. foliosa, S. patens,* and *D. spicata,* was determined to be inhibited by higher free-water salinity levels in silt clay soils (3.2% total salts) than in clay soils (1.8% total salts). *D. spicata* and *S. alterniflora* were determined to be more salt tolerant than *S. foliosa* and *S. patens.* Potassium content of plant tissues for the four species was found to increase significantly (r = 0.89) with increasing sediment salinities.[67] *S. patens* and *S. alterniflora* were found to grow taller and produce more culms when grown in nutrient solution controls than when they were exposed to seawater (2.82% total salts).[68] Seawater caused a reduction in height growth of *S. patens* to 9.3% and culm number to 17.7% of yields from nutrient controls, while reductions in growth of *S. alterniflora* were to 39.9% for height growth and 37.1% for culm number of controls.

Phragmites communis, a species of perennial grass that grows in fresh and brackish water marshes in Japan, was not significantly inhibited in its growth at NaCl concentrations of 300 mM.[69] However, at 500 mM NaCl dry mass was reduced to 0.39 g/plant vs. 0.54

g/plant in controls and 0.55 g/plant at 300 mM NaCl. The Na content of shoots ranged from 0.05% in nonsaline controls to 1.62% in the 500 mM NaCl treatment, while root Na content increased from 0.11 to 4.36%, indicating that Na was sequestered in the roots of *P. communis*. One of the adaptations *P. communis* had to salt stress was to decrease its leaf water content, which reduced the leaf osmotic potential and maintained plant turgor as the soil water potential decreased. Plants were also efficient at excluding Na from the leaves, since even when grown at 250 mM NaCl the leaves contained only 50 mM Na.[69] Uptake experiments indicate that *P. communis* was able to retranslocate Na from the shoots to the rooting medium, which would lower the net uptake of Na in the shoots.

Increasing salinity concentrations of the medium from 0.1 to 500 mM NaCl caused a decrease in the photosynthetic rate of *Distichlis spicata* from 0.97 to 0.57 mg/m^2/s under optimal light (1200 µE/m^2/s) and temperature (35°C day/25°C night) conditions, with the relative growth rate (RGR) decreasing from 0.105 to 0.079 g/g/d.[70] At suboptimal environmental conditions (25°C day/15°C night and 600 µE/m^2/s), the effect of salinity is more dramatic, causing a reduction in photosynthesis from 0.55 to 0.26 mg/m^2/s (500 mM NaCl) and a reduction in the RGR from 0.057 to 0.019 g/g/d. Kemp and Cunningham[70] concluded that dark respiration was not affected by salinity and light and that tolerance to salt stress appeared to be related to the utilization of light energy, possibly in the process of salt excretion and ion compartmentation. Maximal leaf blade length was obtained for *D. spicata* at salinities of 1.5% total soluble salts, with leaf length increasing from about 50% in controls and 2.4% total salts to about 70% in the 1.5% salt treatment.[71] Rhizome growth was responsible for the migration of *D. stricta* from areas of low salinity to high salinity salt pan habitats.

Okusanya and Ungar[72] determined that *Spergularia marina*, *S. rupicola*, and *S. rubra* behaved like true halophytes in that dry mass production was stimulated by 10% sea salt concentrations (Figure 3). Half-strength seawater reduced plant biomass yields to about 50% of the optimum values for the three species. Additions of calcium salts to the 50% sea salt medium greatly enhanced biomass yields for *S. marina* and *S. rupicola* but did not stimulate growth in the dune species *S. rubra*.[72]

Cooper[39] reported that the tolerance of plants to salinity and waterlogging was correlated to their position along a salt marsh gradient. *Salicornia europaea* had higher dry mass production in both drained and waterlogged saline treatments (340 mM NaCl) than in nonsaline treatments, which could explain its success in the lowest portions of the salt marsh. Other species, including *Festuca rubra*, *Juncus gerardii*, and *Armeria maritima*, were inhibited by both salt treatments and waterlogging (Table 7). Some species, such as *Puccinellia maritima*, *Triglochin maritima*, and *Aster tripolium* were inhibited only by the effect of waterlogging or salinity but not both factors. *F. rubra*, *J. gerardii*, and *Armeria maritima* were the least tolerant species to salinity and waterlogging, which agrees with their natural distribution in the upper parts of the salt marsh. *P. maritima* responded to waterlogging and salinity like a pioneer salt marsh invader of the low marsh. The other species studied, *Plantago maritima*, *A. tripolium*, and *T. maritima*, responded to salinity and waterlogging in a manner that correlated well with their distribution on the marsh ecotone.[39]

Salinities up to 150 mM NaCl did not inhibit the growth of the halophyte *Glaux maritima* started from seed collected at Frisian Island (Netherlands), but 300 mM NaCl inhibited both dry mass production and leaf expansion.[73] Stolon production was inhibited by all salinity increments from 60 to 300 mM NaCl. Rozema[73] concluded that the salt tolerance of *Glaux maritima* is based on the efficiency of salt gland excretion of ions and osmotic adjustments by the accumulation of inorganic ions Na, K, and Cl. Seedling growth of *G. maritima* was inhibited by inundation at all salinities tested. Growth and hibernacle (hibernating bud) production was not inhibited significantly at 300 mM NaCl, and a possible stimulation of growth was found at 150 mM NaCl in well-drained treatments. Plants produced a mean of

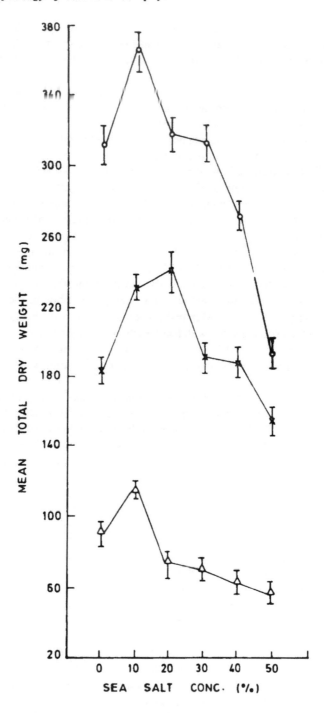

FIGURE 3.   The effect of salinity on the growth of *Spergularia rupicola* (circle), *S. marina* (cross), and *S. rubra* (triangle). Bars represent SE of mean. (From Okusanya, O. T. and Ungar, I. A., *Am. J. Bot.*, 71, 439, 1984. With permission.)

TABLE 7
Summary of an Analysis of Variance of the Effects of Waterlogging and Salinity on Dry Weight Yields of Halophytes

Species	p Values	
	Waterlogging	Salinity
Festuca rubra	0.01	0.01
Juncus gerardii	0.05	0.01
Armeria maritima	0.01	0.05
Plantago maritima	NS	0.01
Aster tripolium	NS	0.01
Triglochin maritima	0.01	NS
Puccinellia maritima	NS	0.01
Salicornia europaea	0.05	0.01

Note: NS = difference is not significant.

From Cooper, A., New Phytol., 90, 263, 1982.

3 hibernacles per treatment, which corresponded to the number of winter hibernacles found under field conditions.[74]

Salinity greater than −0.7 MPa (0.84% NaCl) negatively affected the growth of rosettes of *Centaurium littorale*, a species that occurs along the coasts of northwest Europe, while lower salinity concentrations did not inhibit dry mass production.[75] Plants subjected to treatments with 0.5 meq/l Cl had the highest dry mass in all experiments.

Handley and Jennings[28] determined the effects of specific salts NaCl, KCl, $MgCl_2$, and $CaCl_2$ on the water content of leaves of *Atriplex hortensis*. In older leaves only the monovalent cations caused an increase in leaf water content that was greater than that of control treatments. The 150-mM monovalent chloride salts caused a reduction in yields from 14.6 g/plant in controls to 7.6 g in NaCl and 10.4 g in KCl. Magnesium chloride actually stimulated dry mass yield to 19.0 g/plant.[28]

The effect of salt spray and soil salinity was determined for two coastal dune species *Cakile maritima* and *Salsola kali* and two salt marsh species *A. hastata* and *A. littoralis*.[76] Growth of the four species was not inhibited by salt spray treatments and was actually stimulated in the dune species *Salsola kali*. However, the four species differed in their response to soil salinity, with the dune species showing an inhibition in dry mass productions at 150 and 300 mM NaCl. The two species of *Atriplex* were not significantly inhibited by soil salinity treatments, reflecting their natural ecological position in salt marshes.[76]

VI. COMPARATIVE INVESTIGATIONS WITH HALOPHYTE TAXA

Investigations have been carried out to determine the difference in levels of salt tolerance of species within a plant community, species within a genus, and among populations within a species. Rozema[77] determined the effect of salinity on the growth of *Juncus* species along a salinity gradient: three were halophytic (*J. gerardii*, *J. maritimus*, and *J. alpino-articulatus* ssp. *atricapillus*) and one glycophytic (*J. bufonius* ssp. *bufonius*). All of the species investigated had their highest dry mass production under freshwater conditions. The salt marsh species *J. gerardii* and *J. maritimus* were little inhibited by salinities up to 300 mM NaCl, while *J. bufonius* and *J. alpino-articulatus* ssp. *atricapillus* were very strongly inhibited at 150 mM NaCl in dry mass production, leaf area, rhizome length, and number of shoots per plant. The latter intolerant species died in the 300 mM NaCl treatment, which explains why they do not grow in salt marsh habitats.[77] In a comparison of growth responses of *J. gerardii* with *Agrostis stolonifera*, Rozema and Blom[78] found that a 50% seawater treatment reduced the fresh weight of *A. stolonifera* by 50% when compared to the nonsaline control. Both fresh and dry mass of *J. gerardii* shoots were significantly affected by the 50% seawater treatment. However, root dry mass was not inhibited for *J. gerardii* but was significantly inhibited for *A. stolonifera* in the 50% seawater treatment.[78]

A comparison was made between the growth responses to salinity of the glycophyte *Plantago media* and two coastal halophytes (Netherlands), *P. maritima* and *P. coronopus*. Erdei and Kuiper[79] determined that *P. media* was sensitive to 25 mM NaCl, while *P. coronopus* and *P. maritima* could grow at salinities up to 150 and 300 mM NaCl, respectively. All three species were characterized by accumulation of Na in the shoot and reduced K concentrations when exposed to NaCl treatments. Sodium concentrations in the roots never exceeded 1 meq/g dry weight at the highest salt treatment, while for shoots at 300 mM NaCl, *P. maritima* contained 6 meq Na/g dry weight, at 150 mM NaCl, *P. coronopus* had 4 meq Na/g dry weight, and at 50 mM NaCl, *P. media* contained 3 meq Na/g dry weight. Roots were damaged when they accumulated Na. It was concluded that Na transport was necessary for roots to remain healthy, since cells were damaged with overloading of Na in the roots.[79]

Lasthenia glabrata, a winter annual species, has several adaptations to salt stress, including seed dormancy during the dry summer period when salinity stress is highest, seed germination and rapid growth during the period of reduced salinity (<20% seawater), and reproduction at the time of the year when salinity stress increases.[80] Sensitivity to salt stress proved to be greatest for *L. glabrata* at the earliest stages of development. Populations of the northern subspecies *L. g. glabrata* were exposed to lower salinity levels in the soil than the southern taxa *L. g. coulteri*, indicating that the former subspecies was subjected to less selection pressure. Prolonged periods of salt stress caused an acceleration and condensation of the life cycle of *Lasthenia* taxa, with an early shift to flowering and a high ratio of flowers to vegetative structures. The southern populations were more dormant at high salt concentrations, formed a seed bank, and had lower mortality of mature plants than the northern population. The southern populations also had the ability to restrict Cl uptake compared with the northern subspecies.[80]

The relative salt tolerance of varieties of *Cakile edentula* was studied at different stages of the plant's life cycle.[81] Germination of seeds for the two varieties, *Cakile edentula* ssp. *edentula* var. *lacustris* and var. *edentula*, was uninhibited by salinities up to 1% total salts, with germination percentages ranging from 83 to 100% in the different salt treatments for both varieties. Root growth of seedlings for other Great Lakes variety *(lacustris)* was inhibited by salinities of 0.1% total salts, while the Atlantic Coast variety *(edentula)* was not inhibited by salinities up to 1.0% total salts, and root growth for the latter variety was somewhat stimulated at 0.1% total salts. Similar inhibition of biomass production was found for var. *lacustris* when it was exposed to salt spray, while the coastal variety was uninhibited by up to 90 mg/dm^2/d of salt spray. It was hypothesized that the Great Lakes variety has lost the ability to tolerate both salt spray and soil salinity over the 9000 years that had separated it from the Atlantic Coast variety. Boyd and Barbour[81] concluded that the Atlantic Coast var. *edentula* was better suited for the marine beach environment than the Great Lakes var. *lacustris*.

Comparisons were made of the salt tolerance of *Agrostis stolonifera* plants from maritime and inland populations. Tiku and Snaydon[82] determined that an increase in salinity from 0 to 0.5% NaCl caused a decrease in root growth for plants from all populations. However, the Na content of native soils was highly correlated ($r = -0.73$) with the reduction in biomass yield under NaCl treatments. The Na content of plant shoots was higher for plants

TABLE 8
Live Material Weight as a Percentage of the Plant
Material of New Zealand Coastal Plant Species Grown at
Different Salinites

Species	Salt treatment (%)			
	0.0	0.5	1.0	1.5 or 2.0
Acaena anserinifolia	98.5	0.0	0.0	0.0
Bromus diandrus	97.4	84.5	5.5	0.0
Colobanthus muelleri	66.4	0.0	0.0	0.0
Coprosma acerosa	100.0	92.7	36.3	0.0
Craspedia uniflora	77.8	57.7	25.2	0.0
Cyperus ustulatus	99.0	83.8	76.5	32.0
Desmoschoenus spiralis	84.0	51.2	47.6	4.6
Elymus farctus	85.8	88.8	74.7	63.0
Hydrocotyle novae-zelandiae	96.1	91.6	23.3	0.0
Lagenifera pumila	94.6	43.3	25.7	0.0
Scirpoides nodosa	98.4	95.0	97.0	77.4
Silene gallica	100.0	72.7	27.3	0.0

From Sykes, M. T. and Wilson, J. B., *Acta Bot. Neerl.*, 38, 173, 1989.

collected in low-Na field sites, ranging from 10.6 mg Na/gdw at the Dovey Estuary (200.0 mg Na/100 g soil) to 28.3 mg Na/gdw at Sonning Chalk (15.9 mg Na/100 g soil), indicating that plants from less saline sites were not able to control Na transport to shoots as well as maritime plants.[82]

Seedlings of coastal species were exposed to salt spray from 1982 to 1984.[83] The species, *Myrica pensylvanica* and *Triplasis purpurea*, could tolerate salt spray, but their seedlings were never found in the more seaward grass zone. The grasses *Uniola paniculata* and *Spartina patens* were tolerant of conditions in the more seaward forb zone, but seedlings rarely occurred in that area. Tyndall et al.[83] concluded that other factors besides salt spray were controlling the seaward migration of species on the North Carolina coast.

Sykes and Wilson[84] determined that some New Zealand dune species, such as *Ammophila arenaria*, *Austrofestuca littoralis*, *Scirpoides nodosa*, and *Elymus farctus*, exhibited root salinity tolerance up to 1.5% NaCl (Table 8). Other species, such as *Acaena anserifolia*, *Colobanthus muelleri*, and *Gunnera albocarpa*, were intolerant to salinities of 0.5% NaCl. All 30 species investigated had a reduction in dry mass at some soil salinity level, but the inhibitory salinity level ranged from 0.25% NaCl in *C. muelleri* to 1.5% NaCl in *S. nodosa* (Table 8). However, Sykes and Wilson[85] determined no significant correlation between a species distribution in the field and its relative salt tolerance from laboratory experiments. No significant correlation was found between a species' salt spray tolerance and soil salinity tolerance, with some species being tolerant to both, some to neither, and others to only one of the two types of salt stress.[84,85]

Cartica and Quinn[86] carried out an investigation to determine if there was population differentiation in regard to salt spray responses in the coastal halophyte *Solidago semper-virens*. Bayshore plants responded to salt spray with greater leaf damage than the three other populations investigated, indicating that this population was more sensitive to salt spray than plants from dune areas. The success of this species along dune gradients is related to its ability to develop populations that are tolerant of salt spray, which dune species are exposed to on a daily basis.[86]

VII. MYCORRHIZAL FUNGI

A number of researchers have reported mycorrhizal associations with halophytes in coastal and inland saline environments (Table 9).[87-93] Salt pans in Utah containing over 2.0%

TABLE 9
Presence or Absence of Mycorrhizal Fungi in Halophytes

Species	Present[a] (+)	Absent (−)	Ref.
Armeria maritima		−	87
Aster tripolium	11—20		87,91,93,94
Atriplex confertifolia	+		89
A. hastata		−	87,91
Cenchrus pennisetiformis	39		88
Ceratoides lanata	+		89
Cochlearia anglica	+		87
Distichlis spicata	34.2		86,90,92
Festuca idahoensis	+		92
Glaux maritima	1—10		87,91
Glyceria maritima	+		87,93
Halimione portulacoides	0—10		87,91
Haloxylon griffithii		−	88
Hordeum jubatum	+		97
Juncus gerardi	+		91,93
J. maritimus		−	87,91
Limonium vulgare	0—15	−	87,91
Plantago maritima	16—20		91
Puccinellia maritima	11—15		91
Salicornia brachystachya	0—15		91
S. europaea		−	87,93
S. pacifica		−	90
S. rubra		−	90
Salsola kali		−	88
Scirpus maritimus		−	87
Spartina anglica		−	91
S. townsendii		−	87
Spergularia marginata		−	87,93
Sporobolus arabicus	60		88
Suaeda fruticosa		−	88
S. maritima	1—10		91
S. nudiflora		−	88
Tamarix dioica		−	88
Triglochin maritima		−	87,93

[a]   Percent of roots infected.

total salts did not contain vegetation that was infected by mycorrhizal fungi.[90] However, a high percentage of mycorrhizal roots of *Distichlis spicata* were found in areas of reduced salinity (0.05 to 0.10% total salts). The percentage of VAM roots decreased to nearly zero at 1.0% total soil salinity. A number of halophytic species from Pakistan have been reported to have VAM fungal associations.[88] A survey of halophytes in the Bergen-op-Zoom salt marsh, Netherlands indicates that some species are highly mycorrhizal (*Aster tripolium, Limonium vulgare, Festuca rubra, Plantago maritima,* and *Glaux maritima*).[91] Other species, such as *Spartina anglica, Triglochin maritima, Juncus gerardii,* and *Atriplex hastata,* were not infected by VAM fungi (Table 9). The K and P content of *Aster tripolium* was not affected by the presence of mycorrhizal fungi.[91] However, the Na content of shoots was lower for mycorrhizal than for nonmycorrhizal plants when they were grown at 300 mM NaCl. Benefits to mycorrhizal plants in increased water uptake are evidenced by increased leaf elongation and smaller diurnal changes in leaf thickness. Mycorrhizal plants may have less loss in turgor during the day when the stomata are open.[91]

Mason[93] reported that the halophytes which possess VAM mycorrhiza on British salt marshes include *Plantago coronopus, P. maritima, Aster tripolium, Glaux maritima, Armeria maritima, Cochlearia anglica, Agrostis alba*, and *Glyceria maritima* (Table 9). Vesicular arbuscular mycorrhizae were not observed for the roots of *Salicornia europaea, Triglochin maritima, Juncus maritima, J. gerardi*, and *Spergularia marginata*.

Fries[94] observed mycorrhizal fungi on salt marsh species from the west coast of Sweden. Several halophytes were found to have VAM, including *Plantago maritima, Matricaria maritima, Aster tripolium, Glaux maritima*, and *Armeria maritima*.

Boullard[87] reported that several species of halophytes from the Normandy coast (France) have VAM mycorrhizal fungi, including *Halimione portulacoides, Aster tripolium, Glyceria maritima*, and *Plantago maritima* (Table 9). Other coastal halophytes were found not be infected by mycorrhizal fungi, including *Spartina towensendii, Salicornia herbacea, Triglochin maritima, Scirpus maritimus, Juncus maritimus, Atriplex hastata, A. tornabeni, Spergularia marginata*, and *Limonium vulgare*. Seasonal variation in the amount of infection and presence of vesicles occurred.[95] The infection of halophytes followed a regular seasonal pattern for plants that normally contained VAM. However, those that were only occasionally infected did not show any seasonal pattern, and infection was random.

Cooke and Lefor[96] reported VAM mycorrhizal associations for *Spartina patens, Distichlis spicata, Iva frutescens*, and *Limonium carolinianum*, but not for *S. alterniflora* and *Phragmites australis* on the Indian River salt marsh (Connecticut). Species used for salt marsh restorations 10 years earlier did not contain VAM (*S. patens, S. spicata*, and *S. alterniflora*), and the authors concluded that this might account for the poor establishment of vegetation locally.[96]

The perennial grass *D. spicata* had a mycorrhizal association with the mycorrhizal fungus *Glomus fasciculatum*.[97] An experiment was carried out to determine whether mycorrhizae would alleviate the detrimental effects on growth of NaCl. No general trends were found in terms of enhancement of growth in saline conditions when mycorrhizal and nonmycorrhizal *D. spicata* plants were grown at salinities ranging from 0 to 2000 mg NaCl/kg dry soil. In a laboratory investigation with the moderately salt tolerant grass, *Hordeum jubatum*, it was determined that plants inoculated with the mycorrhizal fungus *G. fasciculatum* had about a tenfold increase in dry weight compared to nonmycorrhizal plants.[98] However, spore germination of the fungus was inhibited with increased salinity, but a mean of 14.3% of the spores germinated at 0.65% NaCl compared with 28.6% at 0.25%. Lowest spore germination was at 30°C and highest at 22°C. The fungal inoculum level in saline soils was found to be high in summer and winter, decreasing sharply in the spring when colonization of the host *H. jubatum* occurs. Host growth and fungal colonization of roots are essentially completed by late spring before soil salinity levels rise sharply, soil moisture levels decrease, and soil temperatures increase in midsummer.[98] Two halophytic grasses in the Alvord desert, Oregon, *Festuca idahoensis* and *Distichlis spicata*, were found to contain mycorrhizal fungi on their roots.[92] Spores of VAM fungi *G. mosseae* and *G. macrocarpum* were found in the soil, but their numbers were negatively correlated ($r = -0.902$, $p < 0.05$) with the soil Na concentration.

VIII. PHENOLOGY

The phenological growth pattern of halophytes varies considerably among species. Some perennial species such as the upper marsh species *Plantago maritima, Limonium vulgare*, and *Triglochin maritima* are reported to have a large portion of their above-ground biomass produced early in the growing season prior to the hypersaline soil conditions that developed

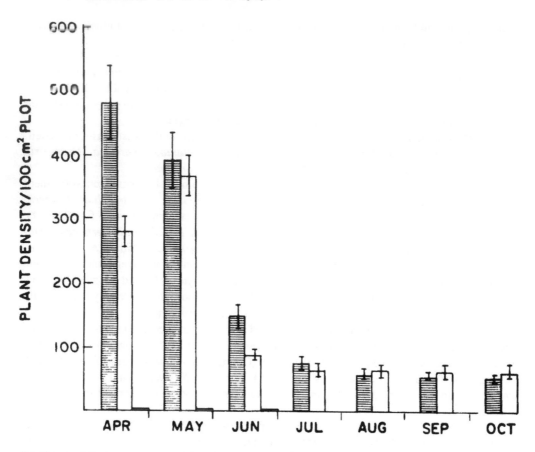

FIGURE 4.   Monthly measurements of *Salicornia europaea* densities (plants/100 cm²) in three locations on a salt marsh. Field (diagonal bar), edge (open bar), pan (solid bar). Lines represent SE of mean. (From McGraw, D. C. and Ungar, I.A., *Ohio J. Sci.*, 81, 109, 1981. With permission.)

in midsummer.[47] Other perennials, including the lower marsh species *Aster tripolium*, *Puccinellia maritima*, and *Halimione portulacoides*, were found to have a regular increase in biomass production throughout the growing season. Annual halophytes in the genera *Salicornia*, *Suaeda*, *Atriplex*, and *Spergularia* tended to demonstrate maximal growth later in the growing season.[49,99-101]

McGraw and Ungar[49] monitored the survival of plants in *Salicornia europaea* populations and determined that from 62 to 100% of the seedlings did not survive to maturity within the three zones investigated (Figure 4). Plants from the less saline zone had shoot dry mass production that was 75% greater than for plants in the more saline zone. Dry mass production was low from early to midsummer, but in late summer and fall, plant dry mass increased to eightfold that of the value prior to August. Low soil water potentials from May through July were limiting dry mass production in these field populations, while the period of maximal growth occurred just prior to reproductive maturation.[49]

Although these patterns may be genetically determined, continuous growth in *Atriplex triangularis* and *Salicornia europaea* was found to be limited by hypersaline conditions in certain zones on an inland salt pan.[50,100] Plants growing in the different zones on the marsh were found to have different periods of active growth. In the most saline areas, *S. europaea* biomass increase stopped in July, while in the least saline zone a large growth increment occurred between July and August (Figure 2).[50] *A. triangularis* was found growing in three distinct zones along a salinity gradient at a Rittman, OH salt marsh, and biomass yields were directly related to the level of environmental stress.[100] In the low-salt zone, growth

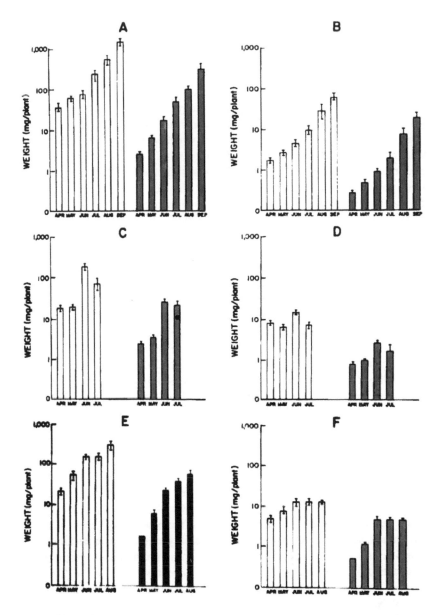

FIGURE 5.   Seasonal pattern of root and shoot fresh mass (open bar) and dry mass (closed bar) for *Atriplex triangularis*. (A) *Hordeum-Atriplex* zone shoots; (B) *Hordeum-Atriplex* zone roots; (C) *Atriplex* dry zone shoots; (D) *Atriplex* dry zone roots; (E) *Atriplex* wet zone shoots; (F) *Atriplex* wet zone roots. (From Riehl, T. E. and Ungar, I. A., *Acta Oecol. Oecol. Plant.*, 4, 27, 1983. With permission.)

continued gradually throughout the growing season until September, while in the more wet high-salt habitats, growth was gradual to August, and then plants died without reproducing, while in the dry high-salt zone growth terminated in June, and plants died in July prior to reaching reproductive maturity (Figure 5). These data for *S. europaea* and *A. triangularis* indicate that environmental stress has a strong influence on the phenological development and growth pattern of halophytes in salt marsh habitats.

Jefferies et al.[47] reported that populations of *S. europaea* on the Stiffkey salt marsh, Norfolk, England had a different growth pattern from one another, with the plants in the lower marsh showing a gradual growth increment response from May to September, while the upper marsh population had little growth early in the growing season and made most of its biomass production in August. When plants from the two habitats were reciprocally transplanted, they tended to maintain the growth pattern of plants from the parental population, indicating that the differences in growth pattern for the plants from the two different populations were genetically determined.[102] Gene flow is reduced in these populations, because inbreeding predominates as the reproductive strategy in this species. Electrophoretic variability was studied for material from the two *Salicornia* populations, and it was found that the upper and lower salt marsh plants had patterns which were different from one another.[103] These data indicated a lack of genetic variability within the two populations of diploid *Salicornia*, indicating that reproductive isolation and inbreeding were genetically differentiating these populations. Similar responses were obtained with field transplants of *S. europaea* on an inland salt pan at Rittman, OH, with short form and tall form transplants tending to resemble their parental populations in growth parameters.[104] However, greenhouse uniform garden treatments indicated that plants from each population had a great deal of phenotypic plasticity and varied in response depending on the time of transplantation. The plants transferred from tall zones tended to produce much greater biomass (293 mg/plant) than did plants from the short *Salicornia* zone (172 mg/plant) in the early transplant, while plants from the very tall zone produced significantly greater biomass (586 mg/plant) than those from the other zones (averaging 193 mg/plant) in the later transplant.[104] The limited data available indicate that both environmental and genetic factors may be influencing the phenological development and the rate of growth of *Salicornia* populations under field conditions. Smythe and Hutchinson[108] determined that *Carex lyngbyei* was also phenotypically plastic. Shoot height, shoot biomass, and above-ground biomass of transplants from various salt marsh locations (Strait of Georgia, Canada) resembled plants in their new habitat after a 6-month period. Phenotypic plasticity could be of adaptive value to plants in the unpredictable environment of salt marshes.[108]

IX. CONCLUSIONS

The effect of salinity on the growth of halophytes differs from glycophytes in that salt marsh and salt desert species can tolerate higher salinities than plants from nonsaline habitats. Many halophytes show growth inhibition with salt increments, but they have adaptations such as salt glands, salt hairs, succulence, and the ability to adjust osmotically, which allows them to survive and grow in saline habitats. Dry mass production of some halophytes is definitely stimulated by relatively high salinity concentrations (100 to 200 mM NaCl). However, most halophytes make optimal growth in nonsaline conditions or with micronutrient levels of NaCl (Table 1, Chapter 2).

Significant questions remain concerning the factors which inhibit growth when plants are exposed to salinity stress. It has been suggested that reduced turgor may play a significant role,[105] but in some cases reduced cell wall extensibility may play an important role in inhibiting plant growth. Other factors that must be considered are nutrient unavailability, specific ion toxicity, energy required for ion uptake in osmotic adjustment, ion antagonisms, energy required to control ion uptake in root cells, and salt pumps in glands. Other problems are related to the difference between the salt tolerance of cells vs. whole plant responses to salinity.

REFERENCES

1. **Chapman, V. J.**, *Salt Marshes and Salt Deserts of the World*, Stechert-MacMillan, Pennsauken, NJ, 1974.
2. **Waisel, Y.**, *Biology of Halophytes*, Academic Press, New York, 1972.
3. **Osmond, C. B., Bjorkman, O., and Anderson, D. J.**, *Physiological Processes in Plant Ecology*, Springer-Verlag, Berlin, 1980.

4. Flowers, T. J., Hajibagheri, M. A., and Clipson, J. W., Halophytes, *Q. Rev. Biol.*, 61, 313, 1986.
5. Munns, R., Greenway, H., and Kirst, G. O., Halotolerant eukaryotes, in *Encyclopedia of Plant Physiology*, Lange, O. L., Nobel, P. S., Osmond, C. B., and Ziegler, H., Eds., Springer-Verlag, Berlin, 1983, 59.
6. Wainwright, S. J., Plants in relation to salinity, *Adv. Bot. Res.*, 8, 221, 1980.
7. Lagerwerff, J. V., Osmotic growth inhibition and electrometric salt-tolerance evaluation of plants, *Plant Soil*, 31, 77, 1969.
8. Billard, J.-P., Binet, P., Boucaud, J., Coudret, A., and Le Saos, J., Halophilie et resistance au sel rflexions sur l'halophilie et quelques-uns de ses aspects physiologiques, *Bull. Soc. Bot. Fr.*, 123, 39, 1976.
9. Partridge, T. R. and Wilson, J. B., Salt tolerance of salt marsh plants of Otago, New Zealand, *N.Z. J. Bot.*, 25, 559, 1987.
10. Clough, B. F., Growth and salt balance of the mangroves *Avicennia marina* (Forsk.) Vierh. and *Rhizophora stylosa* Griff. in relation to salinity, *Aust. J. Plant Physiol.*, 11, 419, 1984.
11. Burchett, M. D., Clarke, C. J., Field, C. D., and Pulkownik, A., Growth and respiration in two mangrove species at a range of salinities, *Physiol. Plant*, 75, 299, 1989.
12. Brun, A., Effets de NaCl sur une halophyte tropicale (*Canavalia obtusifolia* DC.): croissance, disribution de Na^+ et K^+, etat hydrique, *Acta Oecol. Oecol. Plant.*, 9, 173, 1988.
13. Hocking, P. J., Effects of sodium and potassium chlorides on the growth and accumulation of mineral ions by *Cyperus involucratus* Rottb., *Aquatic Bot.*, 21, 201, 1985.
14. Barbour, M. G., Salt tolerance of five California salt marsh plants, *Am. Midl. Nat.*, 84, 262, 1970.
15. Barbour, M. G., Is any angiosperm an obligate halophyte?, *Am. Midl. Nat.*, 84, 105, 1970.
16. Watkins, C. B., Brown, J. M. A., and Dromgoole, F. I., Salt-tolerance of the coastal plant, *Tetragonia trigyna* Banks et Sol. ex Hook, (climbing New Zealand spinach), *N.Z. J. Bot.*, 26, 153, 1988.
17. Ashby, W. C. and Beadle, N. C. W., Salinity factors in the growth of Australian saltbushes, *Ecology*, 38, 344, 1957.
18. Zid, E. and Boukhris, M., Quelques aspects de la tolerance de l'*Atriplex halimus* L. au chlorure de sodium, *Oecol. Plant.*, 12, 351, 1977.
19. Matoh, T., Watanabe, J., and Takahashi, E., Effects of sodium and potassium salts on the growth of a halophyte *Atriplex gmelini*, *Soil Sci. Plant Nutr.*, 32, 451, 1986.
20. Black, R. F., Effect of NaCl in water culture on the ion uptake and growth of *Atriplex hastata* L., *Aust. J. Biol. Sci.*, 9, 67, 1956.
21. Black, R. F., Effects of NaCl on the ion uptake and growth of *Atriplex vesicaria* Heward, *Aust. J. Biol. Sci.*, 13, 249, 1960.
22. Greenway, H., Growth stimulation by high chloride concentrations in halophytes, *Isr. J. Bot.*, 17, 169, 1968.
23. Uchiyama, Y., Salt tolerance of *Atriplex nummularia*, *Tech. Bull. Trop. Agric. Res. Ctr. Japan*, 22, 1, 1987.
24. Aslam, Z., Jeschke, W. D., Barrett-Lennard, E. G., Setter, T. L., Watkin, E., and Greenway, H., Effects of external NaCl on the growth of *Atriplex amnicola* and the ion relations and carbohydrate status of the leaves, *Plant Cell Environ.*, 9, 571, 1986.
25. Mozafar, A., Goodin, J. R., and Oertli, J. J., Na and K interactions in increasing the salt tolerance of *Atriplex halimus* L., *Agronomy J.*, 62, 478, 1970.
26. Priebe, A. and Jager, H. J., Einfluss von NaCl auf wachstum und ionengehalt unterschiedlich salztoleranter Pflanzen, *Angew. Bot.*, 52, 331, 1978.
27. Storey, R. and Wyn Jones, R. G., Responses of *Atriplex spongiosa* and *Suaeda monoica* to salinity, *Plant Physiol.*, 63, 156, 1979.
28. Handley, J. F. and Jennings, D. H., The effect of ions on growth and leaf succulence of *Atriplex hortensis* var. *cupreata*, *Ann. Bot.*, 41, 1109, 1977.
29. Pigott, C. D., Influence of mineral nutrition on the zonation of flowering plants in coastal salt-marshes, in *Ecological Aspects of the Mineral Nutrition of Plants*, Rorison, I. H., Ed., Blackwell Scientific, Oxford, 1969, 25.
30. Yeo, A. R. and Flowers, T. J., Salt tolerance in the halophyte *Suaeda maritima* L. Dum.: Evaluation of the effect of salinity upon growth, *J. Exp. Bot.*, 31, 1171, 1980.
31. Baumeister, W. and Schmidt, L., Uber die Rolle des Natriums im pflanzlichen Stoffwechsel, *Flora*, 152, 24, 1962.
32. Hekmat-Shoar, H., Analyse des relations phylogeniques entre trois *Suaeda* par culture en milieux diversement sales, *Bull. Soc. Bot. Fr.*, 125, 293, 1978.
33. Eshel, A., Response of *Suaeda aegyptiaca* to KCl, NaCl and Na_2SO_4 treatments, *Physiol. Plant*, 64, 308, 1985.
34. Eshel, A., Effects of NaCl and KCl on growth and ionic composition of the halophytic C_4 succulent chenopods *Salsola kali*, *Suaeda monoica* and *Suaeda aegyptiaca*, *Aust. J. Plant Physiol.*, 12, 319, 1985.
35. Williams, M. D. and Ungar, I. A., The effect of environmental parameters on the germination, growth, and development of *Suaeda depressa* (Pursh) Wats., *Am. J. Bot.*, 59, 912, 1972.

36. **Boucaud, J. and Ungar, I. A.,** Halophile et resistance au sel dans le genre *Suaeda* Forsk, *Bull. Soc. Bot. Fr. Actualites Bot.*, 3/4, 23, 1978.

37. **Grouzis, M.,** Exigences ecologiques comparees d'une salicorne vivace et d'une salicorne annuelle; germination et croissance des stades jeunes, *Oecol. Plant.*, 8, 367, 1973.

38. **Hagege, D., Kevers, C., Boucaud, J., and Gasnar, T.,** Activites peroxydasiques, production d'ethylene, lignification et limitation de croissance chez *Suaeda maritima* cultive en l'absence de sel, *Plant Physiol. Biochem.*, 26, 609, 1988.

39. **Cooper, A.,** The effects of salinity and waterlogging on the growth and cation uptake of salt marsh plants, *New Phytol.*, 90, 263, 1982.

40. **Halket, A. C.,** The effect of salt on the growth of *Salicornia, Ann. Bot.*, 29, 143, 1915.

41. **van Eijk, M.,** Analyse der Wirkung des NaCl auf die Entwicklung, Sukkulenz und Transpiration bei *Salicornia herbacea* sowie Untersuchungen uber den Einfluss der Salzaufnahme auf die Wurzelatmung bei *Aster tripolium, Rec. Bot. Trav. Neerl.*, 36, 559, 1939.

42. **Webb, K. L.,** NaCl effects on growth and transpiration in *Salicornia bigelovii*, a salt marsh halophyte, *Plant Soil*, 24, 261, 1966.

43. **Langlois, J.,** Influence du rythme d'immersion sur la croissance et le metabolisme proteique de *Salicornia stricta* Dumort, *Oecol. Plant.*, 6, 227, 1971.

44. **Austenfeld, F.-A.,** Untersuchungen zum Ionen haushalt von *Salicornia europaea* L. unter besonderer Beruchtsichtung des Oxalats in Abhangiigkeit von der Substratsalinitat, *Biochem. Physiol. Pfl.*, 165, 303, 1974.

45. **Ungar, I. A.,** The effects of salinity and hormonal treatments on growth and ion uptake of *Salicornia europaea, Bull. Soc. Bot. Fr. Actualites Bot.*, 124, 95, 1978.

46. **Caldwell, M. M.,** Physiology of desert halophytes, in *Ecology of Halophytes*, Reimold, R. J. and Queen, W. H., Eds., Academic Press, New York, 1974, 379.

47. **Jefferies, R. L., Davy, A. J., and Rudmik, T.,** The growth strategies of coastal halophytes, in *Ecological Processes in Coastal Environments*, Jefferies, R. L. and Davy, A. J., Eds., Blackwell Scientific, Oxford, 1979, 243.

48. **Ungar, I. A.,** Inland halophytes of the United States, in *Ecology of Halophytes*, Reimold, R. J. and Queen, W. H., Eds., Academic Press, New York, 1974, 235.

49. **McGraw, D. C. and Ungar, I. A.,** Growth and survival of the halophyte *Salicornia europaea* under saline field conditions, *Ohio J. Sci.*, 81, 109, 1981.

50. **Riehl, T. E. and Ungar, I. A.,** Growth and ion accumulation in *Salicornia europaea* under saline field conditions, *Oecologia*, 54, 193, 1982.

51. **Suehiro, K.,** On the optimum growth of *Salicornia herbacea* L. under the various seawater concentrations in soil, *Mem. Fac. Educ. Kagawa Univ.*, 26, 105, 1976.

52. **Grouzis, M., Heim, G., and Berger, A.,** Croissance et accumulation de sels chez deux salicornes annuelles du littoral mediterranean, *Oecol. Plant.*, 12, 307, 1977.

53. **Kawasaki, H., Takada, H., and Kamisaka, S.,** Requirement of sodium chloride for the action of gibberellic acid in stimulating hypocotyl elongation of a halophyte, *Salicornia herbacea* L., *Plant Cell Physiol.*, 19, 1415, 1978.

54. **Flowers, T. J.,** Salt tolerance in *Suaeda maritima* (L.) Dum., *J. Exp. Bot.*, 23, 310, 1972.

55. **Boucaud, J.,** Action de la salinite, de la composition du milieu et du pretraitement des semences sur la croissance de *Suaeda maritima*, var. *flexilis* Focke en cultures sans sol., *Bull. Soc. Linn. Norm.*, 101, 135, 1970.

56. **Boucaud, J.,** Auto-ecologie et etude experimentale des exigences eco-physiologiques de *Suaeda maritima* (L.) Dum., var. *macrocarpa* Moq. et var. *flexilis* Focke, *Oecol. Plant.*, 7, 99, 1972.

57. **Boucaud, J.,** Action de la salinite de la composition du milieu et du pretraitement des semences de *Suaeda maritima* var. *macrocarpa* Moq. en cultures sans sol, *Bull. Soc. Linn. Norm.*, 8, 92, 1967.

58. **Boucaud, J. and Ungar, I. A.,** Influence of hormonal treatments on the growth of two halophytic species of *Suaeda, Am. J. Bot.*, 63, 694, 1976.

59. **LeSaos, J. and Binet, P.,** Halophilie et alimentation calcique chez *Suaeda maritima* (L.) Dum. variete *macrocarpa* Moq., *Acta Oecol. Oecol. Plant.*, 8, 127, 1987.

60. **McNulty, I.,** Some effects of chloride salts on greasewood *Sarcobatus vermiculatus* (Hook) Torr., *Utah Acad. Sci. Arts Lett.*, 40, 185, 1963.

61. **Phleger, C. F.,** Effect of salinity on growth of a salt marsh grass, *Ecology*, 52, 908, 1971.

62. **Mahall, B. E. and Park, R. B.,** The ecotone between *Spartina foliosa* Trin. and *Salicornia virginica* L. in salt marshes of northern San Francisco Bay, *J. Ecol.*, 64, 421, 1976.

63. **Haines, B. L. and Dunn, E. L.,** Growth and resource allocation responses of *Spartina alterniflora* Loisel. to three levels of NH_4-N, Fe, and NaCl in solution culture, *Bot. Gaz.*, 137, 224, 1976.

64. **Tremblin, G. and Binet, P.,** Halophilie et resistance au sel chez *Halopeplis amplexicaulis* (Vahl) Ung., *Acta Oecol. Oecol. Plant.*, 5, 291, 1984.

65. **Neales, T. F. and Sharkey, P. J.,** Effect of salinity on growth and on mineral and organic constituents of the halophyte *Disphyma australe* (Soland.) J. M. Black, *Aust. J. Plant Physiol.*, 8, 165, 1981.
66. **Parrando, R. T., Gosselink, J. G., and Hopkinson, C. S.,** Effects of salinity and drainage on the growth of three salt marsh grasses, *Bot. Gaz.*, 139, 102, 1978.
67. **Smart, R. M. and Barko, J. W.,** Influence of sediment salinity and nutrients on the physiological ecology of selected salt marsh plants, *Coast. Mar. Sci.*, 7, 487, 1978.
68. **Halvorson, W. L. and Singer, A. C.,** Growth responses of *Spartina patens* and *Spartina alterniflora* analyzed by means of a two-dimensional factorial design, *Am. Midl. Nat.*, 91, 444, 1974.
69. **Matoh, T., Matsushita, N., and Takahashi, E.,** Salt tolerance of the reed plant *Phragmites communis, Physiol. Plant,* 72, 8, 1988.
70. **Kemp, P. R. and Cunningham, G. L.,** Light, temperature and salinity effects on growth, leaf anatomy and photosynthesis of *Distichlis spicata* (L.) Greene, *Am. J. Bot.*, 68, 507, 1981.
71. **Hansen, D. J., Dayanadan, P., Kauffman, P. B., and Brotherson, J. D.,** Ecological adaptations of salt marsh grass *Distichlis spicata* (Gramineae), and environmental factors affecting its growth and distribution, *Am. J. Bot.*, 63, 635, 1976.
72. **Okusanya, O. T. and Ungar, I. A.,** The growth and mineral composition of three species of *Spergularia* as affected by salinity and nutrients at high salinity, *Am. J. Bot.*, 71, 439, 1984.
73. **Rozema, J.,** An eco-physiological investigation into the salt tolerance of *Glaux maritima* L., *Acta Bot. Neerl.*, 24, 407, 1975.
74. **Rozema, J.,** Population dynamics of *Glaux maritima* and ecophysiological adaptations to salinity and inundation, *Oikos*, 30, 539, 1978.
75. **Freijsen, A. H. J.,** Growth-physiology, salt-tolerance and mineral nutrition of *Centaurium littorale* (Turner) Gilmour: adaptations to its olygothrophic and brackish habitat, *Acta Bot. Neerl.*, 20, 577, 1971.
76. **Rozema, J., Frits, B., Duck, T., and Wesselman, H.,** Salt-spray stimulated growth in strand-line species, *Physiol. Plant,* 56, 204, 1982.
77. **Rozema, J.,** An ecophysiological study on the response to salt of four halophytic and glycophytic *Juncus* species, *Flora,* 165, 197, 1976.
78. **Rozema, J. and Blom, B.,** Effects of salinity and inundation on the growth of *Agrostis stolonifera* and *Juncus gerardii, J. Ecol.*, 65, 213, 1977.
79. **Erdei, L. and Kuiper, P. J. C.,** The efffect of salinity on growth, cation content, Na^+-uptake and translocation in salt-sensitive and salt-tolerant *Plantago* species, *Physiol. Plant,* 47, 95, 1979.
80. **Kingsbury, R. W., Radlow, A., Mudie, P. J., Rutherford, J., and Radlow, R.,** Salt stress responses in *Lasthenia glabrata*, a winter annual composite endemic to saline soils, *Can. J. Bot.*, 54, 1377, 1975.
81. **Boyd, R. S. and Barbour, M. G.,** Relative salt tolerance of *Cakile edentula* (Brassicaceae) from lacustrine and marine beaches, *Am. J. Bot.*, 73, 236, 1986.
82. **Tiku, B. L. and Snaydon, R. W.,** Salinity tolerance within the grass species *Agrostis stolonifera* L., *Plant Soil*, 35, 421, 1971.
83. **Tyndall, R. W., Teramura, A. H., Mulchi, C. L., and Douglass, L. W.,** Effects of salt spray upon seedling survival, biomass, and distribution on Currituck Bank, North Carolina, *Castanea,* 52, 77, 1987.
84. **Sykes, M. T. and Wilson, J. B.,** The effect of salinity on the growth of some New Zealand sand dune species, *Acta Bot. Neerl.*, 38, 173, 1989.
85. **Sykes, M. T. and Wilson, J. B.,** An experimental investigation into the response of some New Zealand sand dune species to salt spray, *Ann. Bot.*, 62, 159, 1988.
86. **Cartica, R. J. and Quinn, J. A.,** Responses of populations of *Solidago sempervirens* (Compositae) to salt spray across a barrier beach, *Am. J. Bot.*, 67, 1236, 1980.
87. **Boullard, B.,** Les mycorhizes des especes de contact marin et de contact salin, *Rev. Mycol.*, 23, 282, 1958.
88. **Khan, A. G.,** The occurrence of mycorrhizas in halophytes, hydrophytes and xerophytes, and of endogone spores in adjacent soils, *J. Gen. Micro.*, 81, 7, 1974.
89. **Miller, R. M., Moorman, T. B., and Schmidt, S. K.,** Interspecific plant association effects on vesicular-arbuscular mycorrhiza occurrence in *Atriplex confertifolia, New Phytol.*, 95, 241, 1983.
90. **Kim, C. K. and Weber, D. J.,** Distribution of VA mycorrhiza on halophytes on inland salt playas, *Plant Soil*, 83, 207, 1985.
91. **Rozema, J., Arp, W., Diggelin, J. V., Esbroek, M. V., Broekman, R., and Punte, H.,** Occurrence and ecological significance of vesicular arbuscular mycorrhiza in the salt marsh environment, *Acta Bot. Neerl.*, 35, 457, 1986.
92. **Ho, I.,** Vesicular-arbuscular mycorrhizae of halophytic grasses in the Alvord Desert of Oregon, *Northwest Sci.*, 61, 148, 1987.
93. **Mason, E.,** Note on the presence of mycorrhiza in the roots of salt marsh plants, *New Phytol.*, 27, 193, 1928.
94. **Fries, V. N.,** Beobacchtungen uber die thamniscophage mykorrhiza einiger halophyten, *Bot. Notisera,* 2, 255, 1944.

95. Boullard, B., Halophytes et mycorhizes, *Rev. Mycol.*, 10, 292, 1964.
96. Cooke, J. C. and Lefor, M. W., Comparison of vesicular-arbuscular mycorrhizae in plants from disturbed and adjacent undisturbed regions of a coastal salt marsh in Clinton, Connecticut, USA, *Environ. Manage.*, 14, 131, 1990.
97. Allen, E. B. and Cunningham, G. L., Effects of vesicular-arbuscular mycorrhizae on *Distichlis spicata* under three salinity levels, *New Phytol.*, 93, 227, 1983.
98. Ballard, B., The mycorrhizae of some inland halophytes, Doctoral dissertation, Ohio University, Athens, 1988.
99. Ungar, I. A. and Capilupo, F., An ecological life history study of *Suaeda depressa* (Pursh) Wats., *Adv. Front. Plant Sci.*, 23, 137, 1969.
100. Riehl, T. E. and Ungar, I. A., Growth, water potential and ion accumulation in the inland halophyte *Atriplex triangularis*, *Acta Oecol. Oecol. Plant.*, 4, 27, 1983.
101. Ungar, I. A., Benner, D. K., and McGraw, D. C., The distribution and growth of *Salicornia europaea* on an inland salt pan, *Ecology*, 60, 329, 1981.
102. Jefferies, R. L., Davy, A. J., and Rudmik, T., Population biology of the salt-marsh annual *Salicornia europaea* agg, *J. Ecol.*, 69, 1, 1981.
103. Jefferies, R. L. and Gottlieb, L. D., Genetic differentiation of the microspecies *Salicornia europaea* L. *(sensu stricto)* and *S. ramosissima*, J. Woods, *New Phytol.*, 92, 123, 1982.
104. Ungar, I. A., Population characteristics, growth and survival of the halophyte *Salicornia europaea*, *Ecology*, 68, 569, 1987.
105. Neumann, P. M., Volkenburgh, E. V., and Cleland, R. E., Salinity stress inhibits bean leaf expansion by reducing turgor, not wall extensibility, *Plant Physiol.*, 88, 233, 1988.
106. Drake, D. R. and Ungar, I. A., Effects of salinity, nitrogen, and population density on the survival, growth, and reproduction of *Atriplex triangularis* (Chenopodiaceae), *Am. J. Bot.*, 76, 1125, 1989.
107. Badger, K. S., The salinity tolerance and the distributional ecology of *Hordeum jubatum* L., Ph.D. dissertation, Department of Botany, Ohio University, Athens, 1989.
108. Smythe, S. R. and Hutchinson, I., Ecological plasticity in *Carex lyngbyei*: evidence from transplant experiments, *Can. J. Bot.*, 67, 3618, 1989.

Chapter 4

PHOTOSYNTHESIS

I. INTRODUCTION

The effect of salinity on the net photosynthetic rates of halophytes has been investigated both under field conditions and in the laboratory to determine if the degree of salt tolerance of a species is reflected in its photosynthetic response. Recently, researchers have attempted to determine the causes for reductions in photosynthetic activity, and these can be broken down into the following general categories; (1) reduction in stomatal conductance, (2) reduction in carboxylase activity, (3) unavailability of substrate, and (4) inhibition of the light reaction mechanism.[1-12] Data available concerning the biochemical responses of plants to salinity have been reviewed recently by Long and Baker.[13] They pointed out that primary production of halophytes would also be affected by a reduction in leaf area when plants were exposed to salinity stress. If salinity stress caused a reduction in leaf area, the biomass production of plants would be reduced even though the net assimilation rates remained constant.

II. EFFECTS OF SALINITY ON ASSIMILATION

Halophytes differ among species in their net assimilation rate response to increased salinity, but most species will have a decrease in photosynthetic activity at some salinity level (Table 1). Kleinkopf and Wallace[14] determined that the net assimilation rate and transpiration rates of *Tamarix ramosissima* changed very little with increased salinity from 10 to 200 mM, indicating that the 50 to 60% decrease in biomass yield at higher salinities was due to energy loss through increased respiration or salt pumping.[14] Tiku[15] compared the net photosynthetic rates of the C_3 succulent *Salicornia rubra* with the C_4 grass *Distichlis stricta* and found contrasting responses in the two species. All concentrations of NaCl up to -3.2 MPa stimulated CO_2 uptake of *Salicornia*, with the greatest increase in CO_2 uptake occurring between -0.8 and -1.6 MPa. Net photosynthetic rates in nutrient control plants were 4.3 mg/g/h in comparison to 6.5 mg/g/h in the -1.6 MPa NaCl treatment. All concentrations of NaCl -0.8 to -3.2 MPa were found to inhibit photosynthesis in *Distichlis*, with plants in nonsaline treatments having rates as high as 19.7 mg/g/h, as compared with rates as low as 2.5 mg/g/h for -1.6-MPa NaCl treatments.[15] Although salinity stimulated photosynthesis in *Salicornia*, the chlorophyll content was reduced about 75% with salt increments, while in *Distichlis* the chlorophyll content of leaves remained unchanged. Similar salt stimulation of photosynthesis and growth and reduced chlorophyll content was observed by Chavan and Karadge[16] in *Sesbania grandiflora*, and they concluded that photosynthetic rates were not limited by chlorophyll content. Kemp and Cunningham[4] found similar results with *Distichlis* and concluded that reductions in CO_2 uptake were caused primarily by increased stomatal resistance which was triggered by salinity increments. Dark respiration of *Distichlis* was not significantly affected by salinity increments.[4] High light treatments interacted synergistically with salinity to produce optimum net photosynthetic rates at 40 klx and -0.8 MPa for *S. rubra*, while the optimum for *D. stricta* was reached in the nonsaline control.

The effect of salinity on net photosynthetic rates of three California perennial marsh species was determined by Pearcy and Ustin.[7] Species varied in their response to salinity stress; net photosynthesis of *Salicornia virginica* was not inhibited at salinities of 450 meq/l, *Spartina foliosa* was inhibited at salinities above 300 meq/l, and *Scirpus robustus* was

TABLE 1
Percent Change in Net Photosynthesis Rates (P) from Control for Halophytes Grown at Different Salinities

| Species | Control (P) | Salinity (mM) | | | Ref. |
		150—250	300—400	450—550	
Abronia maritima	2.8 nM/cm²/s	71.4	—	35.7	2[a]
Alternanthera philoxeroides	1.5 nM/cm²/s	86.7	33.3	—	8[a]
Atriplex californica	1.0 nM/cm²/s	120.0	—	70.0	2[a]
A. halimus	3.3 cm³/cm²/s	90.9	84.9	60.6	1[a]
A. leucophila	2.8 nM/cm²/s	82.1	—	71.4	2[a]
Avicennia marina	15 µM/m²/s	100.0	66.7	46.7	6[a]
Distichlis spicata	1.0 mg/m²/s	68.0	—	59.0	4
Limonium californicum	7.5 mg/dm²/h	—	86.7	53.3	29
Plantago maritima	17.2 µM/m²/s	94.2	79.1	—	12
Puccinellia nuttalliana	22 µM/m²/s	90.9	72.7	—	11[a]
Scirpus robustus	32 µM/m²/s	100.2	56.3	15.6	7[a]
Spartina foliosa	40 µM/m²/s	100.3	100.0	50.0	7[a]

[a] Data taken from graph.

inhibited at above 150 meq/l in artificial sea salt treatments (Table 1). Pearcy and Ustin[7] concluded from analysis of leaf conductance and intercellular CO_2 partial pressures that inhibition of photosynthesis at high salinities was primarily because of reduced photosynthetic capacity of the mesophyll and secondarily due to a reduction in leaf conductance (Figure 1). Temperature optima for photosynthesis in these three species were 25°C in *Salicornia* (>20 µmol/m²/s), 30°C in *Scirpus* (30 µmol/m²/s), and 35°C in *Spartina* (40 µmol/m²/s). At the highest salt stress (450 meq/l), *Spartina* net photosynthetic rates were reduced about 50% and *Scirpus* about 80% from that of their peak rates.

A comparative study of photosynthetic and respiratory responses of the C_4 grasses *Spartina foliosa* and *Distichlis spicata* and C_3 succulents *Salicornia europaea and Batis maritima* from California salt marshes indicated that net assimilation rates of the C_4 grasses were more inhibited at −2.5 MPa than were the succulents (Table 2). Kuramoto and Brest[17] reported that respiration of *Spartina* was also significantly inhibited at all salinities, but the other species were not significantly affected (Table 2). Storage of salts in the vacuole or a dilution effect protected the chloroplasts of succulents, while the exposure of chloroplasts to high salt concentrations in the grasses might account for inhibition of the photosynthetic process.[17] The succulent halophytes had low PEP carboxylase activity (0.24 to 0.33 abs/min/mg chl) when compared with the graminoid taxa (0.79 to 1.06 abs/min/mg chl), which behaved like C_4 plants.

Mallott et al.[18] reported that *Spartina anglica* plants treated with seawater had a greater capacity to maintain high net assimilation at high temperatures than plants grown in nonsaline conditions. Field observations indicated that *S. townsendii* had highest net assimilation rates in midsummer, with reduced rates in spring and fall.[19] However, *Spartina* was found to be well adapted to maintaining the C_4 carboxylation pathway under low temperatures.[20] A model was created by Long and Incoll[19] that could fairly accurately predict net assimilation rates of *S. townsendii* using the light and temperature conditions at different periods of the growing season. Under high light conditions 1800 µmol/m²/s *Spartina alterniflora* maintained similar net assimilation rates in nutrient controls and 3.0% NaCl treatments, while under reduced light 500 µmol/m²/s photosynthesis was reduced to about 50% of the control at the same salinity.[21] At low illumination stomatal resistance increased about 280% while in high light it increased about 30% in the high salt treatment when compared with the control. Photosynthesis was limited by low light and increased stomatal resistance, and

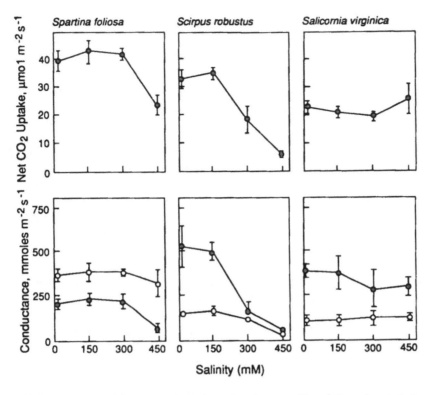

FIGURE 1. Effects of salinity on CO_2 uptake, leaf conductance (filled circle), and mesophyll conductance (open circle) of *Spartina foliosa*, *Scirpus robustus*, and *Salicornia virginica* (From Pearcy, R. W. and Ustin, S. L., *Oecologia*, 62, 68, 1984. With permission.)

TABLE 2
Rates of Photosynthesis and Respiration in Four Salt Marsh Species Exposed to Three Water Potentials

Species	Water potential ($-MPa$)		
	0	1.25	2.54
Photosynthesis			
Salicornia europaea	7.59 ± 0.64 a	5.99 ± 0.27 a,b	4.75 ± 0.12 b
Batis maritima	5.64 ± 0.85 a	5.52 ± 1.78 a	4.13 ± 0.70 a
Spartina foliosa	15.53 ± 2.40 a	16.40 ± 0.74 a	5.57 ± 1.68 b
Distichlis spicata	31.95 ± 3.34 a	8.53 ± 2.66 a	−2.81
Respiration			
Salicornia europaea	6.02 ± 0.10 a	4.84 ± 0.69 a	4.48 ± 0.60 a
Batis maritima	9.93 ± 0.95 a	7.68 ± 0.82 a	4.48 ± 0.57 a
Spartina foliosa	13.84 ± 2.18 a	10.67 ± 2.01 a,b	8.02 ± 0.25 b
Distichlis spicata	11.13 ± 1.21 a	11.08 ± 0.94 a	11.88 ± 2.70 a

Note: Photosynthetic rates are expressed as mg CO_2 dm^{-2}h^{-1}. Respiratory rates are expressed as mg CO_2 gm dry wt/h. All values are expressed as the mean ± 1 SE. Different letters in a row indicate significantly different means at $p = 0.05$.

From Kuramoto, R. T. and Brest, D. E., *Bot. Gaz.*, 140, 295, 1979. With permission.

Longstreth and Strain[21] concluded that salinity would rarely limit photosynthesis in *S. alterniflora*.

The effects of short-term exposure to salinity of the mangrove *Avicennia marina* var. *australasica* were investigated by Ball and Farquhar.[6] At 50 m*M* NaCl, the CO_2 assimilation rate was 14.9 μmol/m²/s at a mean intercellular CO_2 concentration of 177 μl/l. Assimilation rates decreased with increases in salinity over 100 m*M* NaCl, with a 37% decrease at salinity levels above 250 m*M* NaCl (9.1 μmol/m²/s at 500 m*M* NaCl). Ball and Farquhar[6] determined that both stomatal conductance and intercellular CO_2 concentrations decreased with the decline in assimilation rate. The rate of dark respiration was not affected by salinity, maintaining a mean of 1.5 μmol/m²/s in controls and 1.4 μmol/m²/s in leaves of plants receiving salinity treatments. The decline in photosynthetic rate was rapidly reversed when plants were returned to 50 m*M* NaCl from the 500 m*M* salt treatments. Short-term salinity treatments led to a strategy of stomatal response that allowed for limited water loss relative to the assimilation rate, while photosynthetic capacity was also decreasing. Ball and Farquhar[6] concluded that stomatal conductance and reduced photosynthetic capacity were colimiting the assimilation rate of *Avicennia*. Similar results were obtained with relatively long term investigations with the mangroves *A. marina* and *Aegiceras corniculatum*, where reductions in both stomatal conductance and photosynthetic capacity caused a decline in the CO_2 uptake under conditions of increasing medium salinity.[5]

Three California beach species, *Atriplex leucophylla*, *A. californica*, and *Abronia maritima* had reduced photosynthesis and decreased leaf conductance when exposed to artificial seawater at high salinity levels (−2.0 MPa).[2] Water use efficiency (WUE) generally increased for all three species with increased salinity of the medium, with the C_4 species *Atriplex leucophylla* having a higher WUE than the two C_3 species. These gas exchange data for *Atriplex* species corroborate the results found by Gale and Poljakoff-Mayber[22] for *A. halimus*. *A. patula* had an increase in both stomatal and mesophyll resistance when exposed to increased salinity.[23] Net assimilation was reduced about 50% for *A. patula* in 400 m*M* NaCl when compared with nonsaline controls. DeJong[24] reported that CO_2 uptake for the C_4 *A. leucophylla* was saturated at lower CO_2 pressures than the C_3 species.

Net photosynthesis rates were gradually reduced in the salt-tolerant Australian shrub *A. nummularia* at salinities ranging from 0 to 4.0% NaCl.[25] The reduction in net photosynthetic rate at 4.0% NaCl was 85% of that in controls, while at 1.0% NaCl the reduction was 33% of the controls. Transpiration rates dropped sharply at 1.0% NaCl to 40% of the controls, and this level was maintained in up to 4.0% NaCl treatments. The gradual decrease in net photosynthesis for *A. nummularia* indicates that net photosynthesis was being limited by mesophyll resistance as well as by stomatal closure.

A. halimus plants exposed to high NaCl or Na_2SO_4 salinity had an increase in mesophyll resistance and stomatal resistance, causing reductions in both transpiration and photosynthetic rates.[1] Gale et al.[26] found that net primary production of *A. halimus* plants exposed to salt stress was reduced less at a high relative humidity than at a low relative humidity. *A. halimus* had an increase in respiration at 0.75 MPa, and Schwarz and Gale[27] concluded that decreases in growth at low salinities may be caused by a shift from anabolic to catabolic processes. The assimilation rate of the mangroves *Aegiceras* and *Avicennia* was higher for plants at various salinities when the evaporation rate was reduced, indicating that there is a negative interaction between salt stress and reduced relative humidity on the net assimilation rates of plants.[5]

Abdulrahman and Williams[28] reported that the net photosynthesis of *Salicornia fruticosa* was optimal at 171 m*M* NaCl, and the peak salinity for biomass production and lowest assimilation was at 855 m*M* NaCl. Dark respiration was reduced about 50% at the higher temperatures and highest salinity (855 m*M* NaCl), but values were not significantly different from controls in a salinity range of 17 to 342 m*M* NaCl at any temperature. Both mesophyll

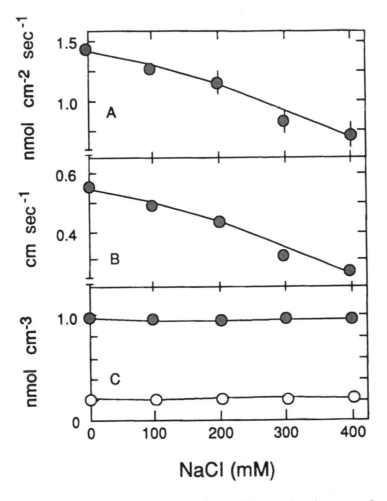

FIGURE 2. Mean (A) net photosynthesis rate, (B) stomatal conductance, and (C) intercellular CO_2 concentration (open circle) and compensation point (closed circle) for *Alternanthera philoxeroides* grown at five salinities. (From Longstreth, D J., Bolanos, J. A., and Smith, J. E., *Plant Physiol.*, 75, 1044, 1984. With permission.)

resistance and stomatal resistance limit assimilation, but the various salt treatment results indicated that mesophyll resistance was more limiting.[28]

Intercellular CO_2 concentrations for *Alternanthera philoxeroides* remained fairly constant at a salinity range of 0 to 400 mM NaCl, while stomatal conductance and assimilation decreased in a parallel pattern with salt increments (Figure 2). Stomatal conductance and net photosynthesis were significantly linearly related ($r^2 = 0.997$), but since intercellular CO_2 concentrations did not vary with salinity treatment the nonstomatal characteristics may account for the 70% decline in cellular metabolism over this salinity range. Longstreth et al.[8] concluded that on a mesophyll area basis, soluble protein levels were relatively unchanged in the salinity treatments, but total chlorophyll decreased with each salinity increment. Reductions in dry matter accumulation for *Alternanthera* were positively correlated with reductions in the plant's net photosynthesis responses.

The photosynthetic capacity of some halophytes has been reported by Flanagan and Jefferies[12] to be directly under the control of stomatal conductance, but the relative effects of stomatal vs. nonstomatal control of photosynthesis in halophytes exposed to salinity stress appears to vary between taxa. *Plantago maritima* plants were exposed to salinities ranging from 50 to 350 mM NaCl, and it was determined that photosynthesis was not limited

TABLE 3
Effect of Salinity on Gas Exchange Characteristics and Chlorophyll Content in *Plantago maritima* L.

Characteristic	NaCl concentration (mol m^{-3})			Significance level
	50	200	350	
Assimilation, A (μmol m^{-2} s^{-1})	17.2 $\pm$ 3.0	16.2 $\pm$ 2.0	13.6 $\pm$ 2.9	NS
Leaf conductance (mmol H$_2$O m^{-2} s^{-1})	369.4 $\pm$ 84.7	227.4 $\pm$ 71.9	172.4 $\pm$ 49.1	*
Intercellular CO$_2$ (μmol mol^{-1})	269.6 $\pm$ 14.8	225.0 $\pm$ 29.2	215.0 $\pm$ 16.9	*
Water-use efficiency A/E (mmol mol^{-1})	5.68 $\pm$ 1.12	8.07 $\pm$ 1.99	8.45 $\pm$ 1.32	*
Chlorophyll (μg cm^{-2})	78.2 $\pm$ 8.2	76.7 $\pm$ 9.1	75.4 $\pm$ 9.1	NS
Dark respiration (μmol m^{-2} s^{-1})	1.83 $\pm$ 0.22	2.12 $\pm$ 0.25	2.05 $\pm$ 0.50	NS
CO$_2$ compensation point (μmol mol^{-1})	48.7 $\pm$ 4.8	54.0 $\pm$ 4.0	57.1 $\pm$ 11.9	NS

Note: Values are the mean and one standard deviation, n = 5. Measurement conditions were 600 μmol photons m^{-2} s^{-1}; leaf temperature, 20°C; leaf to air vapor pressure difference, 1.0 kPa: 350 μmol mol^{-1} CO$_2$, 21% O$_2$.

* $p < 0.05$.

From Flanagan, L. B. and Jefferies, R. L., *Plant Cell Environ.*, 11, 239, 1988. With permission.

biochemically, either by intercellular responses to CO$_2$ or by RUBP carboxylase activity (Table 3). The main limitations for the halophyte *Plantago* were reduced leaf conductance and low intercellular CO$_2$ concentrations.[12] Leaf net photosynthetic rates were reduced from 17.2 μmol/m^2/s at 50 mM NaCl to 13.6 μmol/m^2/s at 350 mM NaCl. Both leaf conductance (369.4 to 172.4 mmol/m^2/s) and intercellular CO$_2$ concentrations (269.6 to 215.0 μmol/M) were significantly affected by increased salinity (50 to 350 mM NaCl). Dark respiration was not significantly affected by salinity increases. Water use efficiency increased from 5.7 to 8.5 with increased salinity (Table 3), reflecting a reduction in water loss because of reduced stomatal conductance. *Plantago* may be more efficient in the compartmentation of excess ions than the halophytes studied by Ball and Farquhar[5] and Pearcy and Ustin,[7] and, thus, ionic stress may not limit the mesophyll capacity for photosynthesis as was found with these other halophytes.[12]

Woodell and Mooney[29] determined the difference in effect of sudden osmotic stress on a glycophyte *Phaseolus vulgaris* and a halophyte *Limonium californicum*. Seawater had no significant effect in reducing the photosynthetic rate or respiration rate of *Limonium*. After 17 h of seawater treatment, the net photosynthesis rate for the halophyte was still 85% of the initial rate. *Phaseolus* photosynthesis was significantly inhibited over time, dropping in 19 h to about 10% of the initial net photosynthetic rate under nonsaline conditions. Further experiments were tried with *Limonium* by increasing the salinity of the media to 1.5 and 2 × that of seawater, and it was found that even at twice seawater concentrations, photosynthesis was maintained at 55% of the normal rate. Woodell and Mooney[29] suggested that *Limonium* could maintain its high photosynthetic rates under seawater conditions because it possessed salt glands that maintained the osmotic adjustment of leaves. They hypothesized that *Limonium* might accumulate ions in the vacuole and thus reduce chloroplast ionic content, or its enzymes may be insensitive to hypersaline conditions.

Some species of halophytes have higher biomass production when exposed to saline conditions than they have in nonsaline controls. Tremblin and Coudret[30] found that *Halopeplis amplexicaulis* had optimal assimilation at 150 mM NaCl under high light conditions, with lower rates occurring at 10 and 200 mM NaCl in both low and high light regimes (150 and

TABLE 4

Percentage Change in Net Photosynthesis Rate (P) from Control of Halophytes
Grown at Different Salinities

Species	Control (P)	Salinity (mM)				Ref.
		100—200	300—400	500—600	700—1000	
Halopeplis amplexicaulis	149 μg/gdw/d	238.3	—	—	—	30
Mesembryanthemum crystallinum	16 mg/mg chl/12 h	125.0	93.8	—	—	31[a]
Salicornia fruticosa	10.5 mg/gdw/h	109.5	104.8	71.4	52.4	28[a]
S. rubra	1.5 mg/g/h	226.7	68.2	—	68.2	15
Spartina alterniflora	39.6 mg/dm²/h	100.5	94.4	—	—	21
Sporobolus airoides	120 μg/gfw/min	116.7	—	66.7	21.0	32[a]
Suaeda maritima	14 mg/mg chl/12 h	128.6	132.1	—	—	31[a]
Tamarix ramosissima	16 mg/gdw/h	187.5	—	—	—	14[a]

[a] Data taken from graph.

TABLE 5

Net Photosynthesis, RUBISCO Activity, and Protein Content in *Beta vulgaris* Plants
Exposed to 180 mM NaCl for 11 d (N = 10)

	Monraic		Kawemegapoly	
	Control	Salt	Control	Salt
CO_2 fixation plant (mg CO_2 dm^{-2} h^{-1})	108 ± 15.1	132 ± 13.9	202 ± 11.3	141 ± 6.97
RUBISCO activity (mole CO_2 mg^{-1} protein h^{-1})	0.97 ± 0.08	1.14 ± 0.12	0.86 ± 0.05	0.52 ± 0.05
Protein content (mg dm^{-2})	2.19	2.06	3.94	4.03

From Heuer, B. and Plaut, Z., *J. Exp. Bot.*, 40, 437, 1989. With permission.

300 μmol/m²/s).[30] *Suaeda maritima* was found to make optimal biomass production at 400
mM NaCl.[31] If assimilation was expressed on a chlorophyll content basis (milligrams of
CO_2 per milligram chlorophyll), Winter[31] determined that photosynthesis increased with
increases in salinity up to 400 mM NaCl, which agreed with the biomass production results
for *S. maritima*. When assimilation was calculated on a fresh weight or dry weight basis,
CO_2 uptake apparently decreased. Respiratory rates remained relatively constant in the
different salinity treatments.[31] Other researchers have found similar salt stimulation of photosynthesis in *Atriplex*, *Salicornia*, *Suaeda*, and *Sporobolus* (Table 4).[2,9,15,28,31,32]

III. BIOCHEMICAL INHIBITION

Curtis et al.[33] found that net photosynthetic rates of *Hibiscus cannabinus* were not
inhibited by increased salt stress. It was concluded from a study of leaf carbohydrate content
that nonosmotically active carbohydrate content accumulated in expanding leaves under salt
stress, and growth was, therefore, limited not by the unavailability of carbohydrates but by
the plant's inability to use these materials for osmotic adjustment and growth.

The net photosynthesis rate for *Beta vulgaris* cultivar Kawemgapoly was reduced from
that in controls in 180 mM NaCl, but cultivar Monriac was not affected by increased salinity.
Heuer and Plaut[34] concluded that RUBISCO activity was inhibited by either a lack of the
substrate RUBP or by a specific ion toxicity, because enzyme levels were the same in
controls and salt-treated plants (Table 5).

Webb and Burley[35] determined that the rate of CO_2 assimilation by *Spartina alterniflora*
was reduced by a treatment with 1% NaCl. Smaller pools of malate were found in the NaCl-
treated plants vs. controls, indicating that NaCl was inhibiting the C_4 photosynthetic pathway.

Oxygen evolution from chloroplasts of *Suaeda maritima* was stimulated by salinities up to 310 mM NaCl.[9] In 10% of the cells, Na and Cl concentrations of chloroplasts of *Suaeda* equaled 257 mM Na and 212 mM Cl, while 90% of the cells had levels of 80 mM for both of these ions. Hajibaghori et al.[9] concluded that either ion compartmentation occurred in chloroplasts or that the total ion content of chloroplasts could be high. Electron transport of chloroplast thylakoid membranes of *Aster tripolium* was inhibited at all concentrations of NaCl.[10] Low concentrations of Na stimulated photosystem I activity in *Aster* grown at higher salinities, while higher concentrations (>300 mM NaCl) were inhibitory. Wignarajah and Baker[10] determined that the salinity of the culture medium in which plants were grown had no effect on photosystem II responses in *Aster* chloroplasts, with salinity concentrations above 100 mM being inhibitory to all treatments.

Ball and Anderson[36] determined that the rate of oxygen evolution was reduced linearly in thylakoids isolated from chloroplasts in both the halophyte *Avicennia marina* ($r = -0.93$) and the glycophyte *Pisum sativum* ($r = -0.93$) when salinity was increased from 0 to 500 mM NaCl. The rate of oxygen evolution at 500 mM NaCl was 50% of that measured at 10 mM NaCl. Accumulation of inorganic ions in chloroplasts exposed to light was concluded to result in rapid damage to photosystem II, possibly by causing the loss of proteins from the membranes. The results of salinity stress could be reversed by returning the thylakoids to a low salt concentration. Both a halophyte and glycophyte were equally salt sensitive in their PS II response to increased salinity.[36]

IV. INDUCTION OF CHANGE IN CARBOXYLATION PATHWAY

Young leaves of the grass species *Aeluropus litoralis* used the C_3 carbon fixation pathway under nonsaline conditions, but when plants were exposed to 100 mM NaCl leaves showed an increased activity of PEP carboxylase and the production of aspartate. Shomer-Ilan and Waisel[37] suggested that NaCl treatments caused a shift from the C_3 to C_4 pathway which is of adaptive value, because C_4 plants are more efficient at assimilation under high light and high temperature conditions. However, Downton and Torokfalvy[38] in a quantitative analysis of fixation products of *Aeluropus* found similar levels of malate and aspartate and high PEP carboxylase activity in both nonsaline controls and 100 mM NaCl treatments, indicating that this grass is generally using the C_4 carboxylation pathway (Table 6). The CAM carboxylation pathway was induced in *Mesembryanthemum crystallinum* when the NaCl content of the medium was increased, while the C_3 pathway was used in the nonsaline controls.[39] A similar shift to the CAM pathway was found for the halophyte *Carpobrotus edulis* when the salinity level was raised to 400 mM NaCl.[40] Other forms of stress that produced a leaf water deficit, such as high light intensity and low relative humidity, were also found by Winter[41] to induce the CAM pathway in *Mesembryanthemum*. Vernon et al.[42] reported that *Mesembryanthemum* plants stressed with 500 mM NaCl were induced to CAM metabolism, but when salts were flushed from the soil the PEP carboxylase mRNA levels dropped by 77%. The expression of the CAM pathway in *Mesembryanthemum* is reversible, and it is regulated by transcription of PEP carboxylase mRNA. Exposure of the succulent halophyte *Disphyma clavellatum* to increased salinity, 0 to 400 mM NaCl, did not result in a change in the diurnal pattern of CO_2 assimilation or accumulation of malate which would be evidence for a switch to the CAM pathway.[43] Therefore, although salinity may induce the CAM pathway in some species, one cannot assume that all succulent halophytes will switch to the CAM pathway when they are exposed to salt stress.

V. STABLE CARBON ISOTOPE RATIOS

Haines[44] determined that stable carbon isotope ratios for Sapelo Island, GA halophytes were -13.6‰ for the tall form of the C_4 grass *Spartina alterniflora* and -12.7‰ for the

TABLE 6
Photosynthetic Characteristics of Low Salt and High Salt
Aeluropus litoralis Leaves

	Treatment	
Parameter	−NaCl	+NaCl
Leaf anatomy	"Kranz" type	"Kranz" type
Photosynthesis (μmol CO_2/min/mg Chl)		
In 21% O_2	1.59 ± 0.15	1.49 ± 0.03
In 2% O_2	1.62 ± 0.14	1.49 ± 0.03
CO_2 compensation point (μl/l)	2	3
Carboxylating enzymes (μmol CO_2/min/mg Chl)		
PEP carboxylase	12.15 ± 0.15	12.56 ± 0.14
RUDP carboxylase	3.32 ± 0.09	3.28 ± 0.31
$^{14}CO_2$ fixation products (%)		
Aspartate	53.4 ± 1.0	59.3 ± 0.6
Malate	38.2 ± 0.8	28.5 ± 0.8
Phosphorylated compds.	8.5 ± 1.9	12.1 ± 0.3

From Downton, W. J. S. and Torokfalvy, E., *Z. Pflanzenphysiol.*, 75, 143, 1975. With permission.

TABLE 7
Carbon Isotope Ratios for Selected Halophytes

Species	Carbon isotope ratio (%)	Photosynthetic pathway	Habit	Ref.
Aegiceras corniculatum	−28.7	C_3	Tree	48
Atriplex patula	−25.0	C_3	Leaf succulent	59
Avicennia marina	−21.9	C_3	Tree	48
Batis maritima	−26.6	C_3	Leaf succulent	45
Borrichia frutescens	−24.5	C_3	Leaf succulent	45
Disphyma clavellatum	−20.6	C_3	Leaf succulent	43
Distichlis spicata	−13.1	C_4	Graminoid	44
Juncus roemerianus	−22.8	C_3	Graminoid	44
Puccinellia nuttalliana	−25.0	C_3	Graminoid	3
Salicornia europaea	−27.0	C_3	Stem succulent	59
S. europaea var *rubra*	−26.0	C_3	Stem succulent	3
S. virginica	−26.2	C_3	Stem succulent	45
Sesuvium maritimum	−26.0	C_3	Leaf succulent	59
Spartina alterniflora	−12.7	C_4	Graminoid	44
Sporobolus virginicus	−13.3	C_4	Graminoid	44
Suaeda linearis	−25.0	C_3	Leaf succulent	59

short form. The C_4 grasses *Distichlis spicata* and *Sporobolus virginicus* had carbon isotope values of −13.1 and −13.3‰. The C_3 succulents *Borrichia frutescens* and *Salicornia virginica* had carbon isotope values of −25.0‰, while the C_3 rush *Juncus roemerianus* had a value of −22.8‰ (Table 7). The carbon isotope ratios of two halophytes, *Salicornia rubra* and *Puccinellia nuttalliana*, native to inland saline areas of Alberta, Canada were found by Guy et al.[3] to be less negative when plants were grown at high salinities than in low salt treatments. Correlations between medium water potential and carbon isotope ratio were r = 0.94 for *Salicornia* and r = 0.98 for *Puccinellia*. Carbon isotope ratios range from −34.9‰ in nonsaline controls to about −25.0‰ at −4.0 MPa. The authors suggested that a shift in the carboxylation pathway from RUBP carboxylase to PEP carboxylase could account for the carbon isotope changes in the halophytes *Salicornia* and *Puccinellia*.[3]

Gas exchange observations and other data indicated that the succulent halophytes *Borrichia frutescens*, *Batis maritima*, and *Salicornia virginica* used the C_3 pathway for photosynthesis.[45] Carbon isotope ratios for the three species were −24.5‰, −26.6‰, and −26.2, respectively (Table 7). There was a negligible change <0.0009 meq NaOH/gfw in titratable acid from night to day, indicating that dark fixation was not occurring.[45]

Luttge et al.[46] determined from an analysis of carbon isotope ratios that two succulent halophytes from the northern coast of Venezuela, *Batis maritima* (−26.4‰) and *Sesuvium portulacastrum* (−25.8‰) carried out the C_3 pathway of photosynthesis. In both *S. portulacastrum* and *B. maritima* the transpiration rates, leaf conductances, and net photosynthetic rates were higher in the rainy season than in the dry season. Dilution of salts during the rainy season and better moisture relations probably account for higher net photosynthetic rates at this time compared with periods of increased drought stress and salt stress in the dry season.[46]

Beer[47] recently reviewed the literature concerning the photosynthesis of submerged marine angiosperms. He concluded that many of the seagrasses were found to have low light compensation and saturation points. The CAM pathway has not been reported for any of the seagrasses studied. Of the ten species reported on up to now, nine have the C_3 pathway: *Halodule uninervis*, *Syringodium isoetifolium*, *S. filiforme*, *Thalassodendron ciliatum*, *Thalassia testudinum*, *T. hemiprichii*, *Halophila stipulacea*, *H. spinulosa*, and *Halodule wrightii*. Only one species, *Cymodocea nodosa*, was reported to have the C_4 pathway of photosynthesis.[47]

The carbon isotope ratio was measured for two mangrove species, *Aegiceras corniculatum* and *Avicennia marina*, and a glycophyte, *Phaseolus vulgaris*, that were grown at salinities from 50 to 500 m*M* and relative humidities from 30 to 90%.[48] Carbon isotope ratios became more positive with an increase in environmental stress, and Farquhar et al.[48] concluded that this fractionation change was related to reduced intercellular CO_2 levels because of stomatal closure under stress. Similar conclusions were reached by Guy and Reid[11] in a study of *Puccinellia*, indicating that reduced CO_2 conductance because of stomatal closure could cause an increase in the carbon isotope ratio because of a difference in fractionation. Net assimilation rates for *P. nuttalliana* and stomatal conductance were both reduced with salinity increments ranging from −0.4 to −1.5 MPa. RUBP carboxylase activity decreased about 33% from controls in the −2.5 MPa NaCl treatment. Dark respiration did not increase with increased salinity. Guy and Reid[11] concluded that photosynthesis in *Puccinellia* was colimited by decreased stomatal conductance and photosynthetic capacity, but the former factor was more significant. Carbon isotope ratios became more positive with increased salinity. At ambient CO_2 in −1.5 MPa NaCl it increased 2.43‰ from the nonsaline control, while at high CO_2 concentrations the increase was 6.35‰. Also, no increase was found in PEP carboxylase activity, indicating that the original suggestion by Guy et al.[3] of a shift in carboxylation pathway was not correct. It was concluded that salt stress resulted in less CO_2 uptake by the leaf because of stomatal closure, which limited fractionation by RUBP carboxylase.[11]

VI. FIELD INVESTIGATIONS

The primary production of the tall and short forms of the dominant halophyte *Spartina alterniflora* of Atlantic coastal marshes of the U.S. has been investigated by a number of researchers.[49-55] The tall form of *S. alterniflora* had higher net assimilation rates during the growing season than the short form.[49] Pezeshki and Delaune[50] reported a seasonal average of 31% lower stomatal conductance and 35% lower net photosynthetic rates in the short vs. tall form of *Spartina* in a Louisiana Gulf Coast marsh. Giurgevich and Dunn[49] determined that metabolic factors were found to be more significant than stomatal conductance in determining the reduced rate of CO_2 uptake in the short form. Water use efficiency during the growing season was usually higher in the tall form than in the short form of *Spartina*. A number of edaphic factors have been suggested to explain reduced net photosynthetic rates in the short form, including higher soil salinity, lower nitrogen availability, greater soil waterlogging, increased anaerobic root respiration, sulfide accumulation in the reducing environment, and restricted nutrient uptake.[50-53] The effect of sediment anaerobiosis on photosynthetic rates of *S. alterniflora* was measured, and Pezeshki et al.[52] determined that CO_2 uptake was reduced by 35% for plants exposed to anaerobic conditions. Plants had a mean net assimilation rate of 8.0 $\mu mol/m^2/s$ in aerobic medium and 5.2 $\mu mol/m^2/s$ in the anaerobic treatment. Mean stomatal conductance was reduced 35% in the anaerobic treatment, from 80 $mmol/m^2/s$ under aerobic conditions to 49 $mmol/m^2/s$. Dry mass production of plants in anaerobic treatments was 27% less than in aerobic treatments, which could be reflecting the decline in CO_2 uptake.[52] Partial stomatal closure, as is indicated by reduced leaf conductance under anaerobic conditions, could be the reason for reduced assimilation in *Spartina*.

Seasonal photosynthetic rates for *Juncus roemerianus* varied in rate from 12.0 mg $CO_2/dm^2/h$ in September to 21.5 mg $CO_2/dm^2/h$ in March.[53] The photosynthetic response of *Juncus* was plastic, and temperature optima shifted toward prevailing daytime temperatures, indicating that this species could easily adapt to seasonal changes in environmental conditions. Blum et al.[54] reported that *Juncus* productivity was highest in the spring when respiration rates were reduced. Internal resistances to CO_2 uptake were found to be larger than that of stomatal resistance. WUE remained constant in all seasons, except when the highest temperatures were reached, and then stomatal resistance played an important role in controlling water loss.

Measurements of community carbon dioxide exchange were made by Drake[55] to determine the effect of light on net photosynthesis in two salt marsh communities, grass = *Spartina patens* and *Distichlis stricta*, and mixed = *Scirpus olneyi*, *Spartina*, and *Distichlis* on the Kirkpatrik Marsh, Maryland.[54] The leaf area index for these two communities ranged from 2.4 to 4.3. Mean values for photosynthesis from June 1 to August 15 were 20.3 $\mu mol/m^2/s$ for the mixed grass and 23.0 $\mu mol/m^2/s$ for the grass communities respectively. Maximum net assimilation values occurred in June for the mixed grass community, 34.3 $\mu mol/m^2/s$, and in July for the *Distichlis-Spartina* community, 39.1 $\mu mol/m^2/s$. The principal limiting factor for photosynthesis in these communities was found to be light.[55] The two plant communities were not light saturated even at >1500 $\mu mol/m^2/s$ PPFD. Dark respiration was determined at midday, and it accounted for about 20% of the gross photosynthetic measurement (mixed 16.9 $\mu mol/m^2/s$, grass 20.4 $\mu mol/m^2/s$). Field measurements of *Salicornia fruticosa* community net photosynthetic rates indicated that reduced gas exchange occurred during the summer, and this was related to dry conditions at that time of the year.[56] Salinity increments from 0.5 to 5% NaCl resulted in a total reduction in the net assimilation rate of *Salicornia* by 61%. The sharpest reductions in assimilation occurred between 3 and 4% NaCl (26% decrease) and between 4.0 and 5.0% NaCl (41% decrease), while between the broader range 0.5 to 3.0% NaCl, there was only an 18% decrease in net assimilation.

VII. CONCLUSIONS

Photosynthetic responses to salinity are species specific, and the chief difference between halophytes and glycophytes is that the former can tolerate higher salinity levels than the latter. The reduction in photosynthetic rates because of salt stress is controlled by several factors, including reduced stomatal conductance, mesophyll resistance in RUBP carboxylase or PEP carboxylase activity, reduction in available RUBP and PEP substrates, and inhibition of thylakoid membranes.[13]

A review of the literature by Drake[31] indicated that C_3 halophytes had photosynthetic rates in the laboratory ranging from 0.3 to 3.5 mM CO_2/cm²/s, and C_4 halophytes had a range from 0.4 to 5.7 µM CO_2/cm²/s. The mean values for the C_3 and C_4 species did not differ significantly from each other, averaging 1.6 and 1.4 mM/cm²/s, respectively. Both the C_3 and C_4 pathways are found in the graminoid halophytes, while most of the succulent halophytes in North America have been reported to use the C_3 pathway. The CAM pathway has not been reported for succulent halophytes in North America.[44,45,57] However, the CAM pathway has been reported for the succulents *Mesembryanthemum crystallinum* and *Carpobrotus edulis*, and CAM was induced when these species were exposed to increased salinity.[39,40]

A number of researchers have reported a decrease in the net assimilation rate of halophytes when salt levels were increased in the medium (Table 1).[2,4,12,15,22,25,58] Other investigators have found that optimal photosynthetic activity occurs at between 100 to 200 mM NaCl in the medium for specific halophytes (Table 5).[15,16,28,32,40]

Most data concerning the photosynthetic rates of plants have been collected from individual leaves or shoots of plants. More data are needed concerning whole plant and community photosynthetic responses of halophytes. The precise inhibitory effect of stressful NaCl concentrations is still unknown and may vary with individual species. In the literature, high salt concentrations were reported to decrease leaf chlorophyll content, reduce leaf area, increase stomatal resistance, decrease carboxylation enzyme levels, decrease substrate levels, damage thylakoid membranes, and inhibit enzymatic activity.[1-13]

REFERENCES

1. Gale, J. and Poljakoff-Mayber, A., Interrelations between growth and photosynthesis of salt bush (*Atriplex halimus*) grown in saline madia, *Aust. J. Biol. Sci.*, 23, 937, 1970.
2. DeJong, T. M., Comparative gas exchange and growth responses of C_3 and C_4 beach species grown at different salinities, *Oecologia*, 34, 59, 1978.
3. Guy, R. D., Reid, D. M., and Krouse, H. R., Shifts in carbon isotope ratios of two C_3 halophytes under natural and artificial conditions, *Oecologia*, 44, 241, 1980.
4. Kemp, P. R. and Cunningham, G. L., Light, temperature and salinity effects on growth, leaf anatomy and photosynthesis of *Distichlis spicata* (L.) Greene, *Am. J. Bot.*, 68, 507, 1981.
5. Ball, M. C. and Farquhar, G. D., Photosynthetic and stomatal responses of two mangrove species, *Aegiceras corniculatum* and *Avicennia marina*, to long term salinity and humidity conditions, *Plant Physiol.*, 74, 1, 1984.
6. Ball, M. C. and Farquhar, G. D., Photosynthetic and stomatal responses of the grey mangrove, *Avicennia marina*, to transient salinity conditions, *Plant Physiol.*, 74, 7, 1984.
7. Pearcy, R. W. and Ustin, S. L., Effects of salinity on growth and photosynthesis of three California tidal marsh species, *Oecologia*, 62, 68, 1984.
8. Longstreth, D. J., Bolanos, J. A., and Smith, J. E., Salinity effects on photosynthesis and growth in *Alternanthera philoxeroides* (Mart.) Griseb, *Plant Physiol.*, 75, 1044, 1984.
9. Hajibagheri, M. A., Harvey, D. M. R., and Flowers, T. J., Photosynthetic oxygen evolution in relation to ion contents in the chloroplasts of *Suaeda maritima*, *Plant Sci. Lett.*, 34, 353, 1984.
10. Wignarajah, K. and Baker, N. R., Salt induced responses of chloroplasts activities in species of differing salt tolerance. Photosynthetic electron transport in *Aster tripolium* and *Pisum sativum*, *Physiol. Plant*, 51, 387, 1981.
11. Guy, R. D. and Reid, D. M., Photosynthesis and the influence of CO_2 enrichment on ¹³C values in a C_3 halophyte, *Plant Cell Environ.*, 9, 65, 1986.

12. **Flanagan, L. B. and Jefferies, R. L.,** Stomatal limitation of photosynthesis and reduced growth of the halophyte, *Plantago maritima* L., at high salinity, *Plant Cell Environ.,* 11, 239, 1988.

13. **Long, S. P. and Baker, N. R.,** Saline terrestrial environments, in *Photosynthesis in Contrasting Environments,* Baker, N. R. and Long, S. P., Eds., Elsevier, Amsterdam, 1986, Chap. 3.

14. **Kleinkopf, G. E. and Wallace, A.,** Physiological basis for salt tolerance in *Tamarix ramosissima, Plant Sci. Lett.,* 3, 157, 1974.

15. **Tiku, B. L.,** Effect of salinity on the photosynthesis of the halophyte *Salicornia rubra* and *Distichlis stricta, Bot. Gaz.,* 37, 23, 1976.

16. **Chavan, P. D. and Karadge, B. A.,** Growth, mineral nutrition, organic constituents and rate of photosynthesis in *Sesbania grandiflora* L. grown under saline conditions, *Plant Soil,* 93, 395, 1986.

17. **Kuramoto, R. T. and Brest, D. E.,** Physiological response to salinity by four salt marsh plants, *Bot. Gaz.,* 140, 295, 1979.

18. **Mallott, P. G., Davy, A. J., Jefferies, R. L., and Hutton, M. J.,** Carbon dioxide exchange in leaves of *Spartina anglica* Hubbard, *Oecologia,* 20, 351, 1975.

19. **Long, S. P. and Incoll, L. D.,** The prediction and measurement of photosynthetic rate of *Spartina townsendii (sensu lato)* in the field, *J. Appl. Ecol.,* 16, 879, 1979.

20. **Thomas, S. M. and Long, S. P.,** C_4 photosynthesis in *Spartina townsendii* at low and high temperatures, *Planta,* 142, 171, 1978.

21. **Longstreth, D. J. and Strain, B. R.,** Effects of salinity and illumination on photosynthesis and water balance of *Spartina alterniflora* Loisel, *Oecologia,* 31, 191, 1977.

22. **Gale, J. and Poljakoff-Mayber, A.,** Interrelations between growth and photosynthesis of salt bush (*Atriplex halimus* L.) grown in saline media, *Aust. J. Biol. Sci.,* 23, 937, 1970.

23. **Longstreth, D. J. and Nobel, P. S.,** Salinity effects on leaf anatomy, *Plant Physiol.,* 63, 700, 1979.

24. **DeJong, T. M.,** Comparative gas exchange and growth responses of C_3 and C_4 beach species grown at different salinities, *Oecologia,* 36, 59, 1978.

25. **Uchiyama, Y.,** Salt tolerance of *Atriplex nummularia, Tech. Bull. Trop. Agric. Res. Ctr.,* 22, 1, 1987.

26. **Gale, J., Naaman, R., and Poljakoff-Mayber, A.,** Growth of *Atriplex halimus* L. in sodium chloride salinated culture solutions as affected by the relative humidity of the air, *Aust. J. Biol. Sci.,* 23, 947, 1970.

27. **Schwarz, M. and Gale, J.,** Maintenance respiration and carbon balance of plants at low levels of sodium chloride salinity, *J. Exp. Bot.,* 32, 933, 1981.

28. **Abdulrahman, F. S. and Williams, G. J.,** Temperature and salinity regulation of growth and gas exchange of *Salicornia fruticosa* (L.) L., *Oecologia,* 48, 346, 1981.

29. **Woodell, S. R. J. and Mooney, H. A.,** The effect of sea water on carbon dioxide exchange by the halophyte *Limonium californicum* (Boiss.) Heller, *Ann. Bot.,* 34, 117, 1970.

30. **Tremblin, G. and Coudret, A.,** Salinite, transpiration et echanges de CO_2 chez *Halopeplis amplexicaulis* (Vahl.) Ung., *Acta Oecol. Oecol. Plant.,* 7, 417, 1986.

31. **Winter, K.,** Wachstum und Photosyntheseleistung der Halophyten *Mesembryanthemum nodiflorum* L. und *Suaeda maritima* (L.) Dum. bei variierter NaCl-Salinitat des Anzuchtmediums, *Oecologia,* 17, 317, 1974.

32. **El-Sharkawi, H. M. and Michel, E.,** Effects of soil salinity and air humidity on CO_2 exchange and transpiration of two grasses, *Photosynthetica,* 9, 277, 1975.

33. **Curtis, P. S., Zhong, H. L., Lauchli, A., and Pearcy, R. W.,** Carbohydrate availability, respiration, and the growth of kenaf (*Hibiscus cannabinus*) under moderate salt stress, *Am. J. Bot.,* 75, 1293, 1988.

34. **Heuer, B. and Plaut, Z.,** Photosynthesis and osmotic adjustment of two sugarbeet cultivars grown under saline conditions, *J. Exp. Bot.,* 40, 437, 1989.

35. **Webb, K. L. and Burley, J. W. A.,** Dark fixation of $^{14}CO_2$ by obligate and facultative salt marsh halophytes, *Can. Bot.,* 43, 281, 1965.

36. **Ball, M. C. and Anderson, J. M.,** Sensitivity of photosystem II to NaCl in relation to salinity tolerance. Comparative studies with thylakoids of the salt-tolerant mangrove, *Avicennia marina* and the salt-sensitive pea, *Pisum sativum, Aust. J. Plant Physiol.,* 13, 689, 1986.

37. **Shomer-Ilan, A. and Waisel, Y.,** The effect of sodium chloride on the balance between the C_3 and C_4 carbon fixation pathways, *Physiol. Plant,* 29, 190, 1973.

38. **Downton, W. J. S. and Torokfalvy, E.,** Effect of sodium chloride on the photosynthesis of *Aeluropus litoralis,* a halophytic grass, *Z. Pflanzenphysiol.,* 75, 143, 1975.

39. **Winter, K. and von Willert, D. J.,** NaCl induzierter Crassulaceensaurestoffwechsels bei *Mesembryanthemum crystallinum, Z. Pflanzenphysiol.,* 67, 166, 1972.

40. **Winter, K.,** NaCl-induzierter Crassulaceensaurestoffwechsel bei einer weiteren Aizoacee: *Carpobrotus edulus, Planta,* 115, 187, 1973.

41. **Winter, K.,** CO_2-Fixierungsreaktionen bei der Salzpflanze *Mesembryanthemem crystallinium* unter variierten Aussenbedingungen, *Planta,* 114, 75, 1973.

42. **Vernon, D. M., Ostrem, J. A., Schmitt, J. M., and Bohnert, H. J.,** PEPcase transcript levels in *Mesembryanthemum crystallinum* decline rapidly upon relief from salt stress, *Plant Physiol.,* 86, 1002, 1988.

43. Neales, T. F., Fraser, M. S., and Roksandic, Z., Carbon isotope composition of the halophyte *Disphyma clavellatum* (Haw.) chinnock (Aizoaceae), as affected by salinity, *Aust. J. Plant Physiol.*, 10, 437, 1983.

44. Haines, E. B., Stable carbon isotope ratios in the biota, soils and tidal water of a Georgia salt marsh, *Est. Coast. Mar. Sci.*, 4, 609, 1976.

45. Antlfinger, A. E. and Dunn, E. L., Seasonal patterns of CO_2 and water vapor exchange of three salt marsh succulents, *Oecologia*, 43, 249, 1979.

46. Lüttge, U., Popp, M., Medina, E., Cram, W. J., Diaz, M., Griffiths, H., Lee, H. S. J., Shafer, C., Smith, J. A. C., and Stimmel, K.-H., Ecophysiology of xerophytic and halophytic vegetation of a coastal alluvial plain in northern Venezuela, *N. Phytol.*, 111, 283, 1989.

47. Beer, S., Photosynthesis and photorespiration of marine angiosperms, *Aquatic Bot.*, 34, 153, 1989.

48. Farquhar, G. D., Ball, M. C., von Cammerer, S., and Roksandic, Z., Effect of salinity and humidity on ^{13}C value of halophytes—evidence for diffusional isotope fractionation determined by the ratio of intercellular/atmospheric partial pressure of CO_2 under different environmental conditions, *Oecologia*, 52, 121, 1982.

49. Giurgevich, J. R. and Dunn, E. L., Seasonal patterns of CO_2 and water vapor exchange of the tall and short height forms of *Spartina alterniflora* Loisel in a Georgia salt marsh, *Oecologia*, 43, 139, 1979.

50. Pezeshki, S. R. and DeLaune, R. D., Carbon assimilation in contrasting streamside and inland *Spartina alterniflora* salt marsh, *Vegetatio*, 76, 55, 1988.

51. Giurgevich, J. R. and Dunn, E. L., A comparative analysis of the CO_2 and water vapor responses of two *Spartina* species from Georgia coastal marshes, *Est. Coast. Shelf Sci.*, 12, 561, 1981.

52. Pezeshki, S. R., Delaune, R. D., and Lindau, C. W., Interaction among sediment anaerobiosis, nitrogen uptake and synthesis of *Spartina alterniflora*, *Physiol. Plant*, 74, 561, 1988.

53. Giurgevich, J. R. and Dunn, E. L., Seasonal patterns of CO_2 and water vapor exchange of *Juncus roemerianus* Scheele in a Georgia salt marsh, *Am. J. Bot.*, 65, 502, 1978.

54. Blum, U., Seneca, E. D., and Stroud, L. M., Photosynthesis and respiration of *Spartina* and *Juncus* salt marshes in North Carolina: some models, *Estuaries*, 1, 228, 1978.

55. Drake, B. G., Light response characteristics of net CO_2 exchange in brackish wetland plant communities, *Oecologia*, 63, 263, 1984.

56. Eckardt, F. E., Dynamique de l'ecosysteme, strategie des vegetaux, et echanges gazeux: cas des enganes a *Salicornia fruticosa*, *Oecol. Plant.*, 7, 333, 1972.

57. Drake, B. G., Photosynthesis of salt marsh species, *Aquatic Bot.*, 34, 167, 1989.

58. McNulty, I., The effect of salt concentration on the growth and metabolism of a succulent halophyte, in *Physiological Systems in Semiarid Environment*, Hoff, C. and Riedesel, M., Eds., University of New Mexico Press, Las Cruces, 1969, 255.

59. Martin, C. E., Lubbers, A. E., and Teeri, J. A., Variability in crassulacean acid metabolism: a survey of North Carolina succulent species, *Bot. Gaz.*, 143, 491, 1982.

Chapter 5

GROWTH REGULATORS

I. INTRODUCTION

The growth of even the most salt-tolerant halophytic flowering plants is inhibited by some maximum level of soil salinity.[1] Whether obligate halophytes exist has been questioned by Ungar[2] in his research concerning the distribution of inland halophytes of the prairie region and by Barbour[3] in a comprehensive review of the problem. The decrease in dry mass production induced by increasing salinities may be the result of a change in the balance of growth regulator substances when plants are exposed to salinity, because of either an osmotic stress or a specific ion toxicity. Prisco and O'Leary[4] hypothesized that hormonal synthesis could be restricted at high salinities, thus causing a reduction in growth under osmotic stress. Another possibility is that growth regulators, such as the cytokinins, may either not be transported or their transport is greatly reduced from roots to shoots when the plant is exposed to salt stress. Translocation of cytokinins was reported to be inhibited by osmotic stress in several species of glycophytes, *Helianthus annuus*, *Phaseolus vulgaris*, and *Nicotiana rustica*. The inhibition of cytokinin production was alleviated when osmotic stress was removed.[5]

II. GIBBERELLINS

Exogenous treatments with 0.1 mM GA$_3$ stimulated the growth of the halophytes *Suaeda depressa* and *S. maritima* var. *macrocarpa* at all salinity levels up to 2.0% NaCl (Figure 1).[6] Similar stimulation of growth in *S. maritima* var. *macrocarpa* was determined in the treatments with 0.1 mM kinetin under all salinity treatments, but *S. depressa* dry mass production and shoot elongation were actually inhibited in the kinetin treatments (Figure 1). The highest relative increase in plant biomass was in the nutrient control plants treated with GA$_3$, indicating that salt treatments may be necessary to obtain maximal growth regulator production in *S. maritima* var. *macrocarpa*. Boucaud and Ungar[6] concluded from their investigation that obligate halophytes might require a salt stimulus in order to develop optimal hormonal levels and maximal dry mass production. Salt-stressed plants that were treated with kinetin may develop higher transpiration rates than untreated controls. Kirkham et al.[7] reported that plants which were exposed to salt stress and treated with kinetin had a lower water content than those of controls that were not treated with salt or kinetin. Shoot water content of the two halophytes *S. depressa* and *S. maritima var. macrocarpa* was the lowest or all the salinity × growth regulator combinations in the 2.0% NaCl treatment to which plants were exposed to kinetin, while GA$_3$ treatments did not significantly affect the tissue water content of either species (Table 1).[6] The reduced water content in kinetin-treated plants when compared with saline controls could be caused by the stimulation of stomatal opening. Mizrahi et al.[8] and Kirkham et al.[7] suggested that kinetin stimulated stomatal opening, thus facilitating transpiration and, therefore, magnifying the effect of water stress.

Ungar[9] reported that height growth of the highly salt-tolerant halophyte *Salicornia europaea* was stimulated by treatments with GA$_3$ (0.1 mM) at all salinity levels tested (0 to 3.0% NaCl). Biomass production was optimal, 881 mg/plant in the 2.0% NaCl (340 meq/l) treatment, and with the addition of 0.1 mM GA$_3$ dry mass production increased to 1086 mg/plant, the highest for all experimental treatments (Figure 2). In comparison, nutrient solution controls had a dry mass yield of 338 mg/plant, and plants treated with 3% NaCl had yields of 484 mg/plant.[9] It is possible that endogenous hormonal activity is stimulated

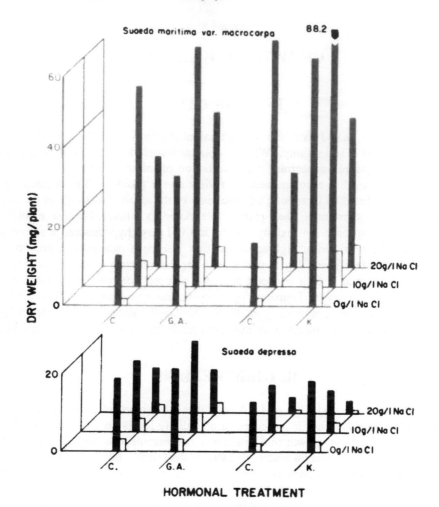

FIGURE 1.   The effect of NaCl and treatments with 10^{-4} *M* GA$_3$ and 10^{-4} *M* kinetin on the dry mass of roots and shoots of *Suaeda macrocarpa* and *S. depressa*. Black bars = shoot dry mass; clear bars = root dry mass; C = control. (From Boucaud, J. and Ungar, I. A., *Am. J. Bot.*, 63, 694, 1976.)

TABLE 1
The Effect of Salinity and Treatment with GA$_3$ (10^{-4} *M*) and Kinetin (10^{-4} *M*) on the Ratio of Tissue Water Content to Dry Mass for *Suaeda maritima* var. *Macrocarpa* and *S. depressa*

Species	NaCl (%)	Control	GA$_3$	Control	Kinetin
S. maritima var. *macrocarpa*	0	7.9:1	11.2:1	8.9:1	9.8:1
	1	12.7:1	12.6:1	13.7:1	11.0:1
	2	10.9:1	10.4:1	11.4:1	9.5:1
S. depressa	0	10.4:1	11.6:1	9.9:1	10.3:1
	1	12.2:1	12.1:1	11.1:1	10.1:1
	2	10.2:1	10.6:1	8.8:1	6.1:1

From Boucaud, J. and Ungar, I. A., *Am. J. Bot.*, 63, 694, 1976. With permission.

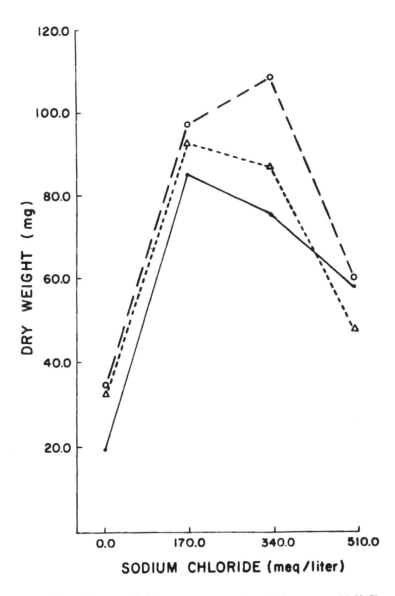

FIGURE 2.   Dry mass of *Salicornia europaea* after a 28-d treatment with NaCl (meq/l); (●) 10⁻⁴ *M* kinetin, and (○) 10⁻⁴ *M* GA₃; (△) = control.

at high salinities in obligate halophytes. However, Nieman and Bernstein[10] determined that GA₃ was only stimulatory to the growth of beans at low salinities, −0 to −1.5 bars, while at −3.0 to −4.5 bars there was no stimulation of dry mass production. They concluded that GA₃ was ineffective at overcoming the suppression of growth that was induced by higher salt concentrations. In contrast to these results, O'Leary and Prisco[11] reported that both fresh and dry mass production of *Phaseolus vulgaris* L. was stimulated by treatments with GA₃ and benzyladenine when plants were grown in a medium containing −4 bars NaCl.

The salt-tolerant woody shrub *Atriplex canescens* was stimulated in both shoot elongation and shoot multiplication when plants were treated with 1.4 μ*M* GA₃.[12] Ke-fu et al.[13] determined that treatments with 100 mg/l GA₃ promoted the growth of the halophyte *Suaeda ussuriensis* at salinities from 0.5 to 2.0% NaCl by 21.6 and 19.4%, respectively. However, at optimal salinity levels for growth at 1.0% NaCl, there was only a 2.2% increase in dry mass production It was concluded that GA₃ could partially counteract the inhibitory effects of salinity on *S. maritimus* at 0.5 and 2.0% NaCl, but had little effect at the optimal salinity concentration for growth. Ke-fu et al.[13] suggested that salt-stressed plants may have a deficiency in GA_3 and that exogenous application of this growth regulator may be overcoming this limitation for growth. Sodium chloride-induced inhibition of seedling growth in *Pennisetum typhoides* was reversed by treatments with 2.89×0.1 mM GA_3.[14] The effect of salinity was moderated by GA_3 treatments, indicating that salinity may be affecting the activity of GA_3 and, thus, limiting growth.

III. CYTOKININS

Cytokinins are produced in the roots of plants and are then transported to shoots. Since the roots are the first organ to be exposed to salinity stress, it is possible that they relay this message via a decrease in the production of cytokinins or in a reduced translocation rate to the shoots. Salt treatments from 0.15 to 0.6% NaCl caused a reduction in the cytokinin content of the root exudates of *Helianthus annuus*.[5] To determine if these results were due to osmotic conditions or a specific ion toxicity, an experiment was carried out using Carbowax 6000, and Itai et al.[5] found that cytokinin levels were also reduced by this osmoticum. The authors concluded that cytokinins provide a communication between roots and shoots of plants, facilitating responses in the shoot to changes in the level of salt stress in the root environment.[5] A reduction in cytokinin levels was measured when leaves of *Eucalyptus occidentalis* plants were exposed to 0.9% NaCl in an investigation by Itai.[15] Cytokinin levels in the leaves were 0.83 μg/g dry weight (52% of controls), while ABA concentration was 14.1 μg/g dry weight (198% of controls).

Waisel[1] reported that a saline pretreatment for two glycophytes (*Phaseolus vulgaris* and *Zea mays*) and four halophytes (*Atriplex halimus, Aeluropus litoralis, Suaeda monoica,* and *Tamarix aphyla*) yielded different results in the two groups. The glycophytes responded to salinity treatments with a decrease in cytokinin levels in their bleeding sap, while for the halophytes the cytokinin level did not decrease when plants were exposed to salt pretreatments (Table 2). For *Aeluropus litoralis,* some stimulation of cytokinin production was indicated at R_f 7-10 in the soybean callus growth bioassay (Table 2).[1]

IV. AUXIN AND ETHYLENE

Reduction of growth in halophytes under saline conditions may be due to the accumulation of inhibitor substance such as ABA under salt stress or the enhanced production of ethylene, which is also inhibitory to plant growth. Plants of *Suaeda fructicosa* collected from saline habitats were found to produce two growth inhibitors. The growth regulator IAA was found to reverse the effects of the inhibitory extracts isolated from *S. fructicosa* leaves by Khan et al.[16] An inhibitor isolated from *S. fructicosa* was reported to be a phenolic compound by Khizar and Khan.[17] The effect of this inhibitor was not reduced at high temperatures. The addition of ACC (1-aminocyclopropane-1-carboxylic acid), an ethylene precursor, in treatments of *Allenrolfea occidentalis* caused a sharp increase in ethylene production in up to 1.4 M NaCl, and all treatments from 0.5 to 4 M NaCl had higher ethylene levels than the distilled water controls (Figure 3).[18,19] In contrast, the glycophyte *Vigna radiata* had its highest ethylene production at 0.2 M NaCl, indicating that glycophytes have different salt optima than halophytes. It may be that the conversion system for the production of ethylene in halophytes is NaCl dependent.[19] Other inorganic salts besides NaCl such as KCl, $MgCl_2$, and $CaCl_2$ and the organic osmoticum PEG 8000 produced similar increases in ethylene production as NaCl.[19] Plant sap of *A. occidentalis* tissue had a NaCl content of 1.2 M, and the osmotica used all stimulated the conversion of ACC to ethylene at this

TABLE 2
The Effect of Salinity on Cytokinin Content of *Aeluropus litoralis* and *Tamarix aphylla,* Using Soybean Callus Growth (g/30 d) as a Bioassay[a]

	Aeluropus litoralis		*Tamarix aphyla*	
R_f	+ NaCl	− NaCl	+ NaCl	− NaCl
1	0.063 ± 0.044	0.045 ± 0.020	0.079 ± 0.025	0.103 ± 0.014
2	0.057 ± 0.030	0.053 ± 0.001	0.114 ± 0.018	0.117 ± 0.025
3	0.060 ± 0.022	0.064 ± 0.008	0.111 ± 0.010	0.108 ± 0.017
4	0.067 ± 0.025	0.053 ± 0.018	0.125 ± 0.002	0.095 ± 0.041
5	0.068 ± 0.026	0.087 ± 0.032	0.085 ± 0.031	0.122 ± 0.033
6	0.084 ± 0.026	0.143 ± 0.028	0.112 ± 0.031	0.070 ± 0.050
7	0.244 ± 0.104	0.053 ± 0.023	0.105 ± 0.031	0.100 ± 0.044
8	0.155 ± 0.015	0.098 ± 0.032	0.103 ± 0.044	0.091 ± 0.025
9	0.136 ± 0.036	0.022 ± 0.006	0.079 ± 0.010	0.121 ± 0.054
10	0.146 ± 0.038	0.032 ± 0.015	0.068 ± 0.020	0.069 ± 0.035

[a] Kinetin 0 ppm = 0.076, 0.001 ppm = 0.130, 0.01 ppm = 0.166, 0.377 ppm = 0.085.

From Waisel, Y., *Biology of Halophytes,* Academic Press, New York, 1972. With permission.

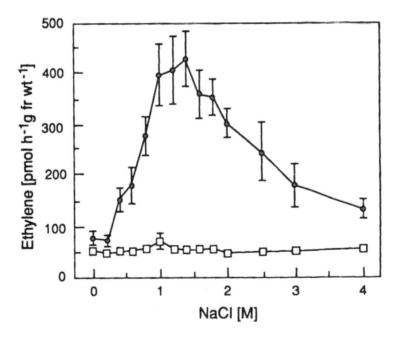

FIGURE 3.   The effect of NaCl on ACC to ethylene conversion in the halophyte *Allenrolfea occidentalis.* Open squares = − ACC; closed circles = + ACC. (From Chrominski, A., Weber, D. J., and Smith, B. N., *Naturwissenschaften,* 73, 274, 1986.)

osmotic concentration. Chrominski et al.[19] concluded from these experiments that the stimulus for ethylene production appears to be osmotic rather than a specific ion toxicity caused by NaCl. The effect of NaCl on the production of ethylene and ethane in *V. radiata* was studied by Chrominski et al.[20] It was determined that NaCl caused an inhibition of ethylene production at higher salt concentrations (0.5 to 1.0 M NaCl), and ethane production was usually from either lipid oxidation or membrane damage.[20]

Experimental data indicated that some halophytes respond to salt with an increase in biomass production.[1,9] The nature of this growth stimulation by relatively high salinity concentrations is not clearly understood. However, Caldwell[21] has suggested that high salinity may provide the rapidly permeating electrolytes necessary for osmotic adjustment during the growth of halophytes. Growth regulators may play a significant role in controlling the permeability of membranes to ion transport. It was determined by Collins and Kerrigan[22] that both ABA and kinetin had strong effects on the movement of ions into the roots of plants. Kinetin increased ion flux into the root and reduced ion and water transport from the root, while ABA increased the flux of ions into roots and also increased the transport of water and ions from the root.

V. CONCLUSIONS

Data regarding the effects of growth regulators on the salt tolerance of halophytes do not present any clear answers at this time and appear to be somewhat inconclusive. Studies using exogenous applications of growth regulators indicate that GA_3 may overcome the inhibitory effects of salinity at lower concentrations but is not effective when plants are treated with salinity levels of seawater.[6,9,13] It may be that salt-stressed plants have a deficiency in GA_3, and the exogenous application of this growth regulator may be overcoming this factor which is limiting plant growth. Plants exposed to salt stress and cytokinins have been reported to have a lower water content than untreated plants.[7,8] Cytokinins influence stomatal movement, and exogenous applications of this growth regulator may cause an increase in the transpiration rate which would put plants under water stress, causing a reduction in plant water content and a decrease in growth. Several researchers have reported reductions in endogenous cytokinin levels when plants were exposed to salt stress.[6,15] The quantity of the growth inhibitors ABA and ethylene have been reported to increase when plants are exposed to an increase in the salinity of the medium in which they are grown.[17,19,20] Increases in ethylene production appear to be caused by an osmotic effect rather than a specific ion toxicity.

Further investigations are necessary to understand the change in the balance of plant growth regulators when plants are exposed to salinity stress. It will be important to determine if the chief effect on growth regulator production is due to an osmotic effect or a specific ion toxicity caused by hypersaline conditions. The relative effects of the various growth regulators on plants under salt stress is not well understood, and future research should be devoted to sort out the responses of individual growth regulators to salt stress and their combined effect on plant growth.

REFERENCES

1. **Waisel, Y.,** *Biology of Halophytes*, Academic Press, New York, 1972.
2. **Ungar, I. A.,** Salt tolerance of plants growing in saline areas of Kansas and Oklahoma, *Ecology*, 47, 154, 1966.
3. **Barbour, M. G.,** Is any angiosperm an obligate halophyte?, *Am. Midl. Nat.*, 84, 105, 1970.

4. Prisco, J. T. and O'Leary, J. W., Enhancement of intact bean leaf senescence by NaCl salinity, *Physiol. Plant,* 27, 95, 1972.

5. Itai, C., Richmond, A., and Vaadia, Y., The role of root cytokinins during water and salinity stress, *Isr. J. Bot.,* 17, 187, 1968.

6. Boucaud, J. and Ungar, I. A., Influence of hormonal treatments on the growth of two halophytic species of *Suaeda, Am. J. Bot.,* 63, 694, 1976.

7. Kirkham, M. B., Gardner, W. R., and Gerloff, G. C., Internal water status of kinetin-treated plants, *Plant Physiol.,* 53, 241, 1974.

8. Mizrahi, Y., Blumenfeld, A., and Richmond, A. E., Abscisic acid and transpiration in leaves in relation to osmotic root stress, *Plant Physiol.,* 46, 169, 1970.

9. Ungar, I. A., The effects of salinity and hormonal treatments on the growth and ion uptake of *Salicornia europaea, Soc. Bot. Fr. Actualites Bot.,* 3-4, 95, 1978.

10. Nieman, R. H. and Bernstein, L., Interactive effects of gibberellic acid and salinity on the growth of beans, *Am. J. Bot.,* 46, 667, 1959.

11. O'Leary, J. W. and Prisco, J. T., Response of osmotically stessed plants to growth regulators, *Adv. Front. Plant Sci.,* 25, 129, 1970.

12. Wochok, Z. S. and Sluis, C. J., Gibberellic acid promotes *Atriplex* shoot multiplication and elongation, *Plant Sci. Lett.,* 17, 363, 1980.

13. Ke-fu, Z., Ming-liang, L., and Jia-yao, L., Reduction by GA$_3$ of NaCl-induced inhibition of growth and development in *Suaeda ussuriensis, Aust. J. Plant Physiol.,* 13, 547, 1986.

14. Huber, W. and Sankhla, N., Eco-physiological studies on Indian arid zone plants. II. Effect of salinity and gibberellin on the activity of the enzyme of amino-acid metabolism in leaves of *Pennisetum typhoides, Oecologia,* 13, 271, 1973.

15. Itai, C., Response of *Eucalyptus occidentalis* to water stress induced by NaCl, *Physiol. Plant,* 43, 377, 1978.

16. Khan, M. I., Khan, M. A., and Khizar, T., Plant growth regulators from species differing in salt tolerance as affected by soil salinity, *Plant Soil,* 45, 267, 1976.

17. Khizar, T. and Khan, M. I., Effect of temperature on the activity of growth inhibitor-A isolated from *Suaeda* leaves, *Biol. Plant.,* 19, 231, 1977.

18. Chrominski, A., Weber, D. J., and Smith, B. N., NaCl-salinity-dependent conversion of ACC to ethylene in the halophyte, *Allenrolfea occidentalis, Naturwissenschaften,* 73, 274, 1986.

19. Chrominski, A., Bhat, R. B., Weber, D. J., and Smith, B. N., Osmotic stress-dependent conversion of 1-aminocyclopropane-1-carboxylic acid (ACC) to ethylene in the halophyte *Allenrolfea occidentalis, Environ. Exp. Bot.,* 28, 171, 1988.

20. Chrominski, A., Khan, M. A., Weber, D. J., and Smith, B. N., Ethylene and ethane production in response to salinity stress, *Plant. Cell. Environ.,* 9, 687, 1986.

21. Caldwell, M. M., Physiology of desert halophytes, in *Ecology of Halophytes,* Reimold, R. J. and Queen, W. H., Eds., Academic Press, New York, 1974, 355.

22. Collins, J. C. and Kerrigan, A. P., Hormonal control of ion movements in the plant root?, in *Ion Transport in Plants,* Anderson, W. P., Ed., Academic Press, New York, 1973, 589.

Chapter 6

PLANT WATER STATUS

I. INTRODUCTION

One of the most significant characteristics of all halophytes is the capacity to adjust their tissue water potentials to a level that is lower than that of the soil water potential in the habitat in which they are growing.[1] Shoot growth in halophytes is dependent on ion transport, because of the ion requirements that are necessary for osmotic adjustment. The high level of ionic accumulation in some halophytes may be related to the high mineral requirement for the maintenance of turgor and osmotic adjustment in salt marsh species.[2] Munns[3] concluded that osmotic adjustment by plants did not directly promote plant growth. In saline soils, most plant species are affected by a decrease in growth, even though turgor regulation takes place due to the uptake of NaCl. Although osmotic adjustment may occur, leaf expansion may be reduced because of salt stress, causing a decrease in plant production.[3] A number of different mechanisms are used by salt-tolerant species to achieve osmotic adjustment, including inorganic ion accumulation, synthesis or accumulation of organic compounds, and water loss.[4-6] In order to avoid the toxic effects of excessive ion accumulation, halophytes have developed a number of mechanisms, including succulence (*Suaeda* — leaf succulence, *Salicornia* — stem succulence), salt glands (*Limonium, Distichlis*), salt hairs (*Atriplex*), dehiscence of plant organs (*Atriplex* — leaves, *Salicornia* — branches), and root exclusion (*Rhizophora* — ultrafiltration, *Atriplex* and *Limonium* — active excretion).[7-13]

II. FIELD MEASUREMENTS OF PLANT WATER POTENTIAL

The relationship between plant zonation and soil water potentials was investigated by Dawson and Bliss[14] on a high-arctic brackish salt marsh. Soil water potentials averaged -0.98 MPa in the zone dominated by *Puccinellia phryganodes* and *Carex ursina*, but measurements of plant water potentials equaled only -0.39 and -0.31 MPa for each of the species, respectively. Dawson and Bliss[14] concluded that the plants in the more saline site must be utilizing a less saline source of water than found in the 0- to 10-cm soil cores sampled, because plant water potentials were not sufficiently low for plants to take up water under these conditions. Freshwater from melting ice may be the water source for these species growing in the seaward portion of the marsh.

Langlois[15] determined the seasonal change in soil water and ionic content and water and ionic content of *Salicornia europaea* plants at the Sallnelles salt marsh on the Orne Estuary (France). The soil water content did not differ significantly from April to August, ranging from 69.2 to 74.3% during the growing season. The soil Na and Cl contents were lower in April than in August, with Na ranging from 0.8 to 1.0% and Cl from 0.8 to 1.0%. Osmotic potential of shoots ranged from approximately -1.5 MPa in April to -2.2 MPa in June. Plant Na and Cl content ranged from 1.8% Cl in April to 3.8% in June, while Na content ranged from 1.2% in April to 2.0% in May on a percent tissue water basis. Langlois[15] reported that the osmotic potential for roots was much less negative than that for shoot segments.

Naidoo[16] found that the soil water potentials in a South African mangrove swamp were lower in the wet hot summer, about -3 MPa, and higher in the cool dry winter, about -2 MPa. Leaf water potentials at midday ranged from -3.8 to -6.0 MPa for *Avicennia marina*, from -3.4 to -5.4 MPa for *Bruguiera gymnorrhiza* and from -3.8 to -4.9 MPa for *Rhizophora mucronata*. Leaf stomatal conductance was reported to be lower for all three

TABLE 1
Seasonal Trends in the Osmotic Potential (−MPa) and Relative
Water Content (% at 0 Turgor) of *Spartina alterniflora*[a]

	June	July	August	September
Relative water content (%)				
Short	70.1	65.4	65.4	63.7
Tall	70.1	65.4	78.2	—
Osmotic potential (MPa)				
Short	−4.23	−4.86	−4.52	−5.04
Tall	−3.06	−3.24	−3.42	—

[a] Data calculated from graph.

From Drake, B. G. and Gallagher, J. L., *Oecologia*, 62, 368, 1984. With permission.

species during the summer, when soil water potentials were more negative than in the winter.[16]

Predawn xylem water potentials were measured for the phreatophyte *Sarcobatus vermiculatus* by Romo and Haferkamp.[17] They determined that xylem water potentials ranged from −1.5 MPa in July to −4.0 MPa in August and September 1982. Osmotic potentials followed a similar pattern and at full turgor ranged from −2.0 MPa from April to July to −3.5 MPa in August. Romo and Haferkamp[17] concluded that *S. vermiculatus* was able to maintain turgor by rooting in soil zones that were saturated with water.

The relationship between tall and short *Spartina alterniflora* and the relative water status of plants were investigated by Drake and Gallagher.[18] Plant height and biomass production declined sharply with increases in interstitial soil salinity concentration. The short form of *S. alterniflora* grew in soils with 2.0 to 5.0% total salts, while tall plants were exposed to a relatively constant 2.0% total salts. Plant osmotic potential was more negative and the pressure potential was lower in the short from of *S. alterniflora* when compared with the tall form (Table 1). The reduction in water potential in both forms of *S. alterniflora* from dawn to midday was caused by a 7 to 24% reduction in RWC (relative water content) of plants. Drake and Gallagher[18] concluded that plant height could be diminished by reductions in leaf water potential and turgor pressure, which would cause a reduction in expansive growth.

Scholander et al.[19] determined that salt-secreting mangroves (*Aegialitis annulata, Avicennia marina, Aegiceras corniculatum*) had higher sap ionic concentrations than did nonsecreting species (*Bruguiera exaristata, Sonneratia alba, Rhizophora mucronata, Lunnitzera littorea*).[19] The nonsecreting species had a xylem sap that was almost salt-free (−0.005 MPa to −0.05 MPa), while salt-secreting species had xylem sap concentrations between −0.1 and −0.2 MPa. Salt-secreting species had secretions which were equivalent to seawater concentrations or higher. Active transport in roots caused a reduction in ion content of the xylem tissue of nonsecreting species.[19]

Scholander et al.[20] determined that sap xylem pressure potential in halophytes ranged from −3.5 to −6.0 MPa when compared with the −2.5 MPa osmotic potential of seawater. The NaCl content of leaf cells was determined to be 100 times greater than that in the xylem sap. Scholander[21] and Scholander et al.[11] concluded that roots act to desalinate the water entering the xylem and can maintain a constant concentration of ions in the xylem which is independent of the rate of transpiration.

Ungar[22,23] determined that the water potential of *Atriplex triangularis* and *Salicornia europaea* plants varied with changing soil salinity during the growing season. The variation in plant water potential paralleled changes in soil water potentials, with less negative values

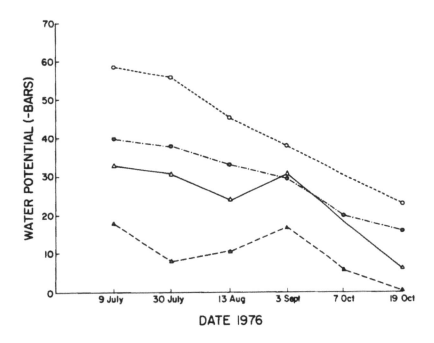

FIGURE 1. Seasonal change in soil water potential (dark symbols) and leaf water potential (open symbols) for *Atriplex triangularis* at low salt (triangle) and high salt (circle) locations. (From Ungar, I. A., *Bot. Gaz.*, 138, 498, 1977. With permission.)

in the spring and more negative water potentials in the summer months when soil salinity levels increased (Figure 1). Field measurements of plant and soil water potentials were made over the growing season in saline location in Ohio.[22] *A. triangularis* leaf water potentials ranged from −0.7 to −6.2 MPa, while soil water potentials ranged from −0.1 to −1.4 MPa. The leaf water potential of *A. triangularis* was highly correlated with soil water potentials (r = −0.90). In an area dominated by *S. europaea*, soil water potentials ranged from −1.6 to −6.0 MPa and shoot water potentials ranged from −6.5 to −9.1 MPa. Shoot water potentials for *S. europaea* were not highly correlated with soil water potentials (r = −0.45). The highly negative shoot water potentials for *S. europaea*, even at soil salinities of −1.0 to −3.0 MPa, could be due to ion accumulation, but may also be in part due to an increase in organic osmotica. Ungar[23] reported that soil water potentials ranged from −0.4 MPa in May to −1.9 MPa in June, averaging −0.9 MPa, in the low salt zone of an Ohio salt pan and from −0.7 to −2.9 MPa, averaging −1.9 MPa, in the high salt zone (Table 2). Leaf water potentials of *A. triangularis* averaged −1.9 MPa in the low salt zone and −2.9 MPa in the high salt zone. Ungar[23] concluded that leaves of *A. triangularis* were able to adjust osmotically to changing soil water potentials during the growing season.

Riehl and Ungar[24] determined tht soil water potentials in an *S. europaea* community decreased from May (−3.0 MPa) to July (< −8.0 MPa). Plant shoot water potentials changed during the growing season and were more negative than the corresponding soil water potentials, indicating that the succulent *S. europaea* was able to adjust osmotically to increasing salinity stress (Figure 2). Changes in plant water potential were closely related to increases in the Na and Cl content of plant shoots.

The leaf water potentials of *A. triangularis* were studied by Riehl and Ungar[25] in three habitats that varied in soil salinity stress. The soils in the *Hordeum-Atriplex* community had Na levels that ranged from 93 μM/ml in May to 400 μM/ml in July. Zones dominated by *Atriplex* had Na levels that ranged from 80 μM/ml in May to 698 μM/ml in July. Plant water potentials in the *Hordeum-Atriplex* zone ranged from −1.0 to −3.0 MPa, while in

TABLE 2
Changes in Soil Water Potential ($-$MPa) and Plant Water Status ($-$MPa) in High-Salt and Low-Salt Sites Containing *Atriplex triangularis*

Month	Low-salt site			High-salt site		
	Soil WP	Plant WP	Plant OP	Soil WP	Plant WP	Plant OP
March	0.6	1.5	—	0.7	1.7	—
April	1.1	1.6	1.9	1.6	3.3	3.6
May	0.4	1.1	1.5	0.7	2.0	2.4
June	1.9	3.3	3.6	2.9	3.3	4.0
August	0.9	2.4	2.7	1.3	2.9	3.5
September	0.5	2.2	2.4	1.6	4.1	4.7
October	1.1	2.0	2.9	1.6	3.0	3.6

From Ungar, I. A., in *Symposium on the Biology of Atriplex and Related Chenopods*, Tiedemann, A. R., McArthur, E. D., Stutz, H. C., Stevens, R., and Johnson, K. L., Eds., Gen. Tech. Rep. INT-172, Forest Service, U.S. Department of Agriculture, Ogden, UT, 1984, 40.

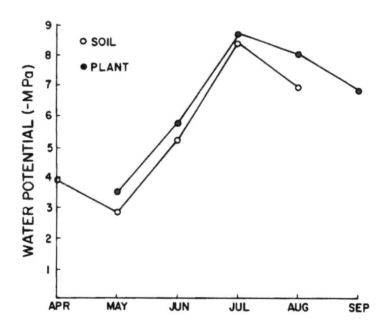

FIGURE 2.   Monthly fluctuations in soil- and shoot-water potentials for *Salicornia europaea* in a highly saline field location. (From Riehl, T. E. and Ungar, I. A., *Oecologia*, 54, 193, 1982. With permission.)

the more saline area the plant water potentials in July were very negative (< -5.0 MPa), but none of these plants survived to reproductive maturity in October (Figure 3). Leaf water potentials of *A. triangularis* were highly correlated ($r = -0.96$) to soil water potentials.

Osmotic potentials were determined for the sap of several halophytes by Kaplan and Gale.[26] Using a cryoscopic technique, they found that for *Atriplex halimus* the osmotic potential of the expressed sap ranged from -2.9 to -4.1 MPa, while *Avicennia marina* had -4.8 MPa and *Reaumuria palaestina* had -2.9 MPa.

Field studies with three coastal halophytes from the Sonoma Creek salt marsh (California)

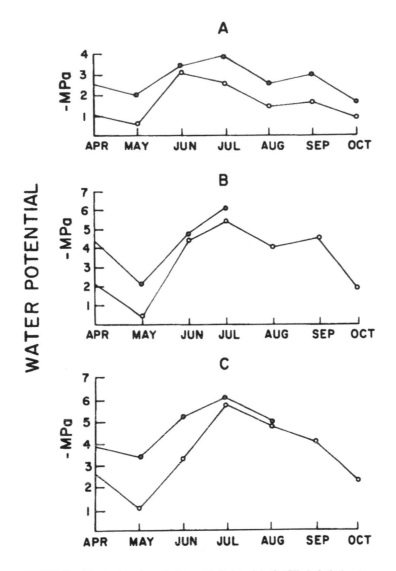

FIGURE 3.   Fluctuations in soil- (open circles) and leaf- (filled circles) water potentials for *Atriplex triangularis* in three zonal communities: (A) *Hordeum - Atriplex*, (B) *Atriplex* dry zone, and (C) *Atriplex* wet zone. (From Riehl, T. E. and Ungar, I. A., *Acta Oecol. Oecol. Plant.*, 4, 27, 1983. With permission.)

indicated that plant water potentials decreased from May to September as the soil water potentials became more negative.[27] The soil water potential was similar in all three zones during the spring (−0.7 to −1.2 MPa), but in the summer months *Spartina foliosa* tended to be higher (−1.7 to −2.3 MPa), *Scirpus robustus* intermediate (−1.8 to −2.0 MPa), and *Salicornia virginica* lower (−1.8 to −3.5). *Spartina*, *Scirpus*, and *Salicornia* all showed a daily rhythm in plant water potentials, with higher values in early morning and evening and lower water potentials from late morning to afternoon. Leaf stomatal conductance in the morning was higher early in the growing season when soil water potentials were higher, but decreased from June through September for *Spartina* and *Scirpus* (Table 3). However little change occurred in the midday values. Ustin et al.[27] determined that plant species were able to lower their water potential to adjust osmotically and reduce leaf stomatal conductance to conserve water as salinity stress increased in the salt marsh during the summer months.

TABLE 3

Seasonal Progression of Morning Maximum and Mean
Midday Leaf Conductance (m/s) during the Growing Season
for *Spartina foliosa* and *Scirpus robustus*[a]

	May	June	July	August	September
Morning (max)					
Scirpus robustus	5.1	11.3	7.4	7.4	3.8
Spartina foliosa	3.5	13.0	7.9	4.8	4.7
Midday (mean)					
Scirpus robustus	2.1	5.3	7.4	6.4	3.2
Spartina foliosa	2.1	4.9	5.3	5.3	4.7

[a] Data calculated from graph.

From Ustin, S. L., Pearcy, R. W., and Bayer, D. E., *Bot. Gaz.*, 143, 368, 1982.
With permission.

TABLE 4

Seasonal Variation in Plant Water Status for Salt Marsh Species

Species, season	Xylem pressure potential (−MPa)	Osmotic potential (−MPa)	Relative water content (%)
Borrichia frutescens (short)			
Summer	3.7 ± 0.22	2.6 ± 0.07	87.8 ± 1.0
Winter	3.1 ± 0.06	2.0 ± 0.28	85.3 ± 2.4
B. frutescens (tall)			
Summer	2.8 ± 0.12	2.3 ± 0.25	89.5 ± 0.7
Winter	1.8 ± 0.24	1.6 ± 0.21	84.6 ± 1.5
Salicornia virginica			
Summer	—	2.9 ± 0.52	93.4 ± 0.6
Winter	—	2.2 ± 0.20	97.7 ± 0.8
Batis maritima			
Summer	4.4 ± 0.22	3.6 ± 0.42	92.1 ± 0.9
Winter	3.0 ± 0.85	2.7 ± 0.40	85.9 ± 5.1
Juncus roemerianus			
Summer	2.2 ± 0.22	—	—
Winter	2.9 ± 0.40	1.7 ± 0.26	68.5 ± 10.1
Spartina alterniflora			
Summer	2.7 ± 0.18	—	91.8 ± 1.5
Winter	2.7 ± 0.17	2.5 ± 0.29	83.9 ± 5.5

From Antlfinger, A. E. and Dunn, E. L., *Am. J. Bot.*, 70, 561, 1983. With permission.

Antlfinger and Dunn[28] found that species growing in higher soil salinities on the Sapelo Island salt marsh in Georgia had more negative xylem pressure potentials than plants growing in less saline areas. Measurements of xylem pressure potential in the halophytes *Borrichia frutescens*, *Salicornia virginica*, *Batis maritima*, *Juncus roemerianus*, and *Spartina alterniflora* ranged from −2.2 to −2.9 MPa for the graminoids and −1.8 to −4.3 MPa for the succulents (Table 4).[28] Tall *Borrichia frutescens* had a xylem pressure potential in the summer of −2.8 MPa, while short plants had a xylem pressure potential of −3.7 MPa. Xylem pressure potentials for *Batis maritima*, *J. roemerianus*, and *S. alterniflora* in the summer were −4.4, −2.2, and −2.7 MPa for each species, respectively (Table 4). Salt content of the succulents ranged from 27 to 48% total salts under interstitial salinity conditions ranging from −1.9 to −2.9 MPa. Salt uptake ranged from 0.6 g NaCl/g C/d in *Borrichia frutescens* to 4.3 g NaCl/g C/d in *Batis maritima*. Antlfinger and Dunn[28] concluded that some mechanism of salt exclusion may exist in the roots, based on the rate of transpiration and potential salt accumulation which was 12 to 30 times the observed values (9 to 54 g NaCl/g C/d).

Stomatal conductance and leaf water potentials of *Cakile edentula* were closely correlated with habitat soil water potentials.[29] The more negative soil water potentials for foredune (< −1.0 MPa in June) vs. beach sites (> −0.3 MPa in June) were associated with reduced stomatal conductance and lowered plant water potential at the foredune sites. Tyndall et al.[29] determined that seedling mortality and growth inhibition on foredunes were probably related to low soil water potentials at this site.

III. PLANT WATER STATUS RESPONSES TO SALT STRESS

The marine angiosperm seagrass *Posidonia australis* grew in a narrow hypersaline salinity range in Botany Bay (Australia), 2.0 to 3.5% total salts, and in Shark Bay (Australia), 3.5 to 5.0% total salts.[30] Laboratory experiments indicated that leaf growth was unaffected by salinities ranging from 1.4 to 5.7% total salts, while net photosynthesis was unaffected by reducing salinity from 3.4 to 1.9% total salts. Turgor pressure varied only between 0.67 to 1.52 MPa over a range of osmotic potentials from −0.83 to −3.89 MPa. Osmotic potentials were adjusted by the accumulation of inorganic ions, Na, K, and Cl in leaves and rhizomes. Tyerman et al.[30] concluded that salt tolerance was not limiting the habitat of *P. australis* to stenohaline conditions.

Karimi and Ungar[31] determined that xylem pressure potentials for *A. triangularis* ranged from −0.75 MPa in −0.06 MPa NaCl to −2.75 MPa in −2.0 MPa NaCl. The xylem pressure potentials were highly correlated (r = 0.77) to medium salinity levels (Figure 4). Leaf water potentials and osmotic potentials were also strongly correlated to medium salinity levels (r = 0.87, r = 0.94), indicating that *A. triangularis* shoots adjusted their water status to soil salinity conditions (Figure 5). Root xylem pressure potentials were also significantly correlated to medium salinity levels (r = 0.69).

Ewing et al.[32] reported that increasing osmotic stress from 0.05 to 2.0% sea salt concentrations caused a corresponding decrease in shoot and root biomass production in three halophytic species, *Puccinellia phryganodes*, *Carex paleacea*, and *Scirpus americanus*. The FW to DW ratio decreased significantly for *Puccinellia* and *Scirpus* as salinity increased, but remained relatively constant for *Carex* (Table 5). Organic substances such as glycine-betaine increased in all three species when they were exposed to salt stress, with *Scirpus* having a fivefold increase. *Puccinellia* had the greatest increase in proline content, nearly fivefold, while the other two species had only small increases.

Poa cookii, a subantarctic maritime tussock grass, responded to increased NaCl concentrations by lowering the water potential in its leaves.[33] Duration of exposure to 300 to 700 m*M* NaCl, external salt concentrations, and leaf age all significantly affected the Na content of plant leaves. The Na content of both young and old leaves increased with increases in media salinity. However, Pammenter and Smith[33] found that metabolically inactive old leaves served as a sink for the accumulation of Na, preventing toxic levels of accumulation of Na in younger leaves.

Jefferies et al.[4] determined that shoot and root water potentials of four species of halophytes *Plantago maritima*, *Triglochin maritima*, *Limonium vulgare*, and *Halimione portulacoides* adjusted to the osmotic potentials of the media in laboratory treatments ranging from −0.05 to −2.4 MPa polyethylene glycol (PEG) and seawater (Table 6). Isotonic concentrations of PEG and seawater produced similar trends, but −2.4-MPa PEG treatments produced consistently lower plant water potentials than seawater in *Plantago maritima* roots

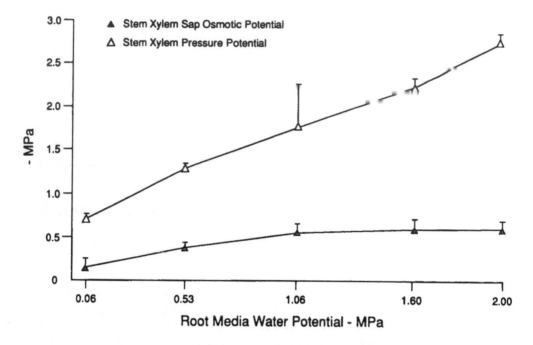

FIGURE 4.   Xylem pressure potential and xylem sap osmotic potential for shoots from *Atriplex triangularis* plants grown at different NaCl levels. (From Karimi, S. H. and Ungar, I. A., in *Symposium on the Biology of Atriplex and Related Chenopods*, Tiedemann, A. R., McArthur, E. D., Stutz, H. C., Stevens, R., and Johnson, K., Eds., Gen. Tech. Rep. INT-172, Forestry Service, U.S. Department of Agriculture, Ogden, UT, 1984, 124.)

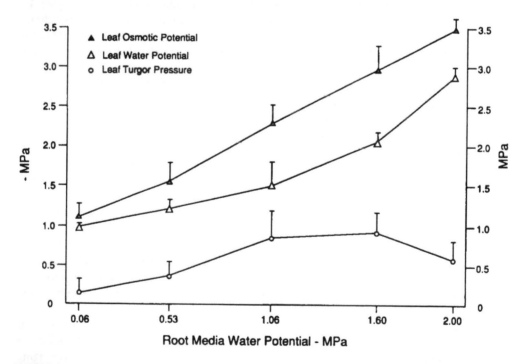

FIGURE 5.   Water potential (− MPa), osmotic potential (− MPa), and pressure potential (MPa) grown at different NaCl levels. (From Karimi, S. H. and Ungar, I. A., in *Symposium on the Biology of Atriples and Related Chenopods*, Tiedemann, A. R., McArthur, E. D., Stutz, H. C., Stevens, R., and Johnson, K., Eds., Gen. Tech. INT-172, Forestry Service, U.S. Department of Agriculture, Ogden, UT, 1984, 124.)

TABLE 5
Adjustment of FW to DW Ratio in Response to Different Levels of Salinity

Species	NaCl (%)				
	0.05	0.5	1.0	1.5	2.0
Puccinellia phryganodes	5.0	—	4.3	—	3.8
Carex paleacea	3.7	3.4	3.7	3.6	3.5
Scirpus americanus	6.4	5.8	5.7	5.6	5.5

From Ewing, K., Earle, J. C., Piccinin, B., and Kershaw, K. A., *Can. J. Bot.*, 67, 521, 1989. With permission.

TABLE 6
Water Potential of Shoots and Roots of Halophytes Grown at Different Concentrations of Seawater and PEG 6000

Species	Seawater (MPa)					PEG (MPa)				
	−0.05	−0.2	−0.5	−1.2	−2.4	−0.05	−0.2	−0.5	−1.2	−2.4
Plantago maritima										
Shoot	−0.7	−0.8	−1.3	−1.9	−2.6	−0.7	−1.1	−1.3	−1.9	−3.9
Root	−0.6	−0.7	−1.3	−1.7	−2.5	−0.6	−1.3	−1.3	−1.7	−3.7
Triglochin maritima										
Shoot	−0.6	−0.8	−1.6	−1.8	−2.5	−0.6	−0.8	−1.8	−2.7	−3.4
Root	−0.6	−0.8	−1.5	−1.6	−2.4	−0.6	−0.7	−1.8	−2.3	−2.6
Limonium vulgare										
Shoot	−0.8	−1.5	−2.6	−3.5	−4.8	−0.8	−1.7	−2.6	−3.5	−4.6
Root	−0.6	−0.7	−1.7	−2.4	−3.3	−0.6	−1.2	−2.2	−3.1	−3.5
Halimione portulacoides										
Shoot	−1.5	−1.6	—	−2.5	−3.6	—	—	—	—	—

From Jefferies, R. L., Rudmik, T., and Dillon, E. M., *Plant Physiol.*, 64, 989, 1979. With permission.

and shoots. *Limonium* and *Halimione* lowered their water potentials more in excess of media water potentials than did *Plantago* and *Triglochin*. All four species were able to adjust osmotically to both PEG and NaCl with shoot osmotic potentials in −2.4-MPa seawater ranging from −4.8 MPa in *Limonium* to −2.5 MPa in *Triglochin*, while osmotic potential values in PEG ranged from −4.6 MPa in *Limonium* to −3.4 MPa in *Triglochin*. Short-term (<14 d), treatments with PEG were not injurious to plants, and the results indicated that the effect of PEG and seawater on plants was due to low osmotic potentials and not to a specific ion toxicity. Jefferies et al.[4] concluded that the chief effect of low water potentials was an osmotic effect rather than a specific ion toxicity induced by NaCl.

Matthews et al.[34] reported that low soil water potentials inhibited leaf expansion in *Helianthus annuus*. Acclimation of leaves allowed growth at leaf water potentials too low for growth in control plants. Growth occurred because leaves adjusted osmotically and maintained turgor in their cells. Acclimated leaves did not recover as well as controls when rewatered because the tissue was less extensible and required more turgor to initiate growth then control plants that were not exposed to low water potentials for extended periods.[34]

Houchi et al.[35] reported that leaf stomatal resistance of *Plantago maritima* decreased from a maximum of 5.2 s/cm in controls of −0.09 MPa to 2.0 s/cm in NaCl treatments of −0.25 MPa. Plants exposed to −0.30 MPa Na_2SO_4 did not differ significantly in stomatal resistance from controls. Water uptake increased in plants treated with NaCl and decreased from control values for *P. maritima* plants treated with Na_2SO_4.

The effect of NaCl and mannitol shock (−0.2 and −0.8 MPa) on transpiration rates of *P. lanceolata* and *P. maritima* was ascertained by Ferard et al.[36] Both species had a decrease in transpiration with osmotica added to the medium. *P. lanceolata* had a 52% reduction in transpiration rate at −0.82 MPa NaCl and a 75% reduction in isotonic solutions of mannitol, while *P. maritima* had only a 25% reduction in its transpiration rate in −0.82 MPa NaCl and a 41% reduction in isotonic solutions of mannitol. At lower levels of salinity shock (−0.21 MPa) the effect of NaCl and mannitol was the same for both species, yielding about a 10% reduction in transpiration. These data indicate that when the species are exposed to salt stress, the stomatal resistance to water loss is better adjusted in the halophyte *P. maritima* than in the glycophyte *P. lanceolata*.[36] Coudret[37] reported that the glycophyte *P. lanceolata* had more difficulty in taking up water and maintaining turgor than did the halophyte *P. maritima* when plants were exposed to 0.25% NaCl. For *P. maritima* an addition of NaCl caused a decrease in the osmotic potential of cells without modifying their state of turgor, while in the glycophyte *P. lanceolata*, retaining turgor was more difficult in the presence of salts. After 3 d of exposure to NaCl, *P. lanceolata* (−1.0 MPa) had a greater change in osmotic potential than did *P. maritima* (−0.5 MPa). Pressure potential increased from 0.96 to 3.19 MPa in *P. lanceolata* and from 0.47 to 1.20 MPa in *P. maritima* when the two species were exposed to NaCl.[37]

Ferard and Coudret[38] determined that the water potential of *P. maritima* adjusted to salt stress ranging from 0 to −1.5 MPa. Data for the adjustment to mannitol solutions correlated with the results obtained for NaCl, with water potential decreasing from −1.2 MPa in a 0-MPa medium to −1.8 MPa for a soil water potential of −1.5 MPa. In the case of mannitol treatments, the pressure potential decreased 0.4 MPa, while for NaCl treatments turgor was not reduced even at −1.5 MPa NaCl, and the pressure potential remained at about 0.9 MPa. Ferard and Coudret[38] suggested that this pressure potential adjustment to low water potentials in NaCl treatments, vs. the reduction in pressure potential in mannitol treatments, was due to osmotic adjustment because of the transport of ions to aerial organs of the plant.

The aquatic plant species *Alternanthera philoxeroides* is capable of growing under a broad range of salinities. Bolanos and Longstreth[39] determined the effect of increasing salinity over time on plant water status. The plant water potential was osmotically adjusted to that of the growth solutions between 0 to 400 mM NaCl. Lowering of the plant osmotic potential at all salinity treatments permitted the maintenance of a water potential gradient between the soil and plant, with plants having a minimum of −0.2 MPa lower water potential than the medium in which they were growing. Pressure potentials increased with increased salinity concentrations of the media. Bolanos and Longstreth[39] concluded that *A. philoxeroides* could develop large driving forces for water uptake when plants were salt stressed and could in this manner maintain a positive water balance. Elasticity of tissues decreased as salinity increased almost 20 times between controls and 400-mM NaCl treatments, making it less difficult for salt-stressed plants to adjust osmotically than for plants at reduced salinities.

IV. CONCLUSIONS

Both field and laboratory investigations indicate that halophytes have adapted to saline habitats by their ability to adjust osmotically to increasing salinity levels.[24,28,31] They are capable of adjusting their water potentials to the soil water potential conditions in the habitats in which they are growing.[24,27,30] The two main mechanisms of adjustment that are apparent include ion accumulation and a reduction in the water content of tissues. Both of these mechanisms lower the plant water potential and permit the plant to take up water as soil salinity levels increase. Plant species growing in both inland and coastal salt marshes have low water potentials. More research is needed to determine precisely how plants are adjusting their ionic content when exposed to salt stress. Are the ions being transported from the soil by the root, or are older senescing leaves transporting ions to the younger leaves? How significant are organic substances in osmotic adjustment in highly saline habitats? Field and controlled laboratory investigations need to be designed to answer these critical questions regarding osmotic adjustment in halophytes and how it differs from the mechanisms in intolerant glycophytes.

REFERENCES

1. **Berger, A.**, L'alimentation en eau en milieu salé, *Soc. Bot. Fr. Actualites Bot.*, 3-4, 159, 1978.
2. **Flowers, T. J., Troke, P. F., and Yeo, A. R.**, The mechanism of salt tolerance in halophytes, *Annu. Rev. Plant Physiol.*, 28, 89, 1977.
3. **Munns, R.**, Why measure osmotic adjustment?, *Aust. J. Plant Physiol.*, 15, 717, 1988.
4. **Jefferies, R. L., Rudmik, T., and Dillon, E. M.**, Response of halophytes to high salinities and low water potentials, *Plant Physiol.*, 64, 989, 1979.
5. **Glenn, E. P. and O'Leary J. W.**, Relationship between salt accumulation and water content of dicotyledonous halophytes. *Plant Cell Environ.*, 7, 253, 1984.
6. **Glenn, E. P.**, Relationship between cation accumulation and water content of salt-tolerant grasses and a sedge, *Plant Cell Environ.*, 10, 205, 1987.
7. **Chapman, V. J.**, *Salt Marshes and Salt Deserts of the World*, J. Cramer, Bremerhaven, 1974.
8. **Waisel, Y.**, *Biology of Halophytes*, Academic Press, New York, 1972.
9. **Osmond, C. B., Bjorkman, O., and Anderson, D. J.**, *Physiological Processes in Plant Ecology*, Springer-Verlag, Berlin, 1980.
10. **Karimi, S. H. and Ungar, I. A.**, Development of epidermal salt hairs in *Atriplex triangularis* Willd., in response to salinity, light intensity, and aeration, *Bot. Gaz.*, 150, 68, 1988.
11. **Scholander, P. F., Bradstreet, E. D., Hammel, H. T., and Hemmingsen, E. A.**, Sap concentrations in halophytes and some other plants, *Plant Physiol.*, 41, 529, 1966.
12. **Jefferies, R. L.**, The ionic relations of seedlings of the halophyte *Triglochin maritima* L., in *Ion Transport in Plants*, Anderson, W. P., Ed., Academic Press, New York, 1972, 297.
13. **Kramer, D.**, The possible role of transfer cells in the adaptation of plants to salinity, *Physiol. Plant*, 58, 549, 1983.
14. **Dawson, T. E. and Bliss, L. C.**, Species patterns, edaphic characteristics, and plant water potential in a high-arctic brackish marsh, *Can. J. Bot.*, 65, 863, 1987.
15. **Langlois, J.**, Evolution de la pression osmotique et des teneurs en eau, en Cl⁻ et Na⁺ au cours de la croissance de l'axe principal chez *Salicornia stricta* Dumort., *Rev. Gen. Bot.*, 75, 377, 1968.
16. **Naidoo, G.**, Seasonal plant water relations in a South African mangrove swamp, *Aquatic Bot.*, 33, 87, 1989.
17. **Romo, J. T. and Haferkamp, M. R.**, Water relations of *Artemisia tridentata* spp. *wyomingensis* and *Sarcobatus vermiculatus* in the steppe of southeastern Oregon, *Am. Midl. Nat.*, 121, 155, 1989.
18. **Drake, B. G. and Gallagher, J. L.**, Osmotic potential and turgor maintenance in *Spartina alterniflora* Loisel., *Oecologia*, 62, 368, 1984.
19. **Scholander, P. F., Hammel, H. T., Bradstreet, E. D., and Hemmingsen, E. A.**, Salt balance in mangroves, *Plant Physiol.*, 37, 722, 1962.
20. **Scholander, P. F., Hammel, H. T., Bradstreet, E. D., and Hemmingsen, E. A.**, Sap pressure in vascular plants, *Science*, 148, 339, 1965.
21. **Scholander, P. F.**, How mangroves desalinate seawater, *Physiol. Plant*, 21, 251, 1968.
22. **Ungar, I. A.**, The relationship between soil water potential and plant water potential in two inland halophytes under field conditions, *Bot. Gaz.*, 138, 498, 1977.
23. **Ungar, I. A.**, Autecological studies with *Atriplex triangularis* in *Symposium on the Biology of Atriplex and Related Chenopods*, Tiedemann, A. R., McArthur, E. D., Stutz, H. C., Stevens, R., and Johnson, K. L., Eds., Gen. Tech. Rep. INT-172, Forest Service, U.S. Department of Agriculture, Ogden, UT, 1984, 40.
24. **Riehl, T. E. and Ungar, I. A.**, Growth and ion accumulation in *Salicornia europaea* under saline field conditions, *Oecologia*, 54, 193, 1982.
25. **Riehl, T. E. and Ungar, I. A.**, Growth, water potential and ion accumulation in the inland halophyte *Atriplex triangularis* under saline field conditions, *Acta Oecol. Oecol. Plant*, 4, 27, 1983.
26. **Kaplan, A. and Gale, J.**, Modification of the pressure-bomb technique for measurement of osmotic potential in halophytes, *J. Exp. Bot.*, 25, 663, 1974.

27. **Ustin, S. L., Pearcy, R. W., and Bayer, D. E.,** Plant water relations in a San Francisco Bay salt marsh, *Bot. Gaz.,* 143, 368, 1982.

28. **Antlfinger, A. E, and Dunn, E. L.,** Water use and salt balance in three salt marsh succulents, *Am. J. Bot.,* 70, 561, 1983.

29. **Tyndall, R. W., Teramura, A. H., and Douglass, L. W.,** Potential role of soil moisture deficit in the distribution of *Cakile edentula, Can. J. Bot.,* 64, 2789 1986.

30. **Tyerman, S. D., Hatcher, A. I., West, R. J., and Larkum, A. W. D.,** *Posidonia australis* growing in altered salinities: leaf growth, regulation of turgor and the development of osmotic gradients, *Aust. J. Plant Physiol.,* 11, 35, 1984.

31. **Karimi, S. H. and Ungar, I. A.,** The effect of salinity on the ionic content and water relations of *Atriplex triangularis,* in *Symposium on the Biology of Atriplex and Related Chenopods,* Tiedemann, A. R., McArthur, E. D., Stutz, H. C., Stevens, R., and Johnson, K., Eds., Gen. Tech. Rep. INT-172, Forestry Service, U.S. Department of Agriculture, Ogden, UT, 1984, 124.

32. **Ewing, K., Earle, J. C., Piccinin, B., and Kershaw, K. A.,** Vegetation patterns in James Bay coastal marshes. II. Physiological adaptation to salt-induced water stress in three halophytic graminoids, *Can. J. Bot.,* 67, 521, 1989.

33. **Pammenter, N. W. and Smith, V. R.,** The effect of salinity on leaf water relations and chemical composition in the subantarctic tussock grass *Poa cookii* Hook F., *N. Phytol.,* 94, 585, 1983.

34. **Matthews, M. A., Volkenburgh, E. V., and Boyer, J. S.,** Acclimation of leaf growth to low water potentials in sunflower, *Plant Cell Environ.,* 7, 199, 1984.

35. **Houchi, R., Morant-Avice, A., Ferard, G., and Coudret, A.,** Effects of NaCl and Na_2SO_4 on water relations of *Plantago maritima* (L.) and *Plantago lanceolata (L.), Biol. Plant,* 30, 457, 1988.

36. **Ferard, G., Coudret, A., and Lasceve, G.,** Action d'un choc osmotique sur les echanges d'eau des *Plantago maritima* L. et des *Plantago lanceolata* L., *C. R. Acad. Sci. Paris,* 294, 775, 1982.

37. **Coudret, A.,** Action du NaCl les contraintes et les relations hydriques dans les parties aeriennes de *Plantago maritima* L. et *Plantago lanceolata* L., *Acta Oecol. Oecol. Plant.,* 2, 111, 1981.

38. **Ferard, G. and Coudret, A.,** Relations entre l'absorption, le transport de l'ion sulfate et les contraintes hydriques chez *Plantago lanceolata et Plantago maritima* soumis a des variations de potentiel osmotique, *Physiol. Veg.,* 20, 703, 1982.

39. **Bolanos, J. A. and Longstreth, D. J.,** Salinity effects on water potential components and bulk elastic modulus of *Alternanthera philoxeroides* Griseb., *Plant Physiol.,* 75, 281, 1984.

Chapter 7

PLANT IONIC CONTENT

I. INTRODUCTION

Numerous papers have reported the ionic status of halophytes under field and laboratory conditions. Attempts have been made to classify halophytes based on their ability to accumulate or exclude particular ions.[1] In an early investigation of osmotic adaptation in plants from coastal marshes in the northeastern U.S., Steiner[2] determined that halophytes such as *Salicornia mucronata, Spartina glabra, S. patens, Distichlis stricta, Atriplex patula var. hastata,* and *Limonium carolinianum* accumulated high concentrations of Na and Cl. He reported that halophytes had three different types of adaptations to hypersaline conditions, including succulence (*Salicornia*), salt secretion by glands (*Spartina* and *Distichlis*), and osmotic adjustment to changing soil salinity levels (*Juncus gerardi*). Certain groups of halophytes may exhibit a syndrome of characteristics that are related to salt tolerance: grasses — salt exclusion by root membranes, dilution by high growth rates, salt glands, decrease in water content; chenopods — succulence, shedding of salt saturated plant parts, salt hairs; mangroves — leaf succulence, salt glands, exclusion of salts at root membranes; general — osmotic adjustment, protoplasmic salt tolerance, ion compartmentation, production of compatible organic solute.[3] One of the factors distinguishing halophytes from glycophytes is their capacity to accumulate selectively large quantities of ions in their cells without disrupting metabolic processes.[4,5] Epstein[4] emphasized the significance of selective ion uptake as an important device by which halophytes could acquire the nutrients necessary for growth in the presence of high concentrations of Na.

In this chapter description of the pattern of ion accumulation in halophytes will be attempted and discussion of some of the mechanisms which may help explain the patterns that exist in different species.

II. IONIC CONTENT OF PLANTS UNDER FIELD CONDITIONS

Surveys of the ionic content of species from different plant communities provide essential data concerning the ionic accumulation by plants (Table 1).[6-18] From this type of investigation it is hoped that we may be able to determine the specific types of adaptations that plants have to resist hypersaline environments, in what organs plants accumulate ions, and to what extent plants can regulate the proportion of various ions in their organs.

In a survey of ion accumulation in halophytes from different parts of Spain, Breckle[6] determined that Na and Cl were generally the most strongly accumulated ions, with some species having relatively high sulfate and K concentrations. Species accumulating high levels of Na and Cl per liter tissue water were *Arthrocnemum glaucum* (700 mM Na and 590 mM Cl), *Suaeda vera* (360 mM Na and 225 mM Cl), *Halimione portulacoides* (290 mM Na and 440 mM Cl), *Arthrocnemum perenne* (1310 mM Na and 770 mM Cl), *Inula crithmoides* (800 mM and 940 mM Cl), *Suaeda brevifolia* (620 mM Na and 220 mM Cl), and *Limonium aragonense* (120 mM Na and 190 mM Cl) (Table 1). Halophytes were classified into four physiotypes: (1) graminoids with very low Na uptake (*Juncus maritimus* and *Lygeum spartum*, 75 mM); (2) Chenopodiaceae which are salt accumulators and have high Na and Cl content (*Arthrocnemum, Suaeda, Halimone*); (3) Brassicaceae with relatively high sulfate uptake (*Lepidium cardamines*); and (4) Plantaginaceae (Plantago spp.) which are high in inorganic ions but low in organic substances.[1] Three general physiological adaptations described for these halophytes from Spain were succulence, salt glands, and ion accumulation, where specific organs or the entire plant accumulated high ionic levels.

TABLE 1
Ionic Content (mM/l Tissue Water) for Selected Halophytes under Saline Field Conditions

Species	Na	K	Ca	Mg	Cl	SO₄	Location	Ref.
Allenrolfea occidentalis	1044	173	0.2	23	526	104	Utah	13
Arthrocnemum glaucum	700	135	6	25	590	20	Spain	6
A. perenne	1310	185	13	105	770	128	Spain	6
Aster tripolium	305	54	11	47	183	50	Austria	1
Atriplex confertifolia	858	409	0.2	56	666	54	Utah	15
A. hastata	303	118	0.8	25	232	23	Austria	1
A. turcomanica	2400	910	—	—	2900	65	Afghanistan	9
Distichlis spicata	410	205	—	21	364	16	Utah	15
Eleocharis palustris	271	107	3.9	35	217	29	Austria	1
Eurotia lanata	681	295	0.1	55	220	14	Utah	15
Halimione portulacoides	290	40	4	20	440	18	Spain	6
Halocnemum strobilaceum	1870	120	—	—	1680	27	Afghanistan	9
Juncus gerardii	137	245	22	34	211	—	Wales	13
Limonium aragonense	120	80	4	110	190	120	Spain	6
Londesia eriantha	1530	310	—	—	430	555	Afghanistan	9
Plantago maritima	224	61	30	71	311	44	Austria	1
Puccinellia maritima	160	159	39	69	229	—	Wales	13
P. peisonis	151	186	6	27	215	12	Austria	1
Salicornia europaea	820	50	19	115	965	—	Wales	13
S. europaea	576	172	0.3	32	535	17	Utah	15
S. utahensis	718	168	1.4	51	606	46	Utah	15
Salsola sclerantha	2470	450	—	—	2200	290	Afghanistan	9
Sarcobatus vermiculatus	665	340	0.1	23	167	19	Utah	15
Scirpus maritimus	1169	111	21	26	186	—	Wales	13
Spartina anglica	346	159	56	38	315	—	Wales	13
Spergularia media	615	70	28	62	481	—	Wales	13
S. nicaeensis	335	70	3	10	280	25	Spain	6
Suaeda brevifolia	710	145	8	110	320	88	Spain	6
S. depressa	1093	243	0.1	24	517	68	Utah	15
S. maritima	451	80	0.4	31	348	64	Austria	1
S. maritima	547	74	38	131	532	—	Wales	13
S. pannonica	450	84	0.6	33	363	47	Austria	1
S. transoxana	960	130	—	—	730	78	Afghanistan	9
S. vera	360	95	5	25	225	35	Spain	6
Triglochin maritima	279	76	32	34	305	—	Wales	13
T. maritima	389	80	38	46	440	27	Austria	1

Sharma and Tongway[7] found a variation in soil salinity at different soil depths for plant communities dominated by *Atriplex vesicaria* and *A. nummularia* at the Falkiner Field Station (Australia). Salts at depths in the soil profile are derived from parent material, but surface accumulations of ions beneath plants are usually due to the decomposition of plant parts. Electrical conductivity in vegetated and unvegetated soils in the *A. nummularia* community indicated that about twofold more ions were found under the vegetation than in bare areas. The NaCl content of leaves ranged from 15.6% in senescing leaves to 23.0% in young leaves, while for *A. vesicaria* the range was from 12.0 to 21.5% in the two leaf age categories, respectively.[7]

Two species of salt desert shrubs (Curlew Valley, UT), *Atriplex confertifolia* and *Ceratoides lanata*, were analyzed to determine their ion content. Bjerregaard et al.[8] reported that new growth shoots of *A. confertifolia* had 8.1 mg Na/g tissue, while shoots of *C. lanata* had 0.9 mg Na/g tissue. The Na content of surface soils (0 to 2 cm) in the *A. confertifolia* community was 0.9 mg Na/g and at 60 to 90 cm depth 7.2 mg Na/g, while in the *C. lanata* community the Na values for soils were lower 0.3 mg/g and 3.3 mg/g for the two soil depths, respectively.[8] The pool of mineral elements in the two plant communities was very similar except for the elevated Na levels in the *A. confertifolia* community.

The ionic content of halophytes from the northern semidesert region of Afganistan was analyzed by Mizrahi and Breckle.[9] Some species had very high Na and Cl content on a dry weight basis, including *Salsola longistylosa* (12.9% Na and 10.5% Cl), *Suaeda altissima* (6.9% Na and 3.6% Cl), *S. transoxana* (9.8% Na and 11.4% Cl), *Halostachys caspica* (10.1% Na and 5.4% Cl), *Halocnemum strobilaceum* (12.0% Na and 17.0% Cl), and *Atriplex turcomanica* (7.9% Na and 15.4% Cl). Other species accumulated relatively high sulfate concentrations: *Salsola dendroides* (11.1% Na and 11.6% sulfate), *Halocharis afghanica* (16.5% Na and 15.2% sulfate), and *Gamanthus commixlus* (12.7% Na and 15.7% sulfate). Further investigations were done in halophytic communities of Iran and Afghanistan, and Breckle[10] classified plants from these salt desert areas into various groups, including stem succulents (*Salicornia*), leaf succulents (*Suaeda*), pseudohalophytes that accumulate ions in the roots (*Juncus*), recreto-halophytes with salt glands (*Limonium, Cressa, Tamarix*) and salt hairs (*Atriplex, Halimione, Mesembryanthemum*), and nonhalophytes. A number of Chenopodiaceae were found to have high Na and Cl content in their shoots: *Salicornia europaea* (5.8% Na and 10.8% Cl), *Halimocnemis mollissima* (11.5% Na and 6.9% Cl), *Halocnemum strobilaceum* (12.2% Na and 17.3% Cl), *Seidlitzia rosmarinus* (13.3% Na and 9.9% Cl), and *Salsola longistylosa* (12.9% Na and 10.5% Cl). Other halophytes had some-what lower ionic content: *Spergularia marina* (3.5% Na and 5.7% Cl), *Aeluropus littoralis* (0.7% Na and 1.2% Cl), *Tamarix karakalensis* (1.1% Na and 3.3% Cl), *Cressa cretica* (1.2% Na and 4.7% Cl), *Glaux maritima* (1.4% Na and 3.1% Cl), and *Limonium reniforme* (1.0% Na and 1.5% Cl).[10]

The ionic content of halophytes was determined for plants growing in saline soils around the Neusiedler Lake in Austria by Albert and Popp.[11] Dramatic differences were found in the capacity of monocotyledons and dicotyledons to accumulate Na and K. The K content was generally higher in monocotyledons and Na content higher in dicotyledons investigated. The following dicotyledon species all had K to Na ratios <1: *Suaeda maritima, S. pannonica, Salicornia prostrata, Atriplex hastata, Chenopodium botryoides, and C. glaucum.* Grami-noids tended to have lower Na content and a preference for K, maintaining a K to Na ratio >1: *Puccinellia distans, Agrostis stolonifera, Phragmites australis, Carex distans,* and *Juncus gerardii.* The species of Chenopodiaceae accumulated large quantities of inorganic ions and were considered to have high protoplasmic tolerance to electrolytes, while most graminoids developed some mechanism to restrict the uptake of Na and Cl.[11] The highest Na concentration was 869 mM/l tissue water in *Salicornia prostrata.* Other dicotyledons had Na contents ranging from 168 mM/l in *Sonchus arvensis* to 770 mM/l in *Lepidium crassifolium.* For monocotyledons such as *Juncus gerardii* and *Carex distans,* the concen-tration of Na was less than 40 mM/l, while for other species higher values were determined, 141 mM/l in *Bolboschoenus maritimus* and 400 mM/l in *Triglochin maritima.*

The ionic content of halophytic species from saline sites in Alberta (Canada) were analyzed by Guy and Reid.[12] Soils in this region were dominated by high sulfate concen-trations. Leaf cation content varied among species growing under various salinity regimes. The percent ash content ranged from a low of 5.6% in the grass species *Puccinellia nuttalliana* to a high of 38.2% in the chenopod succulent *Salicornia rubra.* Other species having intermediate values of ash content were *Hordeum jubatum* (7.4%), *Sonchus arvensis* (11.6%), and *Triglochin maritima* (12.8%) (Table 2). The cation accumulation in shoots of *S. rubra* was 431.9 mM Na/100 g, 56.9 mM K/100 g, and 75 mM Mg/100 g tissue dry mass. *Triglochin maritima* proved to be the second highest Na accumulator with 151.2 mM Na/100 g and 55.1 mg K/100 g tissue dry mass. The grasses *H. jubatum* and *P. nuttalliana* had higher K than Na content in shoots: 46.3 mM/100 g K vs. 16.4 mM/100 g Na in the former and 29.9 mM/100 g K vs. 11.4 mM/100 g dry mass in the latter species.[12]

TABLE 2
Water Content (in Percent of Fresh Weight), Fresh Weight (FW)/Dry Weight (DW) Ratio, Na$^+$, K$^+$, Cl$^-$, SO$_4^{2-}$ from Different Sites in Israel and the Sinai[a]

Species/Site	Water content (% FW)	FW/DW	Na$^+$	K$^+$	Cl$^-$	SO$_4^{2-}$	Malate Dusk	Malate Dawn
Site 1 (Acre)								
Inula crithmoides	91.9	12.3	203	9	392	29	4.4	4.0
Arthrocnemum macrostachyum	89.9	9.9	455	43	490	43	1.6	<1.0
A. cf. fruticosum	89.3	9.4	440	48	537	78	<1.0	2.0
Halimione portulacoides	87.0	7.7	385	34	453	85	—	<1.0
Plantago crassifolia	90.7	10.8	128	22	235	94	1.4	1.2
Site 2 (Tel Aviv)								
Carpobrotus edulis	93.6	15.7	164	5	280	21	5.0	11.2
Mesembryanthemum crystallinum	94.7	19.1	308	10	175	28	9.8	56.8
Pancratium maritimum	89.9	9.8	98	58	132	39	14.4	11.2
Crithmum maritimum	91.3	11.4	226	17	251	12	<1.0	<1.0
Anthemis leucanthemifolia	92.3	12.9	134	20	137	17	5.0	—
Artemisia monosperma	88.1	8.4	212	54	230	30	5.2	1.0
Atractylis flava	88.4	8.6	57	78	99	27	9.0	26.0
Inula crithmoides	91.5	11.8	423	14	540	39	4.6	1.4
Site 3 (Dead Sea)								
Aizoon canariense	86.4	7.4	534	97	135	114	<1.0	<1.0
Mesembryanthemum forsskalii	92.1	12.7	474	74	486	40	16.0	125.6
M. nodiflorum	87.1	7.8	428	60	242	68	8.0	179.0
Arthrocnemum macrostachyum	82.2	5.6	873	41	773	167	<1.0	<1.0
Anabasis setifera	80.6	5.2	902	69	110	123	8.0	21.8
Atriplex halimus	77.3	4.4	487	159	790	111	7.8	9.4
Salsola tetrandra	77.1	4.4	702	164	134	371	2.8	1.6
Suaeda fruticosa	82.1	5.6	1129	20	398	246	—	<1.0
S. monoica	80.6	5.2	733	83	283	188	9.0	3.4
Tamarix nilotica	61.9	2.7	208	50	179	80	5.8	1.0
Nitraria retusa	81.5	5.4	259	27	375	66	<1.0	5.4

[a]   Mineral ion values are expressed as microequivalents per gram fresh weight.

From Winter, K., Troughton, J., Evenari, M., Lauchlli, A., and Luttge, U., *Oecologia*, 25, 125, 1976. With permission.

The ionic composition of salt marsh plants from Ynys Mon (Anglesey, Wales) was analyzed, and Gorham et al.[13] determined that many of the monocotyledon species had relatively high K to Na ratios. Dicotyledons, especially those in the Chenopodiaceae, tended to accumulate high concentrations of Na. The concentrations in millimolar per liter tissue water of Na and K in dicotyledons were as follows: *Suaeda maritima* (547 m*M* Na and 74 m*M* K), *Salicornia europaea* (820 m*M* Na and 50 m*M* K), *Spergularia media* (615 m*M* Na and 70 m*M* K), *Aster tripolium* (360 m*M* Na and 72 m*M* K), and *Glaux maritima* (237 m*M* Na and 103 m*M* K) (Table 1). All of the above species had a K to Na ratio <1. Monocotyledons tended to have relatively high K concentrations, and a number had K to Na ratios of >1: *Juncus gerardii* (137 m*M* Na and 245 m*M* K), *Carex atrubae* (138 m*M* Na and 237 m*M* K), and *Eleocharis uniglumis* (167 m*M* Na and 186 m*M* K). However, some monocotyledons did accumulate relatively high Na levels: *Triglochin maritima* (279 m*M* Na and 76 m*M* K), *Zostera marina* (608 m*M* Na and 186 m*M* K), and *Spartina anglica* (346 m*M* Na and 159 m*M* K). The concept of physiotypes is questioned because the authors suggest that the habitats occupied by the various species may not be equivalent. Species are usually sorted out in definite zonal communities which confuse the issue, and several patterns were discerned from field investigations which indicate the occurrence of differences in physiological responses among species.[13] Therefore, differences in ion content between species could be related to local edaphic conditions rather than specific physiological differences in the way species are responding to salt stress.

Mineral composition of soils and shoots of *Halimione portulacoides* were determined for plants growing at several salt marsh stations on the Oosterschelde Estuary (Netherlands). Field soil salinity values ranged from 0.9 to 5.0% NaCl on a dry soil basis.[14] The Na content for leaves of *H. portulacoides* plants was 129 mg/100 g, Cl 143.3 mg/100 g, and K 34.6 mg/100 g dry mass. The K to Na ratio was <1 for *H. portulacoides*. The salt marsh station with the highest soil salinity contained plants with the highest Na and Cl content. However, Stienstra[14] found no direct correlation between the soil ionic content and leaf ion content. The author concluded that other factors, such as leaf age, prior salinity conditions to which plants were exposed, depth of the water table, and frequency of high soil salinity periods could affect the ion content of leaves.

The ionic content of plant species was determined by Wiebe and Walter[15] for a number of halophytes from the Curlew Valley, UT. Plants growing in the most extremely saline sites had Na concentrations in leaves or shoots ranging from 410 to 1093 mM/l tissue water, Cl content from 364 to 606 mM, and K content from 168 to 243 mM.[15] Species in the Chenopodiaceae had the highest Na content of all plants: *Suaeda depressa*, 1093 mM; *Salicornia rubra*, 576 mM; *Salicornia utahensis*, 718 mM; and *Allenrolfea occidentalis*, 1044 mM (Table 1). The grass *Distichlis spicata* var. *stricta* had 410 mM Na, 205 mM K, and 364 mM Cl. In high sulfate soils of Price, UT the three dominant species were *Eurotia lanata*, *Atriplex confertifolia*, and *Halogeton glomeratus*; they had a Na content of 110, 1000, and 510 mM; K content of 261, 682, and 287 mM; and Cl content of 96, 278, and 113 mM. Sulfate content for the latter three species ranged from 19 to 93 mM, with SO_4 to Cl ratios from 0.19 to 0.49 compared to plants at other sites where the SO_4 to Cl ratio was lower, ranging from 0.04 in *D. spicata* var. *stricta* to 0.22 in *Allenrolfea occidentalis*.[15]

The ionic content of halophytes was determined for plants from Acre, the Dead Sea, and the Red Sea (Israel), and Winter et al.[16] reported that many of the plant species analyzed were characterized by elevated ionic content, chiefly Na and Cl. Most of the species studied had K to Na ratios that were <1, and Na to K ratios in tissues ranged from 10 to 20. Plants growing in coastal marshes tended to have about equivalent concentrations of Na and Cl expressed as millimolar per 1000 g fresh weight: *Inula crithmoides* (203 mM Na and 392 mM Cl), *Arthrocnemum macrostachyum* (455 mM Na and 490 mM Cl), *Halimione portulacoides* (385 mM Na and 453 mM Cl), *Crithmum maritimum* (226 mM Na and 251 mM Cl), and *Cakile maritima* (146 mM Na and 156 mM Cl) (Table 2). Sulfate levels for plants at the Acre coastal site ranged from 29 mM in *Inula crithmoides* to 94 mM in *Plantago crassifolia*. Inland species from the Dead Sea saline region had Na concentrations in their leaves that were often greater than twice that of Cl, but had higher sulfate levels than most of the coastal species: *Aizoon canariense* (534 mM Na, 135 mM Cl, and 114 mM SO_4), *Arthrocnemum macrostachyum* (873 mM Na, 773 mM Cl, and 167 mM SO_4), *Salsola tetrandra* (702 mM Na, 134 mM Cl, and 371 mM SO_4), *Suaeda fruticosa* (1129 mM Na, 398 mM Cl, and 246 mM SO_4), and *Suaeda monoica* (733 mM Na, 283 mM Cl, and 188 mM SO_4) (Table 2).[16] The ratio of Na to Cl varied in the Dead Sea species from 1.8 (*Mesembryanthemum nodiflorum*) to 9.5 (*Seidlitzia rosmarinus*), but these values were reduced to 1.3 and 2.3 when the values for sulfate were added to Cl. Winter et al.[16] classified most of the coastal species as chloride halophytes; many of the inland species with very high Na compared with Cl + SO_4 would be considered alkali halophytes, and some salt-excreting species would be considered sulfate halophytes (e.g., *Reaumeria hirtella*, *Limonium sinuatum*).

An analysis of the mineral content of plants in the field from New Brunswick, Canada by Poulin et al.[17] indicated that *Salicornia europaea* varied in ionic content at different stages of development and in the different organs on a plant. Seeds contained 0.7% Na on a dry weight basis, while young plants had as much as 14.4% Na. More mature plants ranged from 8.0 to 9.4% in Na content. Potassium content varied in plants, ranging from 0.2% in seeds to 2.4% in mature *S. europaea* plants.

Mineral content of soils and halophytes were determined for species growing in saline habitats of Alberta and Saskatchewan, Canada by Redmann and Fedec.[18] Electrical conductivity of soils from Saskatchewan ranged from 13.2 to 49.3 mS/cm and for Alberta 20.7 to 55.0, containing high levels of Na (76 to 850 mM/l), Mg (18 to 87 mM/l), Cl (8 to 137 mM/l), and SO₄ (93 to 526 mM/l) in soils sampled beneath halophytes. Ion content of shoots of *Salicornia rubra* was the highest for all species, 1073 mM total ions per 100 g tissue dry weight, with 515 mM Na/100 g, 277 mM Cl/100 g, and 176 mM SO₄/100 g. Other species with high Na and Cl shoot content were *Suaeda depressa* (467 mM Na/100 g and 140 mM Cl/100 g), *Atriplex patula* (310 mM Na and 140 mM Cl), *Chenopodium rubrum* (213 mM Na and 121 mM Cl), and *Kochia scoparia* (220 mM Na and 44.6 mM Cl/100 g dry weight). Redmann and Fedec[18] concluded from their analyses of plant and soil data that there was a significant positive relationship between the soil ionic content and the total plant tissue ionic content for all species. Plants were classified into two groups: chloride halophytes, containing high Na and Cl concentrations (*Atriplex patula, Chenopodium rubrum, Salicornia rubra, and Suaeda depressa*); and alkali halophytes, containing low Cl and high K, Mg, and Ca concentrations (*Chenopodium rubrum, Kochia scoparia,* and *Salsola pestifer*).

Welch[19] reported that the ash content of *Atriplex canescens* shoots ranged from 13.3 to 16.0% of the dry mass for plants collected from seven locations from the salt desert region of Utah. Soil salinity at these sites ranged from 307 to 4229 mg total salts per kilogram soil. Other investigators have reported relatively high percent ash values for species of *Atriplex: lentiformis,* 31.6%; *polycarpa,* 21.3%[20] *confertifolia,* 21.7%; *nuttallii,* 21.5%; *gardneri,* 23.5 to 25.3%;[19] *vesicaria, nummularia,* and *inflata,* 24.4 to 38.2% (Table 2).[21]

Breckle[22] determined that *Atriplex confertifolia* accumulated both Na and Cl and had a K to Na mean ratio of 0.38, ranging from 0.28 to 0.50, while *Ceratoides lanata* maintained a high K content and had a K to Na ratio of 11.7, ranging from 1.6 to 26 in Curlew Valley, UT during the growing season.[22] Leaf Na ionic content for *A. confertifolia* ranged from 2000 to 2500 mM/l, while for *C. lanata* it ranged from 0.027 to 0.038 mM/l. Potassium content of leaves ranged from 450 to 820 mM/l for *A. confertifolia* and from 420 to 680 mM/l for *C. lanata* at the seven sites investigated. The Cl ion content for *A. confertifolia* ranged from 1200 to 1700 mM/l, while for *C. lanata* it was 120 to 160 mM/l. Sulfate concentrations were also higher for *A. confertifolia,* 170 to 260 mM/l, while for *C. lanata* it ranged from 35 to 75 mM/l.[22]

The relative distribution of Na and K in plant organs is reported for *S. europaea,* and Ernst[23] found that shoots contained >85% of the plants Na, roots about 10%, and seeds a very small amount, <2%. Potassium content varied in plant organs and reached 20% in seeds, 15% in roots, and 60% in shoots late in the growing season. The ion content of shoots of *Salicornia bigelovii* grown in artificial seawater with 3.0% total salts was determined for the various tissue layers: vascular, spongy mesophyll, and palisade.[24] Ash weight was 45% in the palisade tissue, 60% in spongy mesophyll, and 32% in vascular tissue. Sodium content was 13.4 μM/mg ash in palisade, 14.4 μM/mg in spongy mesophyll, and 13.6 μM/mg in the vascular tissue. Stumpf and O'Leary[24] concluded from their data analysis that there was no evidence of Na exclusion from any of the major shoot tissue types. These data differ from the results of Weber et al.[25] where it was reported that Na was excluded from palisade tissue in *Salicornia pacifica* var. *utahensis.* In a study of chlorenchyma tissue of *Suaeda monoica,* Eshel and Waisel[26] reported that Na and Cl were present in high concentrations in both old and young photosynthetic tissues. Energy-dispersive X-ray mi-

TABLE 3
Percentage Ash Content of Selected Halophytes

Species	Organ	Ash content (%)	Ref.
Atriplex canescens	Shoot	16.0	19
A. inflata	Leaf	38.2	22
A. lentiformis	Shoot	14.0	28
A. nummularia	Shoot	22.6	28
A. nummularia	Leaf	35.0	22
A. vesicaria	Leaf	30.8	22
Batis maritima	Leaf	44.0	29
Borrichia frutescens	Leaf	36.9	29
Halodule beaudettei	Leaf	18.2	29
Hordeum jubatum	Leaf	7.4	12
Puccinellia nuttalliana	Leaf	5.6	12
Ruppia maritima	Leaf	18.9	29
Salicornia bigelovii	Shoot	43.2	24
S. rubra	Shoot	38.2	12
S. virginica	Shoot	24.1	29
Salvadora persica	Leaf	28.1	32
Scirpus americanus	Leaf	14.0	29
Sesuvium portulacastrum	Leaf	54.0	30

croanalysis indicated that the Na concentration of shoots was lower in the palisade tissue than in the cortex region of *Salicornia pacifica* var. *utahensis* plants which were collected from a saline area near Goshen, UT.[27] Chloride ion concentration was 50% higher in the cortex than in the palisade region, with the protoplasm of the cortex containing abut 80% of the Cl and the cell wall 20%. Khan et al.[27] determined that a high Cl content was found in all four regions (epidermis, cortex, palisade, and vascular) of the shoot of *S. pacifica* var. *utahensis*.

Field growth experiments were carried out with *Atriplex lentiformis* and *A. nummularia* at Safford, AZ in soils that were irrigated with saline water that had an electrical conductivity of 10.3 mS/cm, 132 mM Na, 118 mM Cl, and 31 mM SO$_4$. Watson et al.[28] determined that shoot ash values were 14% for *A. lentiformis* and 22.6% for *A. nummularia*. Oxalate content was 0.74% and 0.90% of the total dry mass.

The ash content of coastal species from Mississippi was analyzed, and Lanning and Eleuterius[29] determined that some species from salt marsh habitats had high leaf ash content compared with plants growing in nonsaline areas (Table 3). Leaves of the succulent species *Batis maritima* and *Borrichia frutescens* had the highest ash values, 44.0 and 36.9%, respectively. Other species had lower values, such as the aquatic organisms *Halodule beaudettei* (18.2%), *Ruppia maritima* (18.9%), marsh species *Scirpus americanus* (14.0%), and *Salicornia virginica* (24.1%). Species growing in nonsaline habitats had leaf ash contents ranging from 5.7% in *Sorghum halepense* to 18.8% in *Arundinaria gigantea*. For *A. gigantea* the silica content reached 16.1%, while for the succulent halophytes the silica content ranged from 0.02% in *S. virginica* to 0.69% in *B. frutescens*. Ash content of inflorescences of salt marsh plants varied considerably: *Batis maritima* (30.7%), *Salicornia virginica* (12.9%), and *Borrichia frutescens* (8.7%), while species from nonsaline habitats had inflorescence ash contents ranging from 5.4% in *Cyperus polystachyos* to 9.2% in *Juncus polycephalus*.[29]

Sesuvium portulacastrum, a leafy succulent halophyte of tropical salt marshes, grows in soils that have a mean summer electrical conductivity of 109.2 mS/cm, containing 3924 mM Na, 5393 mM Cl, and 57.8 mM K.[30] Ash content of leaves in the summer accounted for 54% of the dry weight. The ion content of leaves was 18.2% Na, 18.0% Cl, and 1% K. High ash contents have been reported for the leaves of a number of other halophytes.[31]

TABLE 4
Daily Mean Inorganic Nutrient Content in Leaves of the Mistletoe
(*Pthirusa maritima*) and Its Two Hosts[a]

Species	Na (mg/g)	K (mg/g)	Ca (mg/g)	Mg (mg/g)
Conocarpus erectus	85.70 ± 59.4	2.2 ± 0.7	12.4 ± 4.5	14.0 ± 1.5
Mistletoe on *C. erectus*	104.80 ± 48.0	26.7 ± 7.6	7.6 ± 4.6	10.9 ± 3.8
Coccoloba uvifera	4.10 ± 0.9	6.1 ± 2.4	16.00 ± 7.9	6.4 ± 1.7
Mistletoe on *C. uvifera*	18.60 ± 7.0	24.5 ± 4.2	16.00 ± 2.4	5.6 ± 1.1

[a]   Values are means ± SE.

From Goldstein, G., Rada, F., Sternberg, L., Burguera, J. L., Burguera, M., Orozco, A., Montilla, M., Zabala, O., Azocar, A., Canales, M. J., and Celis, A., *Oecologia*, 78, 176, 1989. With permission.

The percent ash in young leaves of *Salvadora persica* was 22.9%, while senescent leaves had 28.1% ash (Table 3).[32] Removal of accumulated ions in this species is by shedding of senescent leaves and bark where excess ions are stored. Amonkar and Karmarkar[32] determined that senescing tissue were able to transport K, Mg, and P to other organs, while the bark of *S. persica* served as a sink for Na (426.2 mM/100 g dry tissue) and Cl (339.7 mM/100 g dry tissue).

The hemiparasitic mistletoe *Pthirusa maritima*, a parasite on the mangroves *Conocarpus erectus* and *Coccoloba uvifera*, had a more negative leaf water potential than its host plants.[33] The leaf water potential of the mistletoe was usually −0.5 to −0.7 MPa lower than the water potential of the mangroves. The Na and K content of leaves was higher in the hemiparasite than in the host: 104.8 mg Na/g dry mass for *P. maritima* vs. 85.7 mg Na/g for *C. erectus* (Table 4). Potassium levels were 2.2 mg K/g dry mass vs. 26.7 mg K/g for the host and hemiparasite, respectively. Similar results were reported by Goldstein et al.[33] for the ion content of leaf tissue of *C. uvifera*; the host had fourfold less Na and K than did the hemiparasite (Table 4).

III. SEASONAL AND DAILY CHANGE IN IONIC CONTENT

A few field investigations have been carried out in which researchers have attempted to determine the seasonal change in ionic content of species in salt marsh and salt desert plant communities.[34-44] This type of survey provides essential data concerning the relative change in the abundance of different ions in plants during the year. More research is needed to determine how changes in soil salinity level and other environmental variables are related to changes in the ionic content of salt marsh and salt desert plants during a growing season. Variation in soil salinity content during the year may be reflected in changes in root or shoot ion content at different times during the phenological development of plants.

Harward and McNulty[34] determined that the tolerance of *Salicornia rubra* to excess soil salinity was due to the plant's capacity to absorb Na and Cl ions and in this manner maintain a favorable diffusion gradient for water absorption. Seasonal measurements of Na and Cl in shoots of *S. rubra* indicated that the concentration of these ions in the cell sap increased during the growing season, ranging from 700 mM to a peak of 2 M late in the growing season. Na and Cl accounted for 75 to 93% of the total osmotic pressure of the cell sap, indicating that they were the principle ions accumulated for ionic adjustment in *S. rubra*.[34]

Seasonal measurements were made of ionic content at a site from Goshen, UT which contained a population of the perennial succulent *Salicornia pacifica*.[35] Chloride content of shoots ranged from 14.2% in April to 16.1% in September, and increases in Na were proportional, rising from 8 to 10% of the plant dry mass. Potassium ranged from 4 to 8% of plant dry mass for most of the growing season. Other sites investigated followed similar patterns in relation to ionic content of plants. An analysis of the ionic content of soils at the salt desert playa demonstrated that Na and Cl were the dominant ions, with Cl ranging from 3 to 5% and Na from 2 to 4% in the upper 5 cm of soil throughout the growing season.[35]

Salicornia herbacea was collected at sites from the Aegean Sea (Turkey) with varying seasonal soil salinity content, ranging from 70.6 mS/cm in June to 100.6 mS/cm in August.[36] Plant sap at the above site ranged in electrical conductivity from 113.6 mS/cm in June to 145.4 mS/cm in August. At two other locations the plant sap electrical conductivity ranged from 86.8 to 105.8 mS/cm and 68.6 to 98.8 mS/cm, indicating that ionic content of shoots of *S. herbacea* increased during the summer months. Mert and Vardar[36] determined from an analysis of their results that there was a significant correlation between high tissue salt content and reductions in transpiration rates of plants during each of the summer months: $r = -0.953$ for June, $r = -0.876$ for July, and $r = -0.538$ for August. They concluded that both water and ionic regulation must be taken into account in order to understand the ecophysiology of halophytes.

The seasonal variation in Cl ion content (as a percent dry mass) for halophytes growing in the Pachpadra Salt Basin (India) was determined.[37] *Zygophyllum simplex* had the highest leaf Cl content, 36.2% in July for the six species investigated. Leaf Cl content for *Cressa cretica* ranged from a low of 2.4% of dry mass in September to a high of 10.4% in March. *Suaeda fruticosa* had its lowest Cl concentration in January (7.1%) and highest value (22.4%) in November, but no seasonal pattern was evident. Some species, such as *Trianthema triquetra*, had a relatively small range of Cl change over the year, 4.5 to 8.9% NaCl. Rajpurohit and Sen[37] hypothesized that seasonal variation in leaf Cl content of plant organs was related to number of biological and environmental factors, including leaf age, growth rate, soil ionic content, and osmotic potential of both plants and soils.

Seasonal measurements of change in the ionic content for soils and plants growing at an inland saline site at Rittman, OH were made for *Atriplex triangularis* and *Salicornia europaea* by Riehl and Ungar.[38,39] Soil salinity content ranged from 1.1% total salts to 8.9% total salts at the *S. europaea* sites. These values can be largely accounted for by high concentrations of Na and Cl in the soil, which were measured to range from 200 mM to 1500 mM (Figure 1). It was determined that the Na and Cl content of plants was site dependent, with plants at more highly saline sites having higher ion content than those in less saline habitats (Figure 2). As the soil salinity levels increased from April (1.2%) to July (5.8%) total salts (18.6 to 91.2 mS/cm), *Salicornia europaea* roots and shoots showed an increase in both Na and Cl content, from about 500 to 1500 mM in shoots and from about 200 to 700 mM in roots. A period of heavy precipitation in August caused a reduction in shoot ion content, but it rose to 1800 mM Cl and Na per liter tissue water by September. The concentration of K in roots and shoots ranged from 20 to 120 mM during the growing season (Figure 3).[38]

The effect of seawater salinities from 20 to 50 mS/cm on the accumulation of inorganic ions in shoots of *Salicornia brachiata* was determined by Joshi[40] using both field and laboratory observations. The Na and Cl content of shoots (millimolar per 100 g dry mass) was analyzed for plants grown in the laboratory for 15 weeks, and it was determined that the Na content ranged from about 550 to 678 mM, Cl from 750 to 1000 mM, and K from 25 to 50 mM with increasing seawater concentrations (Table 5). These laboratory data support the measurements of shoot ionic content for field plants, where Na increased from 407 to 742 mM, Cl from 588 to 997 mM, and K from 19 to 37 mM during the growing season. Joshi[40] determined that the highest ionic content in *S. brachiata* plants was during the period of most active growth from October to December (Table 5).

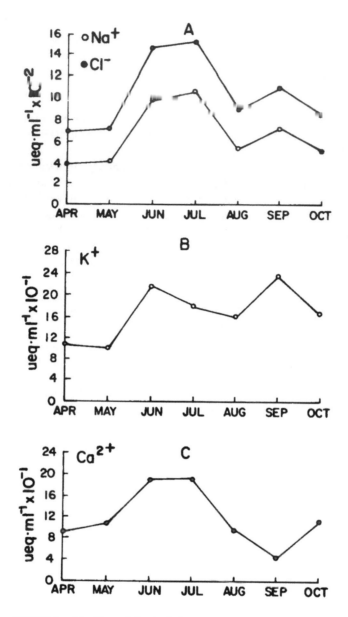

FIGURE 1.   Monthly variation in the ion concentrations (μeq/ml) of soils in the highly saline Edge *Salicornia europaea* zone.

Concentrations of Na, K, and Cl were determined at different times during the growing season for the salt-secreting *Atriplex hymenelytra* from Death Valley, CA.[41] The content of Na in leaves ranged from 730 to 630 mM/l tissue water in January and November, K was 140 and 170 mM, and Cl was 500 and 610 mM. The organic acid oxalate ranged from 130 to 170 mM/l tissue water. Bennert and Schmidt[41] concluded that inorganic ions are used to generate a sufficiently low osmotic potential in plant tissues to ensure water uptake and turgor maintenance.

Seasonal changes in soil salinity were measured in three populations of *A. triangularis*, and Riehl and Ungar[39] and Ungar[42] determined that there was a general increase in soil salinity in these zones until July (Table 6). The low salt zone had a range of salinity from 6 (0.4% total salts) to 20 mS/cm (1.3% total salts), while in the high salt zones conductivity

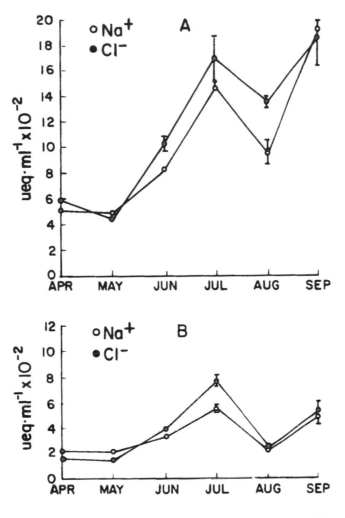

FIGURE 2. Monthly change in the Na and Cl content (μeq/ml) of (A) shoots and (B) roots of *Salicornia europaea* in the Edge *Salicornia* zone. (From Riehl, T. E. and Ungar, I. A., *Oecologia*, 54, 193, 1982. With permission.)

values ranged from 10 (0.64% total salts) to 90 mS/cm (5.8% total salts). The soil extracts contained Na and Cl levels ranging from 40 to 1200 m*M*, which accounted for most of the change in soil salinity during the growing season (Figure 4). Ionic content of plants in all *A. triangularis* zones increased from April to July, with Na and Cl content in the shoots of the low salt zone plants increasing from 100 to 700 mM/l tissue water. In high salt areas Na and Cl concentrations of shoots increased from April to July, from 200 to 2000 mM/l tissue water (Figure 5). Plants in this area died in August and did not reach reproductive maturity. Root Na and Cl content ranged from 100 to 200 m*M* in both high salt and low salt zones, indicating that the roots were transporting most of the Na and Cl to shoots. There were significant correlations ($p < 0.01$) between root and shoot ion content and changes in soil electrical conductivity, soil Na, and soil Cl concentrations during the growing season.[42]

As the soil-water content decreased from September to December in field sites at the Falkiner Field Station (Australia), the Na and Cl content in leaves of *Atriplex vesicaria* and *A. nummularia* increased.[43] The Na content ranged from 5 to 10% on a leaf dry weight basis, and Cl ranged from 5 to 12%, with the lower values for June to August and the higher

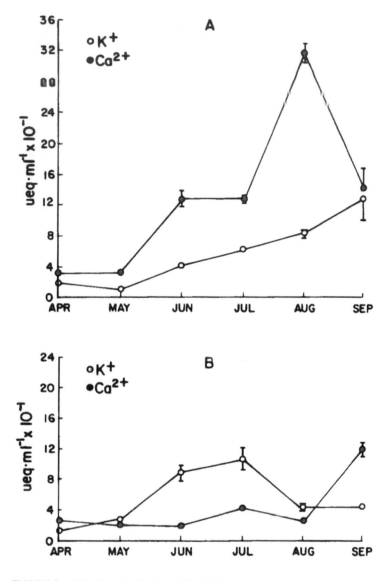

FIGURE 3.   Monthly variation in the K and Ca content (μeq/ml) of (A) shoots and
(B) roots of *Salicornia europaea* plants in the Edge *Salicornia* zone. (From Riehl,
T. E. and Ungar, I. A., *Oecologia*, 54, 193, 1982. With permission.)

values for October to December. Sharma et al.[43] cited a number of factors that contributed
to an increase in the salinity of leaves of *Atriplex* species, including increased soil salinity,
decreased soil water content, and the ion accumulating capacity of plants.

The saltbushes, *Atriplex acanthocarpa* and *A. canescens*, growing in soils with electrical
conductivities ranging from 4 to 12 mS/cm in Zapata County, TX, had leaf Na concentrations
ranging from 7.7 and 2.7% in November to 10.5 and 2.7% in August for the two plant
species, respectively.[44] The K content of leaves remained relatively consistent during the
sample period, ranging from 1.6 and 2.3% to 2.7 and 2.8% for *A. acanthocarpa* and *A.
canescens*, respectively.

The ionic content of *Atriplex hastata* shoot and root material was determined on an area
basis on the Seine Estuary (France) by Binet and Thammavong.[45] The Na content of *A.
hastata* shoots increased during the growing season from a low value of 15 mg/m² in March

TABLE 5
Mineral Ions Content (meq/100 g Dry Wt. ± SE) in *Salicornia brachiata* at Different Growth Stages[a]

Month	Na$^+$	K$^+$	Ca^{2+}	Mg^{2+}	Cl$^-$	SO$_4^{2-}$
October	682 ± 5	31 ± 2	35 ± 3	114 ± 3	867 ± 5	32 ± 1
November	742 ± 5	19 ± 1	31 ± 3	152 ± 5	977 ± 5	41 ± 1
December	731 ± 5	26 ± 3	32 ± 2	99 ± 5	922 ± 6	23 ± 1
January	585 ± 6	37 ± 3	30 ± 2	95 ± 3	675 ± 7	26 ± 2
February	533 ± 9	36 ± 2	29 ± 1	92 ± 2	583 ± 4	22 ± 2
March	523 ± 5	41 ± 1	36 ± 3	102 ± 2	686 ± 5	23 ± 3
April	407 ± 4	28 ± 4	40 ± 1	100 ± 5	586 ± 6	32 ± 2

[a] October to November, vegetative; December, flowering; January to March, post-flowering; April, field-dried plants.

From Joshi, A. J., *J. Plant Physiol.*, 123, 497, 1986. With permission.

TABLE 6
Seasonal Change in the Ionic Content of Soil in a Plant Community Dominated by *Atriplex triangularis*

	Cl (μeq/ml)	Na (μeq/ml)	K (μeq/ml)	Ca (μeq/ml)
April	80	52	0.57	33
May	123	100	0.50	31
June	338	325	0.90	68
July	875	550	1.65	125
August	872	488	1.08	200
September	960	722	1.78	350
October	717	340	0.84	157

From Riehl, T. E. and Ungar, I. A., *Acta Oecol. Oecol. Plant.*, 4, 27, 1983. With permission.

to 1636 mg/m^2 in May. By September the Na content of shoots was 15.7 g/m^2 and K 21.1 g/m^2 for a shoot dry weight of 724 g/m^2 and the Na content was 3.7 g/m^2 and K 6.8 g/m^2 for a root dry weight of 482 g/m^2.[45] Inorganic material made up about 15% of the total dry mass of *A. hastata* plant material throughout the growing season.

The ionic content of *Salicornia europaea* was reported by Von Willert[46] to have a diurnal pattern that is related to the flooding regime at a site. Plants growing on levees that flood infrequently had higher Na and Cl concentrations in their plant sap than plants which were flooded regularly in the tidal zone ditches. The Na and Cl content of levee plants fluctuated daily between a tidal flushing value of 700 to 800 mM/l plant sap at times of the day when plants were flooded by the tides. Ditch plants ranged from about 500 mM/l Na and 600 mM/l Cl during flooding to 650 mM/l Na and 725 mM/l Cl when the tidal action subsided (seawater contains 415 mM Na and 345 mM Cl). Von Willert[46] reported that the ratio of Na to Cl was 1:1.1 in ditch plants and 1:1 for plants growing on the levees (seawater ratio is 1.2:1).

Ruess and Wali[47] reported that the cations Ca, Mg, K, and Na contributed 58% of the osmolality to the xylem sap of *Atriplex canescens*, but the dominate ions were K and Cl for this species from the Red Desert, Wyoming. Osmotic potential measurements of expressed xylem sap indicated that sharp changes occurred over a 48-h period from 15 to 50 mOsm, with highest values during the morning and evening when water movement was lowest.[47] At the highest osmolality the sap contained 14 mM/l K and only 1 mM Na, indicating that there is a selective stimulated uptake of K and decrease in the uptake of Na.

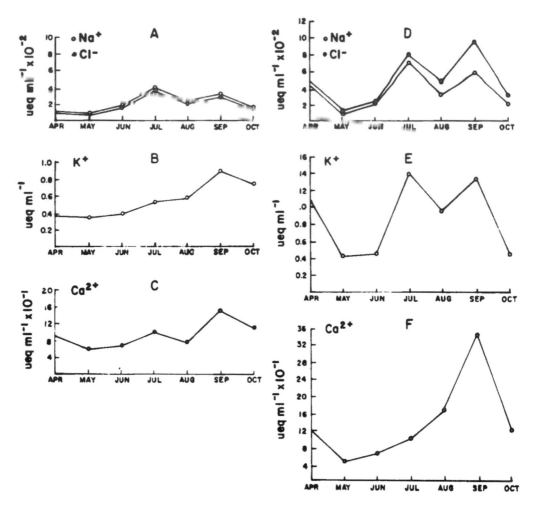

FIGURE 4.   Monthly fluctuation of soil ionic content (μeq/ml) for the *Hordeum-Atriplex* (A,B,C) and *Atriplex* zones (D,E,F) containing *A. triangularis*.

IV. IONIC CONTENT OF PLANTS IN RESPONSE TO SALINITY

The relationship between plant growth responses to salinity and ion uptake have been investigated by a number of researchers.[48-60] The ability of plants to regulate their nutrition has been documented in a review of the subject by Epstein.[48] In the case of halophytes much of the research has been related to the effects of salinity on ion accumulation in plant organs. In recent years studies have been carried out to determine the compartmentalization of ions in plants. Studies of ion distribution in plants and its effect on plant growth have been carried out at several levels of organization: distribution of ions in roots, stems, and leaves; ionic content of particular tissues in plants; and finally with the use of X-ray microprobe analysis the distribution of ions in cells and within cellular compartments. Flowers and Yeo[49] have suggested that a greater concentration of ions are sequestered in the vacuoles of cells than in the cytoplasm. Osmotic balance within cell compartments is therefore hypothesized to be controlled by the production by plants of one or more osmotically compatible solutes, such as proline, glycine-betaine, sugars, glycerol, and mannitol. The accumulation of compatible solutes has been documented by a number of researchers.[49] However, whether the solutes are confined to the cytoplasm as is hypothesized and whether they are present in sufficient concentration in the cytoplasm to balance the high ionic content postulated for the vacuole

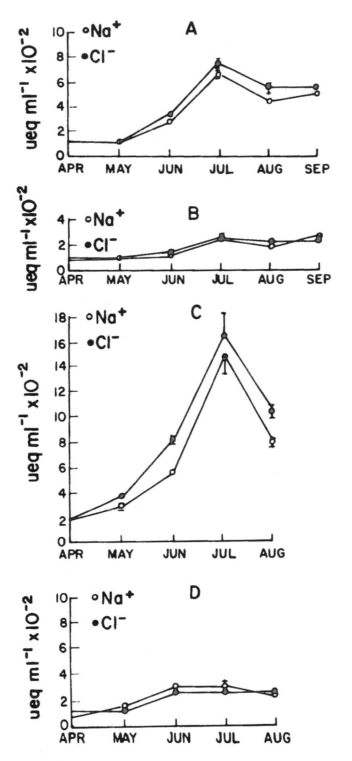

FIGURE 5. Monthly variation in the ion content (μeq/ml) for *Atriplex triangularis* in the *Hordeum-Atriplex* (A = shoot, B = root) and the *Atriplex* (C = shoot and D = root) zones. (From Riehl, T. E. and Ungar, I. A., *Acta Oecol. Oecol. Plant.*, 4, 27, 1983. With permission.)

is still a question open for further investigation. Cheeseman[50] reviewed this subject recently and has come up with some very significant questions regarding the function of organic solutes in determinig the salt tolerance of halophytic flowering plants.

The effect of treatments with 340 mM NaCl and waterlogging on the ion content of eight halophytic species from Northern Ireland salt marshes was determined by Cooper.[51] Plants grown in saline treatments had higher Na content than those grown in nutrient solutions (Table 7). The accumulation of Na in shoots ranged from 9.6 mg/g dry weight in *Festuca rubra* to 110.2 mg/g dry weight in *Salicornia europaea* for plants treated with drained saline solution, while in the nonsaline treatment Na content ranged from 1.2 to 56.8 mg/g dry weight for the two species, respectively. *Juncus gerardi* and *F. rubra* accumulated large amounts of K in the drained saline treatments: 42.1 mg/g K for the latter species and 167.7 mg/g for the former species (Table 7). Waterlogging did not significantly affect the Na content of shoots, whereas for most of the species studied an increase in salinity was correlated with increased Na content of the shoots. The grass *F. rubra* maintained relatively low Na content in shoots in both drained (9.6 mg/g) and waterlogged (6.7 mg/g) saline conditions (Table 7).

Comparisons were made of Na and K content in shoots of several halophytes, and Cheeseman et al.[52] determined that all species had increased Na content in their shoots when salinity was raised to either 0.2 or 0.4 times seawater. In *Spergularia marina* the Na content of shoots increased from 6.7 μM/g fresh weight in controls to 185 μM/g fresh weight in 0.4 times seawater, while root Na content increased from 1.1 to 85 μM/g fresh weight. Potassium content in shoots of *S. marina* decreased from 140 μM/g in nutrient solution controls containing 1 mM K to 88 μM/g fresh weight in the 0.4-times seawater treatment, while root K content did not vary greatly in the treatments, decreasing from 103 to 102 μM/g fresh weight. *Plantago maritima* behaved in a similar manner, with Na content increasing from 1.6 μM/g in controls to 280 μM/g fresh weight in 0.4 times seawater and K decreasing from 179 to 57 μM/g fresh weight in the two treatments, respectively. Four species, *Lycopersicon cheesmanii*, *Aster tripolium*, *Puccinellia maritima*, and *Zea mays*, all had <1 μM/g Na in the controls and increases to 92, 110, 77, and 98 μM/g fresh weight Na in the 0.2-times seawater treatments for each species, respectively. Potassium content of *Lycopersicon cheesmanii* (48 to 60 μM/g), *Puccinellia maritima* (118 to 159 μM/g) and *Zea mays* (119 to 129 μM/g) increased in the 0.2-times seawater treatments when compared with the controls, while in *Aster tripolium* there was a decrease in K content (140 to 104 μM/g) with increased salinity.[52] *Spergularia marina* plants grown at 0.2 times seawater had shoot Na levels that continued to increase for 20 d, from 80 to 170 μM/g fresh weight, while K content decreased from 190 to 80 μM/g.[53] Root tissue of *S. marina* followed a different pattern, with the Na content reaching a steady level 70 μM/g by day 12 and K concentration declined gradually from 100 to 70 μM/g. Transpiration rates for leaves of *S. marina* were reduced by 50% in 0.2-strength seawater vs. the nonsaline controls.[54] Daytime levels of Na and K in the xylem were relatively low (<7 mM) even at moderate salinity concentrations. In *S. marina* more than 80% of the Na and K was in the shoot during the vegetative phase. Cheeseman and Wickens[55] found that even with a 45-fold excess of Na over K in the growth medium, the internal Na and K contents were similar for both roots and shoots.

Okusanya and Ungar[56] compared the effects of salt treatment on the ion content of three species of *Spergularia: S. marina* (salt marsh species), *S. rupicola* (maritime cliff species), and *S. rubra* (glycophytic species of sandy soils). The Na and Cl content of leaves increased in all three species when they were exposed to 0.5 times seawater; Na increased from 1.64 to 4.4 mM/g dry weight in *S. marina*, 1.64 to 4.33 mM/g dry weight in *S. rupicola*, and 1.05 to 2.19 mM/g dry weight in *S. rubra*. The values for Cl followed the same pattern of increase, while the K content decreased in leaves of all three species.[56]

Puccinellia peisonis collected from field sites around the Neusiedler Lake (Austria)

TABLE 7
The Effects of Waterlogging and Salinity on the Shoot Cation Concentrations (mg/g) of the Experimental Species

	Sodium					Potassium					Calcium					Magnesium				
	d	w	ds	ws	D	d	w	ds	ws	D	d	w	ds	ws	D	d	w	ds	ws	D
Festuca rubra	1.2	1.4	9.6	6.7	5.4	32.9	35.0	42.1	34.3	—	1.3	1.5	1.6	1.8	—	3.4	4.6	3.6	3.3	—
Juncus gerardii	13.2	20.2	41.3	66.3	36.6	161.0	191.6	167.7	154.0	—	4.0	4.5	3.7	3.8	—	7.0	8.4	8.2	10.2	—
Armeria maritima	3.8	7.5	21.2	17.7	10.1	44.5	51.9	32.7	32.1	—	1.7	1.7	2.1	1.9	—	10.1	7.2	7.3	10.0	2.9
Plantago maritima	22.2	22.8	58.6	60.8	8.1	27.0	32.9	20.7	20.9	7.4	9.6	5.9	7.3	4.2	2.9	3.7	4.8	8.5	4.3	—
Aster tripolium	11.8	11.3	13.2	23.2	—	53.2	30.7	19.4	17.3	—	2.7	2.5	1.9	2.2	—	2.3	3.0	1.8	2.5	—
Triglochin maritima	25.5	21.6	33.4	32.4	—	64.7	74.6	45.1	59.3	—	7.2	4.8	5.0	3.2	1.8	10.9	11.8	8.9	8.8	—
Puccinellia maritima	5.3	6.4	14.7	10.5	7.3	33.8	50.7	42.5	38.1	—	2.7	1.8	3.4	2.0	—	1.7	1.1	1.8	1.4	—
Salicornia europaea	56.8	49.3	110.2	101.4	17.5	115.4	85.6	57.7	49.2	36.9	2.9	2.7	2.7	1.8	—	5.6	3.1	4.2	2.6	—

Note: d, drained nonsaline; w, waterlogged nonsaline; ds, drained saline; ws, waterlogged saline. D is a confidence interval above which any two treatment means are significantly different.

From Cooper, A., *N. Phytol.*, 90, 263, 1982. With permission.

had higher Na content in the roots than in leaves, whereas leaves and stems had higher K and Cl content than did roots.[57] Plants of *P. peisonis* grown in the laboratory under aerated and unaerated nutrient solutions, containing from 0 to 300 m*M* NaCl, had lower Na and Cl content in the roots than in the shoots. However, leaves from unaerated treatments had about 33% more Na and Cl than did leaves from aerated treatments. Stelzer and Läuchli[57] concluded that exclusion of ions is carried out by the plant controlling Na transport to the stele through activity of the inner cortex and endodermis cells of the root.

Growth of the halophyte *Spartina townsendii* was suppressed by all salt treatments, but was more greatly reduced at salinities above 300 to 400 m*M* NaCl.[58] Shoot ion content per liter tissue water increased in NaCl treatments from 0 to 800 m*M* to levels of 48 to 661 m*M* Na, 29 to 359 m*M* Cl, and declined from 168 to 95 m*M* K. The K to Na ratio in leaves of *S. townsendii* decreased from 3.45 to 0.14 when comparing the 0 and 800 m*M* NaCl treatments. On a dry mass basis at salinities above 200 m*M* NaCl, the levels of Na, Cl, and K remained relatively unchanged. A large percentage of the inorganic ions transported to the leaves might be secreted by salt glands. However, roots of *S. townsendii* had a continuous increase in Na and Cl on a dry mass basis, with increases in medium salinity up to 700 m*M* NaCl. The K to Na ratio in roots at 0 m*M* was 3.75, and at 800 m*M* NaCl it was 0.23. Storey and Wyn Jones[58] concluded that osmotic adjustment at low salinities was due to the uptake of Na and Cl, while at the higher salinities a decrease in leaf water content was more significant in obtaining the water balance for leaves of the grass *S. townsendii*.

Allen and Cunningham[59] reported that mycorrhizal roots of *Distichlis spicata* had higher Na, K, and P content in treatments with 1 and 2 g Na/kg soil than did nonmycorrhizal plants. Leaf Na content did not vary significantly between mycorrhizal and nonmycorrhizal treatments, probably because of the ability of *D. stricta* to excrete ions through salt glands.[59]

Suaeda australis plants were grown in nutrient solutions containing a range of salinities from 0 to 600 m*M* NaCl. Plants grew poorly minus NaCl, and optimum growth occurred at 50 to 150 m*M* NaCl.[60] Sodium content in leaves increased with salt increments to a maximum of 800 m*M*/l tissue water, with Cl reaching 500 m*M*/l tissue water in the 600 m*M* NaCl treatment (Figure 6). Potassium concentration of leaves decreased sharply in 25 to 150 m*M* NaCl treatments and remained relatively constant at about 60 m*M*/l at higher salinities (Figure 6). Concentrations of K in chloroplasts tended to be correlated to leaf K content, but Na and Cl content in the chloroplasts were generally lower than that of leaf extracts, averaging 115 m*M*/l Na and 80 m*M*/l Cl in the 350-m*M* NaCl treatment. Robinson and Downton[60] determined that Na and Cl were accumulated in chloroplasts of *S. australis* at low leaf ionic content, but were excluded when leaf Na and Cl levels were high.[60]

The Na and Cl content of chloroplasts of *Halimione portulacoides* plants grown in nutrient solutions containing 3% NaCl was observed to reach 750 and 450 m*M* for each ion respectively, while in the nonplastid cell fraction Na and Cl concentrations reached 1800 and 1250 m*M* for each ion, respectively. Kappen[61] concluded from these data that *H. portulacoides* had the capacity to regulate the Na and Cl content of different compartments within its cells.

The ionic content of leaf sap and chloroplasts from *Mesembryanthemum crystallinum*, for plants exposed to 0 to 400 m*M* NaCl, was determined by Demmig and Winter.[62] For plants that were grown at 400 m*M* NaCl, the Na content ranged from 156 to 234 m*M* and Cl from 42 to 64 m*M* in the chloroplasts. Total leaf sap Na ranged from 4 m*M* in the nonsaline treatment to 97 m*M* at 20 m*M* NaCl and 498 m*M* at 400 m*M* NaCl, while corresponding Cl values were 1, 69, and 374 m*M* for the three salt treatments, respectively. The change in Na content for chloroplasts was sharp between 0 and 20 m*M* NaCl, rising from 1 to 2 m*M* up to 99 to 148 m*M*. While a 20-fold increase in NaCl (400 m*M*) caused a five-fold increase in tissue Na, there was only a relatively small increase in the chloroplast Na content. The K content of leaf tissue decreased from 151 m*M* in 0 m*M* NaCl to 19 m*M*

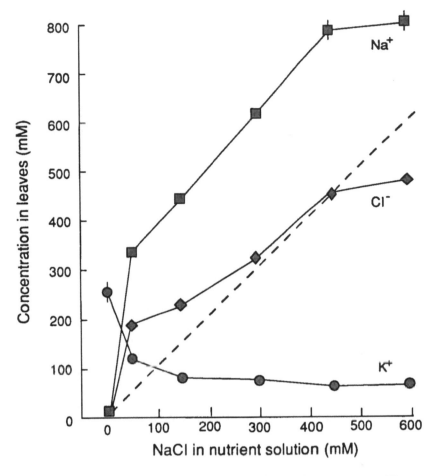

FIGURE 6. Ionic content of leaves from *Suaeda australis* plants that were grown at different levels of NaCl. Dashed line represents equal values in both medium and leaves. (From Robinson, S. P. and Downton, W. J. S., *Aust. J. Plant Physiol.*, 12, 471, 1985. With permission.)

in 400 mM NaCl, while in chloroplasts a similar decrease in K occurred from 110 and 160 mM down to 35 and 54 mM K. Cytoplasmic compartments contained lower concentrations of inorganic salts than the concentration found in vacuoles when *M. crystallinum* was exposed to high salinities.[62]

Greenway et al.[63] reported that growth of *Atriplex nummularia* was optimal at 100 to 200 mM NaCl. Leaf Na content increased from 0.9 to 5.0 mM/g dry weight and Cl increased from 0.4 to 3.6 mM/g in treatments with 1 and 300 mM NaCl. Potassium content of leaves decreased over this NaCl range from 1.8 to 0.6 mM/g dry weight. At optimal growth conditions plants accumulated from 4.1 to 6.4 mM Na/g and 2.0 to 3.0 mM Cl/g dry weight of leaf material. The rate of solute accumulation for *A. nummularia* at high NaCl concentrations may be more rapid than that required for ionic adjustment. Therefore, the pressure potential may be higher at high salinities, causing increased cell expansion and leaf area which would increase plant growth.[64] Similar results were obtained when *A. nummularia* was exposed to salinities ranging from 0 to 1% NaCl, resulting in an accumulation of Na at high salt concentrations and a reduction in K content in leaves. Plants grown in 2% NaCl had 63% of the blade Na content concentrated in the salt hairs (3.6 mg Na/10 cm² of leaf area), while controls contained 0.65 mg Na/10 cm² (49% of the leaf blade Na). The Na accumulation in *A. nummularia* organs could be essential, because Braun et al.[65] concluded from their data analysis that plants were not merely tolerant of NaCl, but required NaCl for their proton-translocating ATPase.

Under conditions of high NaCl concentrations in the medium, Jeschke et al.[66] determined that inadequate osmotic adjustment of roots caused a reduction in growth of *Atriplex am-nicola*. Root tissues had a relatively small amount of Na, even in vacuoles, ranging from 23 mM at 25 mM NaCl to 161 mM Na at 400 mM NaCl in root tips. Chloride ion values followed a similar pattern, ranging from 44 to 181 mM Cl in the two treatments, respectively. Vacuolar content of Na and K was higher than that in the cytoplasm of root cells, with Na ranging from 3 to 41 mM in the cytoplasm and 26 to 163 mM in vacuoles, while K ranged from 131 down to 110 mM in the cytoplasm and 140 to 256 mM in the vacuoles when plants were treated with 25 to 400 mM NaCl.[66] The Na content was determined for isolated vacuoles of *Atriplex gmelini*, and it was found tht the vacuoles contained 569 mM and 260 mM Cl.[67] Protoplasts of *A. gmelini* cells were found to have similar ionic content, 582 mM Na and 254 mM Cl, indicating that most of the ions accumulated in cells were sequestered in the vacuoles.

The accumulation of cations was studied by Aswathappa and Bachelard[68] in species of *Casuarina* with varying degrees of salt tolerance. The two highly salt-tolerant species, *C. glauca* and *C. equisetifolia*, accumulated much smaller quantities of Na and Cl in plant organs than did the moderately tolerant *C. cunninghamiana*, less than 0.8 mM Na/g dry weight for leaves of the two tolerant species and 1.6 mM Na/g dry weight for the latter. Similar results were obtained for Cl accumulation. Older needles of the tolerant species had a significantly greater Na and Cl content than younger leaves, but Aswathappa and Bachelard[68] found no difference between leaves of different ages in *C. cunninghamiana*.

Leaves of a salt-tolerant grass species *Leptochloa fusca* accumulated Na (11 to 185 mM) and Cl (16 to 245 mM) in the 250-mM NaCl treatment when compared with the nutrient controls, but had a 50% decrease (132 to 64 mM) in K.[69] Ionic adjustment was chiefly due to exclusion at the roots, but salt excretion from salt glands in the leaf served as a secondary mechanism. Gorham[69] determined that salt excretion was temperature dependent, with the rate of Na excretion for *L. fusca* plants exposed to 250 mM NaCl increasing from 11.6 nM/m^2/s at 19°C to 258.3 nM/m^2/s at 39°C and Cl excretion ranging from 9.4 to 283 nM/m^2/s for the two temperatures, respectively.

An electron probe was used by Eshel and Waisel[26] to determine the Na and Cl distribution in leaves of *Suaeda monoica*. Maximum Na concentration was found in the outer chlorenchyma where it comprised 20 to 30% of the dry mass compared with 8 to 10% for Cl, while Cl concentrations were highest in the inner chlorenchyma where it made up 15 to 25% of the dry mass compared with 5 to 10% for Na. From these data it was calculated that the Na content in the outer chlorenchyma of *S. monoica* was 1 M and Cl 0.3 M, and in the inner chlorenchyma Na was 0.3 M and Cl 0.6 M.[26]

Measurements were made of ion distribution in root cortical cells of *S. maritima* by use of X-ray microprobe analysis.[70] Plants were grown in the presence of 200 mM NaCl added to the nutrient solution, and it was determined that the Na content in the cytoplasm had a mean value of 118 mM, vacuole 432 mM, and cell wall 95 mM. Chloride ion levels were similar to those of Na in the various compartments: cytoplasm, 90 mM; vacuole, 445 mM; and cell wall, 120 mM.[70]

With an increase in soil salinity from 3.8 to 30 mS/cm electrical conductivity, Bal and Dutt[71] reported that the Na and Cl content of leaf tissue of *Acanthus ilicifolius* increased from 1.6 to 3.0% Na and from 1.0 to 2.0% Cl of leaf dry mass. The Na content of salts excreted from foliar salt glands increased from 14.1 to 30.0 mg/100 g fresh leaf mass, and Cl increased from 29.9 to 54.8 mg/100 g fresh weight with increasing medium salinity.

Dry mass production of *Salicornia europaea* increased with increases in NaCl concentration of the medium from 0 to 250 mM.[72] The Na concentration in shoots increased from

TABLE 8
Effect of 100 mM Salinity Shock on the Concentrations of Selected Free Ions in the Shoot Tips of *Salicornia europaea*

Time (h)	Na$^+$ (mM/l)	K$^+$ (mM/l)	Mg^{2+} (mM/l)	Ca^{2+} (mM/l)	Cl$^-$ (mM/l)	SO$_4$ (mM/l)
0	104.3 ± 13.2	116.9 ± 11.1	102.7 ± 4.1	23.7 ± 4.3	64.8 ± 3.1	91.6 ± 7.4
2	112.3 ± 8.7	119.1 ± 4.5	238.0 ± 6.4	25.5 ± 4.1	93.0 ± 2.4	109.7 ± 5.2
4	127.2 ± 7.0	125.0 ± 6.2	198.8 ± 6.2	28.3 ± 6.7	117.6 ± 2.3	111.3 ± 6.0
6	133.5 ± 14.3	137.0 ± 9.3	202.0 ± 5.3	22.8 ± 7.2	117.2 ± 5.1	106.3 ± 4.3
8	148.3 ± 5.6	104.7 ± 3.5	169.9 ± 4.8	28.7 ± 9.8	149.8 ± 4.8	90.4 ± 4.9
12	154.7 ± 18.4	104.8 ± 4.0	185.0 ± 16.0	27.8 ± 5.8	152.1 ± 3.9	91.7 ± 4.8
16	175.9 ± 11.2	112.2 ± 7.3	159.0 ± 11.4	22.7 ± 6.7	184.4 ± 7.0	84.1 ± 4.7
20	176.7 ± 6.2	100.5 ± 6.8	160.0 ± 5.5	22.8 ± 8.2	194.9 ± 6.8	89.1 ± 10.0
24	208.2 ± 13.0	98.5 ± 3.2	159.0 ± 13.0	21.1 ± 4.3	195.0 ± 4.7	77.4 ± 7.6
48	267.7 ± 24.1	92.6 ± 2.0	139.0 ± 8.9	20.0 ± 3.6	295.9 ± 3.6	67.6 ± 3.4
72	337.6 ± 19.3	73.8 ± 5.3	89.3 ± 7.6	12.3 ± 3.4	371.5 ± 5.0	59.8 ± 4.6

From McNulty, I. B., *Plant Physiol.*, 78, 100, 1985. With permission.

about 100 to 550 mM/100 g dry mass, and K decreased from 155 to 75 mM/100 g with increased salinity. Magnesium concentration of shoot tissue also decreased from 63 to 29 mM/100 g dry mass. Chloride ion content increased from 147 mM in controls to 495 mM in the 250-mM NaCl treatment. Austenfeld[72] concluded that Na and Cl were accumulated by *S. europaea* in proportion to their content in the growth medium.

Growth of seedlings of *S. bigelovii* was optimal in medium containing 3.0% NaCl. Ash content was fivefold higher (45%, or 0.5 g) in the 3.0% NaCl treatment than in the nutrient controls (13%, or 0.1 g) after a 24-d seedling culture treatment.[73] Na and Cl accounted for 85 to 95% of the ash weight in the embryo axis and cotyledons of *S. bigelovii* seedlings. Ion accumulation was considered to be required for osmotic adjustment and to maintain cell turgor. Stumpf et al.[73] concluded that the reason growth was inhibited in the controls is that insufficient Na and Cl were present to generate sufficient turgor for cell enlargement.

Ownbey and Mahall[74] reported that the effect of NaCl salinity (370 to 950 mosmol/l) on salt uptake per gram dry mass of *S. virginica* shoot tissue ranged from 4.5 to 7.6 mosmol/g dry mass.[74] Significant ion uptake by shoots indicate that plants were able to adjust osmotically to the salinity content of the medium. Mean relative growth rates of *S. virginica* were not inhibited when exposed to 950 mosmol/l NaCl. Field investigations indicated that there was a tendency toward salinity-induced growth inhibition at about 1400 mosmol/l. Mahall and Park,[75] concluded that halophytes differed from glycophytes in their increased capacity to avoid salt stress unless extreme soil salinities were present in the growth medium.

McNulty[76] reported from short-term experiments with *S. rubra* that shoots adjusted relatively quickly to treatments with 100 mM NaCl. After a 24-h treatment, plant Na levels increased from 104.3 to 208.2 mM/l extracted sap, and Cl increased from 64.8 to 195.0 mM/l (Table 8). The final measurement at 72 h indicated an increase in Cl to 371.5 mM and Na to 337.6 mM/l. Potassium content of sap decreased from 116.9 to 73.8 mM after the 72-h treatment (Table 8). Ca, Mg, and SO$_4$ concentrations in plants also decreased after 72 h. Compatible organic osmotica did not increase sufficiently to account for the osmotic balance between the vacuole and cytoplasm, indicating that osmotica may have been re-apportioned in the cell compartments during the 72-h experimental period.[76]

Tissue ion concentrations of *A. hastata* var. *salina* were found to increase in culture as external NaCl content was increased, but more Na was taken up than Cl.[77] Plants grown in 100 mM NaCl had 131.5 μM Na/g fresh weight, while plants at 500 mM NaCl had 281.1 μM Na/g fresh weight. In contrast, Cl content ranged from 65.7 to 170.9 μM/g fresh weight at the two salinities, respectively. Anderson et al.[77] concluded that the excess Na accumulated by *A. hastata* var. *salina* plants was balanced by endogenous organic compounds and a hydrogen efflux pump.

Richardson[78] determined that populations of *Atriplex canescens* vary in their ability to accumulate Na. Low Na biotypes and high Na biotypes treated with 4.1- and 18.6-mS/cm irrigation water containing Na_2SO_4 had very different Na and K contents in their leaves, ranging from 2 to 13 m*M* Na/100 g in low Na plants and from 50 to 128 m*M* Na/100 g dry mass in high Na plants in the two irrigation treatments. Potassium content did not vary signficantly in the high Na biotype (71 and 68 m*M*/100 g) for the two irrigation treatments, but for the low Na biotype the range was from 98 to 139 m*M*/100 g in the two treatments, respectively. Accumulation of Na in deciduous leaves could cause an increase in surface soil salinitiy and adversely affect the establishment of herbaceous plant species in the *A. canescens* community.[78]

Leaf dry mass of *A. canescens* and *A. cuneata* was reduced significantly at salinities above 20-mS/cm electrical conductivity.[79] However, both species could tolerate soil salinities as high as 38 mS/cm. The effect of -1.2 MPa Na_2SO_4 and $MgSO_4$ on the ion content of *A. canescens* was determined.[80] Plants in nonsalinized controls had 0.69 m*M*/g leaf dry mass Na, while Na-treated plants had 2.33 m*M*/g Na, and Mg-treated plants had 0.69 m*M*/g Na. Richardson and McKell[79] reported that the K and Ca concentrations of plants were lower in the salt treatments than in the controls — from 0.95 m*M* K and 0.14 m*M* Ca in controls to 0.79 and 0.76 m*M* K and 0.09 m*M* Ca in the Na and Mg salt treatments, respectively — indicating some interference with K and Ca uptake by Na and Mg.

Increases in medium salinity from 0 to 1.0% NaCl resulted in an accumulation of Na and a reduction in K content in the leaves of *A. nummularia*.[81] Plants cultivated in 2.0% NaCl had 63% of the blade Na content concentrated in the vesicular hairs (3.63 mg Na/10 cm^2 leaf area), while controls contained 0.65 mg Na/10 cm^2 leaf area (49% of the leaf blade Na).

Letschert[82] reported that increases in medium NaCl salinity (0 to 80 m*M*/l) resulted in an increase in leaf Na and Cl content for both *A. halimus* and *A. hortensis*. However, *A. halimus* accumulated considerably higher concentrations of Na and Cl in its leaves than did *A. hortensis* under treatments with 80 m*M* NaCl.

Investigations, using the technique of X-ray probe microanalysis, indicated that transfer cells in the roots of *A. hastata* function in selectively taking up K from the medium. Kramer et al.[83] concluded that high Na concentrations in the root tissue indicate that Na is not being excluded by the root, but that Cl may be excluded based on the higher Na to Cl ratio 2:1 in plants grown in a nutrient medium containing a 1:1 ratio.[83,84] The sodium content of roots of *A. hortensis* increased along the length of roots exposed to 1 m*M* Na from 2.3 μM at 0 to 6.25 mm to 10.3 μM/g fresh weight at 50 to 62.5 mm.[85] Potassium content of roots varied in the reverse direction with 50.5 μM/g at 0 to 6.25 mm to 10.7 μM/g fresh weight at 50 to 62.5 mm. Cytoplasmic damage may occur in roots of *A. hortensis* due to a K deficiency or by a toxic accumulation of Na in roots.[85] Inclusion of Na in vacuoles and maintenance of high K levels in the cytoplasm is a fairly general property of higher plant cells.[86] Vacuolar Na/K exchange is indicated when *A. hortensis* plants are exposed to a medium high in Na. Compartmental analysis of ions was done for *A. nummularia* plants treated with from 0 to 50 m*M* NaCl,[87] and it was determined that K content of the cytoplasm decreased from 6.8 to 4.5 μM/g in 50 m*M* NaCl treatments, while Na content increased from 3.1 to 9.9 μM/g in the cytoplasm. Vacuolar Na increased from 57.4 to 108.9 μM/g, while vacuolar K decreased from 171.6 to 139.6 μM/g. Mills et al.[87] concluded that excess ions are either sequestered in the vacuoles of cells or excreted by salt glands.

Some species of grasses were able to accumulate Na in their roots and did not transport high quantities to the leaves. When grown at 224 m*M* NaCl, *Puccinellia distans* was found to accumulate 13.3% Na in its roots and 5.1% in shoots.[88] However at lower salinities, 28 mM NaCl, this pattern was reversed to 1.2% Na and 0.8% Cl in the root and 2.1% Na and 4.7% Cl in shoots.

Microhairs in some species of Gramineae (*Chloris gayana, Cynodon dactylon, Leptochloa digitata, Sporobolus virginicus,* and *Zoysia macrantha*) were reported to contain Cl in leaf washings, while other species with similar hairs with basal cell partitioning membranes, such as *Sporobolus elongatus* and *Eleusine indica,* did not secrete Cl (Table 9).[89] None of the grass species examined without basal cell partitioning membranes were determined to contain high Cl content in their leaf washings or with analyses by X-ray microanalysis. The highest Cl levels were found in *Zoysia macrantha* and *Sporobolus virginicus* leaf washings, averaging 0.49 μM Cl/mg leaf dry weight and 0.24 μM Cl/mg leaf dry weight for the two species, respectively.[89]

An increase in NaCl salinity from 0 to 500 mM caused an increase in both the Na and Cl content of leaf tissue of the succulent halophyte *Disphyma australe,* which grows in saline habitats in southern Australia.[90] The Na content of leaves ranged from 20 mM in nonsaline controls to 465 mM in the 500-mM NaCl treatment, with the Cl concentrations following a similar pattern ranging from 10 to 334 mM in the two treatments, respectively. Other cations (K, Mg, Ca) in the leaf tissue decreased in concentration with increasing medium salinity. Potassium ion content decreased from making up 36% of the total ion content at the lowest salinity to 1.4% at the highest salinity (88.8 to 11.5 mM K). The relative contribution of Na and Cl to total leaf ion content increased from 67% in 25 mM NaCl to 93% in 500 mM NaCl.[90]

The Na and Cl content of roots and shoots of *Salicornia europaea* increased with increments in the NaCl level from 0 to 150 meq/l (Figure 7).[91] Potassium contents of roots and shoots dropped sharply in the 170-meq/l treatment and remained about 50% of the control treatment for plants in the 510-meq/l NaCl treatment (Figure 8).[91] The decrease in K content did not inhibit the growth of *S. europaea* plants, since highest biomass production occurred in the salt treatments between 170 and 340 meq/l NaCl.

Gale et al.[92] found that high relative humidity plus −0.5-MPa NaCl salinity was more inhibitory to dry mass production of *Atriplex halimus* than for plants exposed to low relative humidity. Salinity levels of −0.5 MPa stimulated growth by fourfold in comparison to nutrient controls at low relative humidities, while at high relative humidity an increase in salinity levels inhibited plant growth at all salinities. In contrast to these results, the relative growth rate of *Atriplex spongiosa* was inhibited by increased salinity, but increased relative humidity caused an increase in the growth rate of plants at 100 mM NaCl from 0.16 g/g/d at low humidity to 0.22 g/g/d.[93] At high relative humidity with 100 mM NaCl in the medium, the Na to K ratio was 7.1 compared with 5.8 at low relative humidity. Potassium content of the roots and shoots of *A. spongiosa* decreased, while Na and Cl levels increased with increments of NaCl up to 100 mM/l (Table 10).[93]

Measurements of Na and Cl content were made on leaves of *Halimione portulacoides* for plants grown in nutrient solutions containing 2.0% NaCl. Maier and Kappen[94] reported that Na concentrations ranged from 4.5 to 12.5 mM/g leaf dry mass and Cl from 4 to 12 mM/g leaf dry mass over the year, with ion content increasing in the winter and reaching peak values in February. They found no significant relationship between the freezing tolerance of leaves and the increase in inorganic ion content in the winter. Plants grown in 0% NaCl had similar levels of frost tolerance to those grown in 2% NaCl.[61,94]

Some research has been carried out to determine the effect of high Na content on Ca uptake in halophytes. The presence of NaCl in the growth medium inhibited the translocation of Ca to the leaves of *Cochlearia anglica.*[95] Increasing the concentration of NaCl in the medium from 0 to 170.8 mM caused a sharp decrease in the Ca content of leaves, while the root Ca concentration increased markedly with increasing salinity. The effect of NaCl

TABLE 9
Salt Secretion by Grass Microhairs

Species	Microhair morphology	Presence or absence of basal cell partitioning membranes	Amount of Cl⁻ present in the leaf washings (μ equivalents Cl/mg dry wt leaf)		Presence or absence of crystalline material at the microhair apices (S.E.M.)	X-ray microanalysis of microhair exudates: presence (+) or absence (−) of Na and Cl
			NaCl (100 mM)	Control		
Chloridoideae						
Chloris gayana	Chloridoid	Present	0.2232	0.0017	Present	+
Cynodon dactylon	Chloridoid	Present	0.1485	0.0008	Present	+
Leptochloa digitata	Chloridoid	Present	0.2747	0.0009	Present	+
Zoysia macrantha	Chloridoid	Present	0.4890	0.0152	Present	+
Eleusine indica	Chloridoid	Present	0.0013	0.0006	Absent	−
E. tristachya	Chloridoid	*	0.0008	0.0003	Absent	?
Sporobolus africanus	Chloridoid	*	0.0007	0.0002	Absent	?
S. elongatus	Chloridoid	Present	0.0014	0.0001	Absent	−
S. virginicus	Chloridoid	Present	0.2482	0.0012	Present	+
Bouteloua curtipendula	Chloridoid	Absent	0.0015	0.0013	Absent	−
Eragrostis ciliianensis	Chloridoid	Absent	0.0045	0.0032	Absent	−
E. elongata	Panicoid	Absent	0.0010	0.0008	Absent	−
Enneapogon nigricans	Enneapogon	Absent	0.0005	0.0003	Absent	−
Arundinoideae						
Rytidosperma linkii	Panicoid	Absent	0.0003	0.0015	Absent	−
Panicoideae						
Panicum virgatum	Panicoid	Absent	0.0003	0.0005	Absent	−
Spinefex hirsutus	Panicoid	Absent	0.0006	0.0003	Absent	−
Imperata cylindrica	Panicoid	*	0.0001	0.0002	Absent	−
Pennisetum clandestinum	Panicoid	*	0.0003	0.0000	Absent	−

* Information not available.

From Amarasinghe, V. and Watson, L., *Aust. J. Plant Physiol.*, 16, 219, 1989. With permission.

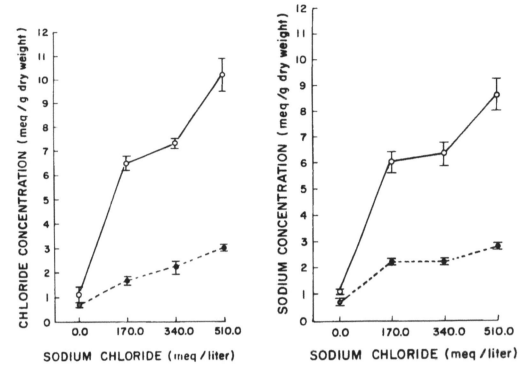

FIGURE 7.   Changes in shoot (O) and root (●) Na and Cl content of *Salicornia europaea* with increased medium NaCl salinity. (From Ungar, I. A., *Soc. Bot. Fr. Actual. Bot.*, 3-4, 95, 1978. With permission.)

in inhibiting Ca transport to leaves is accentuated when plants were not aerated, with Ca uptake decreasing about 50% from aerated controls when exposed to 85.4 and 170.8 mM NaCl.[95]

The growth of *Bromus mollis* from saline grasslands in the Camargue (France) was inhibited by increases in NaCl from 0 to 141 mM. The percent Na in roots and shoots increased with increasing salinity to 4% of the shoot dry weight over 3% of the root dry weight.[96] Addition of $CaCO_3$ to the soil caused a significant decrease in Na accumulation for plants grown at 56 and 141 mM NaCl, but growth responses did not appear to be ameliorated at high salinities by increasing the Ca content of the soil.

The effects of Ca on the uptake of Na and K in roots of *Atriplex gmelini* was determined for plants exposed to various concentrations of artificial seawater, 0.01 to full strength (480 mM Na and 13 mM K).[97] Roots of *A. gmelini* did not concentrate Na at high salinities, and Ca in the medium reduced the root Na content about 50% of that in seawater (394.6 mM) and 0.5 seawater (210.4 mM). Potassium was concentrated by roots, and the roots lost K in the absence of Ca from the medium. It was concluded from this investigation that Ca played a significant role in determining changes in the permeability of membranes.[97]

V. CHANGE IN IONIC CONTENT WITH LEAF AGE

Little research has been carried out to determine the changes in ion content of leaves of different age on a plant. The concentration of ions in leaves of *Jaumea carnosa* subjected to 300 to 500 mM NaCl was determined by St Omer et al.[98] In NaCl treatments, K, Ca, and Mg decreased markedly at all nodes when compared with controls. Leaf Na ion content increased sharply in all NaCl treatments when compared with controls where Na made up

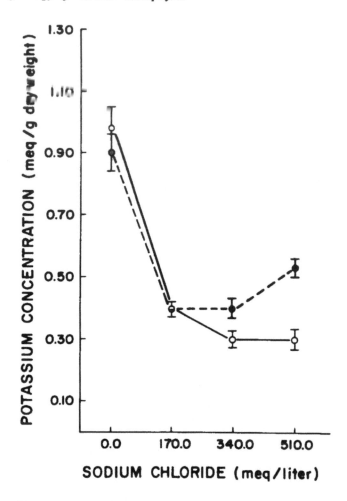

FIGURE 8.   The effect of NaCl salinity on K content of the shoots (O) and roots (●) of *Salicornia europaea*. (From Ungar, I. A., *Soc. Bot. Fr. Actual. Bot.*, 3-4, 95, 1978. With permission.)

1% of the leaf dry weight vs. between 10 and 15% for leaves in all of the NaCl treatments. No significant differences were found in the Na content of *J. carnosa* leaves in any of the treatments that were related to the nodal position of the leaves.[98]

Leaves of the salt marsh species *Atriplex hastata* accumulated fivefold more Na and Cl when they were treated with 200 mM NaCl than in 1-mM NaCl controls.[63] Eleven days after treatment with 200 mM NaCl, old leaves contained 4.2 mM Na and 1.9 mM Cl/g dry mass, while 1-mM NaCl control treatments had 0.87 mM Na and 0.27 mM Cl. On a dry mass basis young leaves had about 10% less Na and Cl than old leaves. Similar results were obtained by Karimi and Ungar[99] with *A. triangularis*, where mature leaves of plants treated with 258 mM NaCl accumulated 5.5 mM Na and 3.2 mM Cl/g dry weight, while young leaves contained significantly lower amounts, 2.6 mM Na and 1.6 mM Cl/g dry weight. Potassium concentration in mature leaves of *A. triangularis* decreased from 171 mM/l tissue water to 30 mM/l in the 517-mM NaCl treatment when compared with the nutrient solution controls, while Na content was 809 mM/l and Cl 679 mM/l tissue water in 517-mM NaCl treatment.[99] Osmond[100] determined that Na uptake by leaf disks of *Atriplex spongiosa* was accompanied by loss of K from the vacuolar compartment.

The ionic content of different tissues of *Suaeda maritima* was measured for plants treated with 50 mM NaCl, and it was determined that the younger apical tissues had a higher K to

TABLE 10

Inorganic Ion Content (mM/l) in the Roots (R) and Shoots (S) of *Atriplex spongiosa* Grown in Different Humidity and NaCl Treatments

NaCl humidity (%)	K		Na		Ca		Mg		Cl	
	R	S	R	S	R	S	R	S	R	S
Low humidity										
NaCl 0	98 ± 7	296 ± 31	1 ± 0	2 ± 0	5 ± 0	30 ± 3	10 ± 1	36 ± 2	1 ± 0	4 ± 0
50	62 ± 5	131 ± 15	76 ± 8	340 ± 36	4 ± 0	12 ± 1	9 ± 1	14 ± 0	22 ± 2	101 ± 10
100	49 ± 5	108 ± 10	79 ± 6	387 ± 29	4 ± 0	12 ± 1	9 ± 1	12 ± 1	26 ± 2	161 ± 13
High humidity										
NaCl 0	91 ± 6	217 ± 23	1 ± 0	4.0 ± 0	5 ± 0	221 ± 23	9 ± 1	15 ± 1	1 ± 0	1 ± 0
50	56 ± 4	97 ± 10	55 ± 7	230 ± 25	5 ± 0	327 ± 33	10 ± 22	8 ± 0	18 ± 3	46 ± 8
100	41 ± 5	69 ± 8	85 ± 7	226 ± 21	4 ± 0	335 ± 31	9 ± 1	9 ± 1	22 ± 2	84 ± 7

From Salim, M., *N. Phytol.*, 113, 13, 1989. With permission.

Na ratio (1.04) than old leaves (0.27), flowers (0.28), and stems (0.40).[101] Sodium content of tissues ranged from 218 mM/g fresh weight in the apex to 433 mM/g in flowers. Plants treated with 150 mM NaCl followed a similar pattern with young leaves (<2 mm) containing 225 mM Na and mature leaves (>20 mm) 490 mM Na, with K to Na ratios ranging from 1.2 to 0.4 in the two leaf sizes, respectively. The analysis of plant material treated with 300 mM NaCl by X-ray microanalysis indicated that leaf primordia had a K to Na ratio of 1.2, young leaf mesophyll vacuoles 0.53, and old leaf mesophyll vacuoles 0.06. The experiments indicated a high K to Na ratio was maintained in young meristematic cells of *S. maritima*, while in older tissue the NaCl is stored in the vacuoles, leading to low K to Na ratios.[101] However, Harvey et al.[102] reported that the cytoplasm and cell wall of mature mesophyll cells have low K to Na ratios (<0.2), indicating that both vacuoles and cytoplasm of mature tissues may have low K to Na ratios.

VI. CONCLUSIONS

It is clear from the above review of field and laboratory investigations that plants have developed a number of mechanisms to cope with hypersaline conditions. Mechanisms such as succulence, salt glands, and salt hairs, regulation of root uptake of ions, and accumulation of ions for osmotic adjustment are all significant factors in determining the level of salt tolerance of plant species. Halophytes may possess more than one mechanism for salt regulation, which may be related to morphological, anatomical, or physiological character-istics of the plant.[3] Some halophytes have the capacity to accumulate high concentrations of Na and Cl and function optimally under these conditions, while other species growing in similar habitats have developed mechanisms to reduce the concentration of inorganic ions reaching the photosynthetic tissues. Salt tolerance may be increased by compartmentation of ions within organs, tissues, or cellular compartments, and the production of organic osmotically compatible solutes. However, at present we really do not know what levels of inorganic ion accumulation are beyond the limits of tolerance for the biochemical processes in plants.[50] Salt tolerance is a complex process involving both cellular and whole plant integrated processes. Currently, we have not been able to isolate a single gene or gene product that has been shown to be directly involved in salt tolerance.[50] One must conclude from these data related to ion accumulation in plants that salt tolerance is a complex process involving several levels of organization in plants and that because of the complex nature of the phenomena, it will be difficult to ascertain the precise genetic differences between salt tolerant and intolerant species. Further investigations are necessary to separate the effects of ions at the cellular levels from problems related to transport and compartmentation of ions in the plant.

REFERENCES

1. **Albert, R. and Kinzel, H.**, Unterscheidung von physiotypen bei halophyten des Neusiedlerseegebietes (Osterreich), *Z. Pflanzenphysiol.*, 70, 138, 1973.
2. **Steiner, M.**, Zur Oklogie der Salzmarschen der nordostilichen Vereinigten Staaten von Nordamerika, *Jahrb. Wissenschaft. Bot.*, 81, 94, 1935.
3. **Albert, R.**, Salt regulation in halophytes, *Oecologia*, 21, 57, 1975.
4. **Epstein, E.**, Mineral metabolism of halophytes, in *Ecological Aspects of the Mineral Nutrition of Plants*, Rorison, I. H., Ed., Blackwell Scientific, Oxford, 1969, 345.
5. **Jefferies, R. L.**, Osmotic adjustment and the response of halophytic plants to salinity, *Bioscience*, 31, 42, 1981.
6. **Breckle, S.-W.**, Ionengehalte halophiler Pflanzen Spaniens, *Decheniana*, 127, 221, 1975.

7. **Sharma, M. L. and Tongway, D. J.**, Plant induced soil salinity patterns in two saltbush (*Atriplex* spp.) communities, *J. Range Manage.*, 26, 121, 1973.

8. **Bjerregaard, R. S., West, N. E., Caldwell, M. M., and Mayland, H. F.**, Standing crops and dynamics of phytomass and minerals in two salt desert shrub communities, *Great Basin Nat.*, 44, 327, 1984.

9. **Mizrahi, N. A. and Breckle, S.-W.** Salzverhaltnisse in Chenopodiaceen Nord-Afghanistans, *Bot. Jahrb. Syst.*, 99, 565, 1978.

10. **Breckle, S.-W.**, Studies on halophytes from Iran and Afghanistan. II. Ecology of halophytes along salt gradients, *Proc. R. Soc. Edinburgh*, 89B, 203, 1986.

11. **Albert, R. and Popp, M.**, Chemical composition of halophytes from the Neusiedler Lake region in Australia, *Oecologia*, 27, 157, 1977.

12. **Guy, R. D. and Reid, D. M.**, Factors affecting $^{13}C/^{12}C$ ratios of inland halophytes. II. Ecophysiological interpretations of patterns in the field, *Can. J. Bot.*, 64, 2700, 1986.

13. **Gorham, J., Hughes, L. L., and Wyn Jones, R. G.**, Chemical composition of salt-marsh plants from Ynys Mon (Anglesey): the concept of physiotypes, *Plant Cell Environ.*, 3, 309, 1980.

14. **Stienstra, A. W.**, Mineral composition of *Halimione portulacoides* (L.) Aellen leaves, *Hydrobiologia*, 52, 89, 1977.

15. **Wiebe, H. W. and Walter, H.**, Mineral ion composition of halophytic species from northern Utah, *Am. Midl. Nat.*, 87, 241, 1972.

16. **Winter, K., Troughton, J., Evenari, M., Lauchli, A., and Luttge, U.**, Mineral ion composition and occurrence of CAM-like diurnal malate fluctuation in plants of coastal and desert habitats of Israel and the Sinai, *Oecologia*, 25, 125, 1976.

17. **Poulin, G., Bourque, D., Eid, S., and Jankowski, K.**, Composition chimique de *Salicornia europaea* L., *Nat. Can.*, 105, 473, 1978.

18. **Redmann, R. E. and Fedec, P.**, Mineral ion composition of halophytes and associated soils in western Canada, *Commun. Soil Sci. Plant Anal.*, 18, 559, 1987.

19. **Welch, B. L.**, Relationships of soil salinity, ash, and crude protein in *Atriplex canescens*, *J. Range Manage.*, 31, 132, 1978.

20. **Goodin, J. R. and McKell, C. M.**, *Atriplex* spp. as a potential forage crop in marginal agricultural areas, *Proc. 11th Int. Grasslands Congr.*, 11, 158, 1970.

21. **Beadle, N. C. W., Whalley, R. D. B., and Gibson, J. B.**, Studies in halophytes. II. Analytic data on the mineral constituents of three species of *Atriplex* and their accompanying soils in Australia, *Ecology*, 38, 340, 1957.

22. **Breckle, S.-W.**, Zur Okologie und zu dem Mineralstoffverhaltnissen absalzender und nicht absalzender Xerohalophyten, *Diss. Bot.*, 35, 1, 1976.

23. **Ernst, W. H. O.**, Okologische Anpassungsstrategien an Bodenfaktoren, *Ber. Dtsch. Bot. Ges.*, 96, 49, 1983.

24. **Stumpf, D. K. and O'Leary, J. W.**, The distribution of Na^+, K^+ and glycinebetaine in *Salicornia bigelovii*, *J. Exp. Bot.*, 36, 550, 1985.

25. **Weber, D. J., Rasmussen, H. P., and Hess, W. M.**, Electron microprobe analysis of salt distribution in the halophyte *Salicornia pacifica* var. *utahensis*, *Can. J. Bot.*, 55, 1516, 1977.

26. **Eshel, A. and Waisel, Y.**, Distribution of sodium and chloride in leaves of *Suaeda monoica*, *Physiol. Plant*, 46, 151, 1979.

27. **Khan, M. A., Weber, D. J., and Hess, W. M.**, Elemental distribution in shoots of *Salicornia pacifica* var. *utahensis* as determined by energy-dispersive X-ray microanalysis using a cryochamber, *Bot. Gaz.*, 147, 16, 1986.

28. **Watson, M. C., O'Leary, J. W., and Glenn, E. P.**, Evaluation of *Atriplex lentiformis* (Torr.) S. Wats. and *Atriplex nummularia* Lindl. as irrigated forage crops, *J. Arid. Environ.*, 13, 293, 1987.

29. **Lanning, F. C. and Eleuterius, L. N.**, Silica and ash in tissues of some plants growing in the coastal area of Mississippi, USA, *Ann. Bot.*, 56, 157, 1985.

30. **Joshi, A. J.**, Amino acids and mineral constituents in *Sesuvium portulacastrum* L., a salt marsh halophyte, *Aquatic. Bot.*, 10, 69, 1981.

31. **Waisel, Y.**, *Biology of Halophytes*, Academic Press, New York, 1972.

32. **Amonkar, D. V. and Karmarkar, S. M.**, Ion regulation in *Salvadora persica*, Linn., *J. Biol. Sci.*, 21, 13, 1978.

33. **Goldstein, G., Rada, F., Sternberg, L., Burguera, J. L., Burguera, M., Orozco, A., Montilla, M., Zabala, O., Azocar, A., Canales, M. J., and Celis, A.**, Gas exchange and water balance of a mistletoe species and its mangrove hosts, *Oecologia*, 78, 176, 1989.

34. **Harward, M. R. and McNulty, I.**, Seasonal changes in ionic balance in *Salicornia rubra*, *Utah Acad. Proc.*, 42, 65, 1965.

35. **Hansen, D. J. and Weber, D. J.**, Environmental factors in relation to the salt content of *Salicornia pacifica* var. *utahensis*, *Great Basin Nat.*, 35, 86, 1975.

36. **Mert, H. H. and Vardar, Y.**, Salinity, osmotic pressure and transpiration relationships of *Salicornia herbacea* in its natural habitat, *Phyton*, 18, 71, 1977.

37. **Rajpurohit, K. S. and Sen, D. N.**, Seasonal variation in chloride ion percentage of plants and soils of Pachpadra salt basin in Indian desert, *Ind. J. Bot.*, 2, 17, 1979.

38. **Riehl, T. D. and Ungar, I. A.**, Growth and ion accumulation in *Salicornia europaea* under field conditions, *Oecologia*, 54, 193, 1982.

39. **Riehl, T. E. and Ungar, I. A.**, Growth, water potential and ion accumulation in the inland halophyte *Atriplex triangularis* under saline field conditions, *Acta Oecol. Oecol. Plant*, 4, 27, 1983.

40. **Joshi, A. J.**, Effects of seawater in amino acids and mineral ions composition in *Salicornia brachiata* Roxb, *J. Plant Physiol.*, 123, 497, 1986.

41. **Bennert, H. W. and Schmidt, B.**, On the osmoregulation in *Atriplex hymemelytra* (Torr.) Wats. (Chenopodiaceae), *Oecologia*, 62, 80, 1984.

42. **Ungar, I. A.**, Autecological studies with *Atriplex triangularis* Willdenow, in Proc. Symp. Biology of Atriplex and Related Chenopods, Tiedemann, A. R., McArthur, E. D., Stutz, H. C., Stevens, R., and Johnson, K. L., Eds., Gen. Tech. Rep. INT-172, Forest Service, U.S. Department of Agriculture, Ogden, UT, 1984, 40.

43. **Sharma, M. L., Tunny, J., and Tongway, D. J.**, Seasonal changes in sodium and chloride concentration of saltbush (*Atriplex* spp.) leaves as related to soil and plant water potential, *Aust. J. Agric. Res.*, 23, 1007, 1972.

44. **Garza, A. and Fulbright, T. E.**, Comparative chemical composition of armed saltbush and fourwing saltbush, *J. Range Manage.*, 41, 401, 1988.

45. **Binet, P. and Thammavong, B.**, Production primaire et accumulation des bioelements au niveau d'une population pure d'*Atriplex hastata*. 1. des rives de l'estuaire de la Seine (France), *Acta Oecol. Oecol. Plant.*, 3, 219, 1982.

46. **Von Willert, D. J.**, Tagesschwankungen des Ionengehalts in *Salicornia europaea* in Abhangigkeit vom Standort und von der Uberflutung, *Ber. Dtsch. Bot. Ges.*, 81, 442, 1968.

47. **Ruess, R. W. and Wali, M. K.**, Daily fluctuations in water potential and associated ionic changes in *Atriplex canescens*, *Oecologia*, 47, 200, 1980.

48. **Epstein, E.**, *Mineral Nutrition of Plants: Principles and Perspectives*, John Wiley & Sons, New York, 1972.

49. **Flowers, T. J. and Yeo, A. R.**, Ion relations of salt tolerance, in *Solute Transport in Plant Cells and Tissues*, Baker, D. A. and Hall, J. L., Eds., Longman Scientific and Technical, Essex, England, 1988, chap. 10.

50. **Cheeseman, J. M.**, Mechanisms of salinity tolerance in plants, *Plant Physiol.*, 87, 547, 1988.

51. **Cooper, A.**, The effects of salinity and waterlogging on the growth and cation uptake of salt marsh plants, *N. Phytol.*, 90, 263, 1982.

52. **Cheeseman, J. M., Bloebaum, P., Enkoji, C., and Wickens, L. K.**, Salinity tolerance in *Spergularia marina*, *Can. J. Bot.*, 63, 1762, 1985.

53. **Cheeseman, J. M. and Wickens, L. K.**, Control of Na^+ and K^+ transport in *Spergularia marina*. III. Relationship between ion uptake and growth at moderate salinity, *Physiol. Plant*, 67, 15, 1986.

54. **Cheeseman, J. M. and Wickens, L. K.**, Control of Na^+ and K^+ transport in *Spergularia marina*. I. Transpiration effects, *Physiol. Plant*, 67, 1, 1986.

55. **Cheeseman, J. M. and Wickens, L. K.**, Control of Na^+ and K^+ transport in *Spergularia marina*. II. Effects of plant size, tissue ion contents and root-shoot ratio at moderate salinity, *Physiol. Plant*, 67, 7, 1986.

56. **Okusanya, O. T. and Ungar, I. A.**, The growth and mineral composition of three species of *Spergularia* as affected by salinity and nutrients at high salinity, *Am. J. Bot.*, 71, 439, 1984.

57. **Stelzer, R. and Lauchli, A.**, Salt- and flooding tolerance of *Puccinellia peisonis*. III. Distribution and localization of ions in the plant, *Z. Pflanzenphysiol.*, 88, 437, 1978.

58. **Storey, R. and Wyn Jones, R. G.**, Salt stress and comparative physiology in the Gramineae. III. Effect of salinity upon ion relations and glycinebetaine and proline levels in *Spartina X townsendii*, *Aust. J. Plant Physiol.*, 6, 831, 1978.

59. **Allen, E. B. and Cunningham, G. L.**, Effects of vesicular-arbuscular mycorrhizae on *Distichlis spicata* under three salinity levels, *N. Phytol.*, 93, 227, 1983.

60. **Robinson, S. P. and Downton, W. J. S.**, Potassium, sodium and chloride ion concentrations in leaves and isolated chloroplasts of the halophyte *Suaeda australis* R. Br., *Aust. J. Plant Physiol.*, 12, 471, 1985.

61. **Kappen, L.**, Widerstandsfahigkeit von Halophyten gagenuber Gefrieren und Salzstress und ihre moglichen biochemischen Ursachen, *Ber. Dtsch. Bot. Ges.*, 92, 55, 1979.

62. **Demming, B. and Winter, K.**, Sodium, potassium, chloride and proline concentrations of chloroplasts isolated from a halophyte, *Mesembryanthemum crystallinum* L., *Planta*, 168, 421, 1986.

63. **Greenway, H., Gunn, A., and Thomas, D. A.**, Plant response to saline substrates. VIII. Regulation of ion concentrations in salt-sensitive and halophytic species, *Aust. J. Biol. Sci.*, 19, 741, 1966.

64. **Greenway, H.,** Growth stimulation by high chloride concentrations in halophytes, *Isr. J. Bot.,* 17, 169, 1968.

65. **Braun, Y., Hassidim, M., Lerner, H. R., and Reinhold, L.,** Studies on H^+-translocating ATPase in plants of varying resistance to salinity, *Plant Physiol.,* 81, 1050, 1986.

66. **Jeschke, W. D., Aslam, Z., and Greenway, H.,** Effects of NaCl on ion relations and carbohydrate status of roots and on osmotic regulation of roots and shoots of *Atriplex amnicola, Plant Cell Environ.,* 9, 559, 1986.

67. **Matoh, T., Watanabe, J., and Takahashi, E.,** Sodium, potassium, chloride, and betaine concentrations in isolated vacuoles from salt-grown *Atriplex gmelini* leaves, *Plant Physiol.,* 84, 173, 1987.

68. **Aswathappa, N. and Bachelard, E. P.,** Ion regulation in the organs of *Casuarina* species differing in salt tolerance, *Aust. J. Plant Physiol.,* 13, 533, 1986.

69. **Gorham, J.,** Photosynthesis, transpiration and salt fluxes through leaves of *Leptochloa fusca* L. Kunth, *Plant Cell Environ.,* 10, 191, 1987.

70. **Hajibagheri, M. A. and Flowers, T. J.,** X-ray microanalysis of ion distribution within root cortical cells of the halophyte *Suaeda maritima* (L.) Dum., *Planta,* 177, 131, 1989.

71. **Bal, A. R. and Dutt, S. K.,** Salt tolerance mechanism in *Acanthus ilicifolius* l., *Indian J. Plant Physiol.,* 30, 170, 1987.

72. **Austenfield, F. A.,** Untersuchungen zum Ionenhaushalt von *Salicornia europaea,* l. unter besonderer Berucksichtigung des Oxalats in Abhangigkeit von der Substratsalinitat, *Biochem. Physiol. Pflanzen,* 165, 303, 1974.

73. **Stumpf, D. K., Prisco, J. T., Weeks, J. R., Lindley, V. A., and O'Leary, J. W.,** Salinity and *Salicornia bigelovii* Torr. seedling establishment. Water relations, *J. Exp. Bot.,* 37, 160, 1986.

74. **Ownbey, R. S. and Mahall, B. E.,** Salinity and root conductivity: differential responses of coastal succulent halophyte, *Salicornia virginica,* and a weedy glycophyte, *Raphanus sativus, Physiol. Plant.,* 57, 189, 1983.

75. **Mahall, B. E. and Park, R. B.,** The ecotone between *Spartina foliosa* Trin. and *Salicornia virginica* L. in salt marshes of Northern San Francisco Bay. II. Soil water and salinity, *J. Ecol.,* 64, 793, 1976.

76. **McNulty, I. B.,** Rapid osmotic adjustment by a succulent halophyte to saline shock, *Plant Physiol.,* 78, 100, 1985.

77. **Anderson, W. P., Willcocks, D. A., and Wright, B. J.,** Electrophysiological measurements on the root of *Atriplex hastata, J. Exp. Bot.,* 28, 894, 1977.

78. **Richardson, S. G.,** High and low sodium biotypes of fourwing saltbush: their responses to sodium and potassium in retorted oil shale, *J. Range Manage.,* 35, 795, 1982.

79. **Richardson, S. G. and McKell, C. M.,** Water relations of *Atriplex canescens* as affected by the salinity and moisture percentage of processed oil shale, *Agronomy J.,* 72, 946, 1980.

80. **Richardson, S. G. and McKell, C. M.,** Salt tolerance of two saltbush species grown in processed oil shale, *J. Range Manage.,* 33, 460, 1980.

81. **Uchiyama, Y.,** Salt tolerance of *Atriplex nummularia, Tech. Bull. Trop. Agric. Res. Ctr.,* 22, 1, 1987.

82. **Letschert, U.,** Zum Mineralstoffhaushalt einiger Chenopodiaceae bei hohen Bor- und Salzangeboten, *Diss. Bot.,* 96, 1, 1986.

83. **Kramer, D., Anderson, W. P., and Preston, J.,** Transfer cells in the root epidermis of *Atriplex hastata* L. as a response to salinity: a comparative cytological and X-ray microprobe investigation, *Aust. J. Plant Physiol.,* 5, 739, 1978.

84. **Kramer, D.,** The possible role of transfer cells in the adaptation of plants to salinity, *Physiol. Plant,* 58, 549, 1983.

85. **Jeschke, W. D. and Stelter, W.,** Measurement of longitudinal ion profiles in single roots of *Hordeum* and *Atriplex* by use of flameless atomic absorption spectroscopy, *Planta,* 128, 107, 1976.

86. **Jeschke, W. D., Stelter, W., Reising, B., and Behl, R.,** Vacuolar Na/K exchange, its occurrence in root cells of *Hordeum, Atriplex,* and *Zea* and its significance for K/Na discrimination in roots, *J. Exp. Bot.,* 34, 964, 1983.

87. **Mills, D., Robinson, K., and Hodges, T. K.,** Sodium and potassium fluxes and compartmentation in roots of *Atriplex* and oat, *Plant Physiol.,* 78, 500, 1985.

88. **Harivandi, M. A., Butler, J. D., and Soltanpour, P. N.,** Effects of soluble salts on ion accumulation in *Puccinellia* spp., *J. Plant Nutr.,* 6, 255, 1983.

89. **Amarasinghe, V. and Watson, L.,** Variation in salt secretory activity of microhairs in grasses, *Aust. J. Plant Physiol.,* 16, 219, 1989.

90. **Neales, T. F. and Sharkey, P. J.,** Effect of salinity on growth and on mineral and organic constituents of the halophyte *Disphyma australe* (Soland.) J. M. Black, *Aust. J. Plant Physiol.,* 8, 165, 1981.

91. **Ungar, I. A.,** The effects of salinity and hormonal treatments on growth and ion uptake of *Salicornia europaea, Soc. Bot. Fr. Actual. Bot.,* 3-4, 95, 1978.

92. **Gale, J., Naaman, R., and Poljakoff-Mayber, A.,** Growth of *Atriplex halimus* L. in sodium chloride salinated culture solutions as affected by the relative humidity of the air, *Aust. J. Biol. Sci.,* 23, 947, 1970.

93. **Salim, M.,** Effects of salinity and relative humidity on growth and ionic relations of plants, *N. Phytol.,* 113, 13, 1989.
94. **Maier, M. and Kappen, L.,** Cellular compartmentalization of salt ions and protective agents with respect to freezing tolerance of leaves, *Oecologia,* 38, 303, 1979.
95. **Le Guen, J.,** Migration du salaium vers les organes aeriens chez un halophyte *Cochlearia anglica* I. Effects du NaCl, *Physiol. Veg.,* 14, 381, 1976.
96. **Bassett, P. A.,** The effect of soil salinity and calcium levels on the growth of *Bromus mollis* in the Camargue, France, *Oikos,* 35, 353, 1980.
97. **Davis, R. F.,** Salinity effects on the electrical and ionic parameters of *Atriplex gmelini,* in *Plant Membrane Transport: Current Conceptual Issues,* Spanswick, R. M., Lucas, W. J., and Dainty, J., Eds., Elsevier/North-Holland, Amsterdam, 1980, 407.
98. **St. Omer, L., Horvath, S. M., and Setaro, F.,** Salt regulation and leaf senescence in aging leaves of *Jaumea carnosa* (Less.) Gray (Asterceae), a salt marsh species exposed to NaCl stress, *Am. J. Bot.,* 70, 363, 1983.
99. **Karimi, S. H. and Ungar, I. A.,** The effect of salinity on the ionic content and water relation of *Atriplex triangularis,* in *Proc. Symp. Biology of Atriplex and Related Chenopods,* Tiedemann, A. R., McArthur, E. D., Stutz, H. C., Stevens, R., and Johnson, K. L., Eds., Gen. Tech. Rep. INT-172, Forest Service, U.S. Department of Agriculture, Ogden, UT, 1984, 124.
100. **Osmond, C. B.,** Ion absorption in *Atriplex* leaf tissue. I. Absorption by mesophyll cells, *Aust. J. Biol. Sci.,* 21, 1119, 1968.
101. **Gorham, J. and Wyn Jones, R. G.,** Solute distribution in *Suaeda maritima, Planta,* 157, 344, 1983.
102. **Harvey, D. M. R., Hall, J. L., Flowers, T. J., and Kent, B.** Quantitative ion localization within *Suaeda maritima* leaf mesophyll cells, *Planta,* 151, 555, 1981.

Chapter 8

WATERLOGGING

I. INTRODUCTION

A number of researchers have suggested that tidal action and the effect of waterlogging play a significant role in determining the zonation of plant species in coastal salt marshes.[1-10] The distribution and growth of halophytes in saline habitats is most probably determined by a number of abiotic and biotic variables. Two factors that have been investigated in some detail are the effects of soil salinity and tidal action.[1,3,4,6] The influence of tidal action can be divided into the direct effect of tidal movement and that of waterlogging of the soil. Tidal silt deposition and tidal abrasion[11] as well as reduced aeration and its ancillary effects on soil conditions could play a significant role in determining the establishment of plant species in coastal salt marshes.

One of the direct effects reported for flooding in salt marshes is a reduction in the level of aeration in the soil which could impede plant growth, since oxygen is necessary for aerobic respiration in roots.[3] The adaptations that plants have developed which allow them to grow in poorly aerated salt marsh soils could play a very significant role in determining the zonation of plants in both inland and coastal salt marshes. Plants growing in areas that are flooded for prolonged periods or frequently exposed to tidal fluxes have either developed adaptations to acquire oxygen in order to survive or must remain dormant during prolonged periods when they are exposed to poorly aerated soils. The availability of some essential nutrients which are necessary for plant growth could be reduced when roots are exposed to poor aeration and more negative redox potentials in waterlogged soils. Some inorganic substances, such as sulfides, may be accumulated in toxic quantities in salt marsh soils. Precise data are needed to sort out the various environmental effects that flooding may have on plants at different stages of plant development.

II. THE EFFECT OF FLOODING ON PLANT DISTRIBUTION AND GROWTH

The relationship between soil aeration and the distribution of plant species was studied on the Humber salt marsh (England), and Armstrong et al.[12] determined that tidal activity could cause significant temporal and zonal differences in the aeration of soils within the salt marsh. In an earlier investigation, Chapman[3] reported that an aerated layer occurred in some salt marshes between the water table and tide during flooding periods, but at the Humber marsh no evidence was found for this aerated layer when soil redox potentials were used as an estimate of soil aeration. *Spartina anglica*, the dominant species of the low marsh, was exposed to reducing conditions that persisted throughout most of the soil profile. However, soils at sites containing *Puccinellia maritima*, *Halimione portulacoides*, and *Elymus pycanthus* could become highly oxidized during periods between spring tides, with *Puccinellia* soils having the shortest periods of oxidation. The higher general salt marsh sites were not affected by high spring tides and had a longer period of oxidation than other sites.[12]

Tidal fluctuations affected species cover on the Oosterwelder salt marsh (Netherlands) over a 15-year period.[9] Inundation by tides reduced plant cover by inducing substrate anaerobiosis, increasing salt stress and deposition of silt. Significant positive or negative correlations were found between inundation frequency and plant species cover at all elevations and during all seasons. Species such as *Festuca rubra*, *Triglochin maritima*, *Juncus maritimus*, and *J. gerardii* were positively affected by inundation, indicating that they were able to migrate into newly exposed areas. Other species such as *Armeria maritima*, *Agrostis stolonifera, Atriplex prostrata,* and *Halimione portulacoides* were negatively correlated with tidal inundation, having a loss of cover because of tidal submergence or the abrasive action of tides. Plant species that produced a seed bank or regenerated vegetatively were rapidly recruited into bare areas. The more salt-tolerant species of halophytes might be better competitors after tidal inundation. Seasonal variation in tidal inundation played a significant role in determining plant establishment, with most positive correlations occurring in the winter and most negative correlations with inundation frequency in the summer.[9]

The depth of the water table was determined to be a critical factor in controlling the distribution of *Limonium diffusum* in French Mediterranean saline soils.[13] The water table level varied seasonally, being highest in the spring and fall. *Limonium* occurred in a transition zone between the highly salt-tolerant halophytes and glycophytes and under saline conditions which glycophytes from the dunes could not tolerate. Soil electrical conductivity measurements ranged from 15 to 73 mS/cm at the eight sites studied. Microtopography and its relationship to the depth of the saline water table was apparently the determining factor for the establishment of *L. diffusum* populations between the higher dune glycophytes and the lower elevation salt marsh species.[13]

Plants may be capable of affecting the level of oxidation of soils in their rhizosphere. Pneumatophores of the mangrove species *Avicennia germinans* increased the oxygen content of soils, and soils beneath this species tended to be more oxidized than soils under *Rhizophora mangle*, which resembled the oxidative state of bare ground.[14] The requirement for oxidized substrate may explain why *A. germinans* usually occurred at higher elevations where its pneumatophores were exposed to air for most of the tidal cycle. The shoots of plants in the salt marsh grass genus *Puccinellia* were determined to be the major source for oxygen that was transferred to roots.[15] Oxygen transfer from shoots to roots was not inhibited by a previously reported high diffusive resistance at the root-shoot junction for *Puccinellia peisonis*.[16] Raskin and Kende[17] proposed that a mass flow of air from the aerial to the submersed plant parts rather than diffusion was the major process for aerating inundated *Oryza sativa* plants. Oxygen consumed by submerged organ respiration and solubilization of respiratory carbon dioxide in the surrounding water caused a reduction in pressure in the air-conducting system of the plant, which was considered to be the driving force for the mass flow of air into submersed plant parts.

By increasing the amount of subsurface drainage in experimental field plots at a Sapelo Island salt marsh (Georgia), it was found that the yields of *Spartina alterniflora* in 1979 and 1980 were >600 g C/m^2 in tiled plots vs. about 425 g C/m^2 in undisturbed controls and <400 g C/m in disturbed controls (trenches dug) that were not tiled (Table 1). Wiegert et al.[18] hypothesized that the increased drainage caused by tiling plots probably reduced sulfide toxicity, increased aeration, and reduced salinity stress at sites containing the intermediate growth form of *S. alterniflora*.

Root oxygen deficiencies along a streamside to inland transect of *S. alterniflora* caused a reduction in height growth and a decrease in plant biomass production. Mendelssohn et al.[19] determined that aerenchyma tissue did not conduct sufficient oxygen to roots for plants to carry on complete aerobic respiration in highly reduced substrates. Redox potentials varied from streamside to inland habitats, remaining relatively high in the more productive streamside habitat and decreasing significantly as one moved along the gradient to the more inland habitat. A shift to anaerobic respiration occurred at sites with reduced interstitial water movement and low substrate oxygen concentrations. Root alcohol dehydrogenase activity, an estimate of alcohol fermentation, was lower at the streamside site where plant biomass production was greatest. The ATP concentration of streamside plants decreased with increased waterlogging, indicating that the supply of ATP could be limiting under conditions of inundation. Another adaptation to waterlogging is related to the accumulation of nontoxic malate as the end product of anaerobic respiration in roots, preventing the accumulation of toxic ethanol but limiting ATP production.[19] Inland populations of *S.*

TABLE 1
Biomass Production (g C/m^2) of
Spartina alterniflora under
Conditions of Increased
Drainage[a]

Treatment	Year	
	1979	1980
Tiled plots	775	630
Undisturbed control	430	425
Disturbed control	200	360

[a] Data taken from graph.

From Wiegert, R. G., Chalmers, A. G., and Randerson, P. F., *Oikos*, 41, 1, 1983. With permission.

alterniflora were under greater environmental stress than streamside forms, as is indicated by the lower biomass yields of plants at the former sites.[20] The adenylate energy charge ratio, which represents the level of metabolically available energy stored as adenylate, was used to estimate sublethal levels of stress in coastal plant communities. It was determined that both the adenylate energy charge ratio (EC ratio) and ATP concentration were higher in streamside populations than for inland populations of *S. alterniflora*. Mendelssohn and McKee[21] concluded from their investigation that the EC ratio could be used as a useful indicator of environmental stress in salt marsh habitats.

McKee and Patrick[10] determined that the vertical distribution of *S. alterniflora* in coastal salt marshes was primarily related to the mean tide range. Differences in tidal amplitude accounted for 70 and 68%, respectively, of the statistical variation for the upper and lower limits of the range for this species over its latitudinal distribution on salt marshes. A number of local edaphic factors could influence the growth and distribution of *S. alterniflora* on salt marshes from Florida to Maine. These include redox potential of the soil, depth and frequency of flooding, nutrient levels, salinity, physical perturbation by tides, and interspecific competition.[10] The effects of waterlogging on die-back of *S. alterniflora* at inland sites of Barataria Bay salt marshes (Louisiana) were investigated to identify the factors associated with the reduced growth and die-back of plants.[21] Reciprocal transplants were made between streamside and inland populations of *S. alterniflora*. Soil redox potential of streamside transplants decreased over time at 1-cm depth in the inland zone until they reached typical inland levels of -100 to -200 mV after 42 d, while transplants from inland soils reached streamside levels of $+50$ to $+100$ mV after 11 d at the 1-cm depth (Table 2). Sulfide levels were ten times higher in the inland (1.0 mM) vs. streamside sites (0.1 mM), with transplant soils reaching site levels after 42 d. Soil salinity concentrations did not differ significantly at the two locations. Alcohol dehydrogenase activity, an indicator of anaerobic respiration, was significantly higher in transplants and controls at the inland site then at the streamside site which had better aeration. After 1 year of transplantation, the inland culms (28 g/pot) had an above-ground biomass equivalent to that of streamside controls (26 g/pot). Although streamside transplants to the inland site had reduced yield (16 g/pot), they achieved about 2.5 times greater biomass production of inland controls (7 g/pot).[21] Physical factors such as soil waterlogging, more negative redox potentials, sulfide accumulation, and increased anaerobic metabolism in plants were related to reduced growth at the inland sites. Die-back may be caused by the increased soil reducing power and sulfide accumulation at inland sites in these Louisiana salt marshes.

TABLE 2
Soil Redox Potentials (mV) at 1-cm Depth for 0 to 42 d after Transplanting for Controls from Streamside (SS) and Inland (II) Sites of *Spartina alterniflora* as well as for Transplants from Inland to Streamside (IS) and Streamside to Inland (SI) Sites.[a]

	Days			
	0	2	11	42
SS	+75	+19	+56	+112
IS	−159	−112	+337	+131
SI	+80	−94	−84	−206
II	−159	−146	−150	−191

[a]   Data taken from graph.

From Mendelssohn, I. A. and McKee, K. L., *J. Ecol.*, 76, 509, 1988. With permission.

Spartina patens was found to have primarily aerobic respiration in dune habitats and high levels of anaerobic respiration in marsh soils. Burdick and Mendelssohn[22] determined that the aerenchyma tissue in shoots was sufficient to supply some oxygen to roots but not at adequate levels to overcome anoxic conditions when plants at either dune or marsh locations were waterlogged. Under waterlogged conditions, plants increased their root alcohol dehydrogenase activity and had low EC ratios. Increased anaerobic respiration, leading to malate and lactate accumulation, occurred in all populations in August when soil waterlogging reached its peak. Development of root aerenchyma tissue for *S. patens* plants that were growing in waterlogged soils was higher than for plants in better-drained soils. It was concluded that anatomical adaptations might not be sufficient to fulfill the oxygen requirements for aerobic respiration in temporally or spatially waterlogged habitats along the gradient studied, causing *S. patens* plants to utilize anaerobic respiration pathways.[22]

III. TOLERANCE TO WATERLOGGING AND RELATED FACTORS

Laboratory experiments were initiated with salt marsh species from Portaferry marshes (Northern Ireland) to determine the effects of salt stress and tidal inundation on salt marsh species. Cooper[8] ascertained that the growth of upper marsh species, such as *Festuca rubra*, *Armeria maritima*, and *Juncus gerardii*, were strongly limited by both soil salinity and waterlogging (Table 3). *Plantago maritima*, *Aster tripolium*, and *Triglochin maritima* had their lowest dry mass production in waterlogged-saline treatments. Growth of lower marsh species was not significantly inhibited by these factors, with *Salicornia europaea* having its maximum biomass production in saline treatments and *Puccinellia maritima* having better growth on waterlogged soils.

The effects of aeration, light intensity, and salinity on the growth of *Atriplex triangularis* was determined by Karimi and Ungar.[23] Lack of aeration caused a reduction in plant dry mass, with the lowest dry mass for all salinity treatments, 0 to 3% NaCl, in the shaded treatment that was unaerated (Figure 1). Highest dry mass yields were in the 0.5% NaCl treatment exposed to high light and aeration. Aerated plants exposed to high light produced

TABLE 3
The Effects of Waterlogging and Salinity on the Mean Dry Mass (mg) of Salt Marsh Species

	Drained nonsaline	Waterlogged nonsaline	Drained saline	Waterlogged saline	D*
Festuca rubra	345.5	138.9	150.8	80.5	148.0
Juncus gerardii	150.5	77.3	59.4	34.8	59.4
Armeria maritima	278.5	158.0	189.9	81.1	126.3
Plantago maritima	723.6	712.8	521.6	499.7	169.8
Aster tripolium	904.2	908.3	740.3	489.0	344.1
Triglochin maritima	154.9	130.5	158.6	92.4	51.6
Puccinellia maritima	369.0	424.1	243.8	282.6	106.2
Salicornia europaea	31.6	19.4	40.7	35.4	17.9

* D is a confidence interval above which any two treatment means are significantly different.

From Cooper, A., *N. Phytol.*, 90, 263, 1982. With permission.

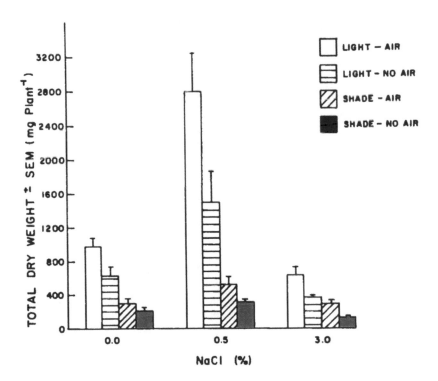

FIGURE 1.   The effect of aeration, salinity, and light intensity on the dry mass production of *Atriplex triangularis*. (From Karimi, S. H. and Ungar, I. A., unpublished data.)

about 2.8 g of total biomass per plant, while unaerated plants had a 50% reduction in yield. Unaerated plants in the shade treatment at 0.5% NaCl produced about 0.3 g/plant total biomass compared with about 0.5 g/plant in the aerated treatment (Figure 1). The number of epidermal trichomes per leaf area was related to the salinity level and aeration treatment.[24] The greatest number of trichomes were found in the high light, aerated treatment in 0.0% NaCl. Unaerated treatments generally had fewer salt hairs per unit leaf area than did aerated treatments. The only exception to this was in the aerated 3.0% NaCl treatments (Figure 2).

The yields of *Puccinellia peisonis* plants in nonsaline-aerated controls were only 30%

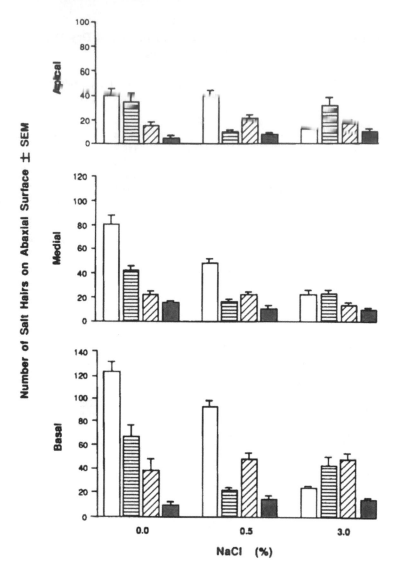

FIGURE 2.   The effect of aeration, salinity, and light intensity on the production of
salt hairs on leaves of *Atriplex triangularia*. Treatments were light-aerated (white
bar), light-unaerated (horizontal line bar), shade-aerated (diagonal line bar), and shade-
unaerated (black bar). (From Karimi, S. H. and Ungar, I. A., *Bot. Gaz.*, 150, 68,
1989. With permission.)

of that for plants in nonsaline-unaerated treatments.[25] At increased salinity concentrations
(200 to 250 mM NaCl), the difference in dry mass production between aerated and unaerated
treatments was not significant. Similar results were found in KCl treatments, but the growth
of plants in the 200- to 300-mM KCl concentrations were more greatly reduced in the aerated
vs. unaerated treatments. Stelzer and Lauchli[26,27] determined that the development of aerenchyma tissue in the root cortical region played a significant role in the ability of *P. peisonis*
to tolerate flooding. The inner cortical cells and the development of a secondary endodermis
external to the primary endodermis were hypothesized to play a significant role as a physiological barrier to the transport of solutes to the stele. Addition of 100 mM NaCl to the
medium caused a significant increase in the respiration rates of roots growing in unaerated
conditions vs. those in aerated treatments, with higher salinity levels causing a reduction in

the respiration rate in both aerated and unaerated treatments.[16] Differences in respiration rates between aerated and unaerated plants were not significant at any of the KCl concentrations, 0 to 300 mM. Oxygen diffusion from the proximal part of the root was greater than when the root-shoot transition zone is flooded, but during long periods of flooding this latter pathway may be the chief source of oxygen.[16]

An experiment was carried out to determine if two genetically distinct forms of *Sporobolus virginicus* varied in their ability to tolerate anaerobic conditions. Donovan and Gallagher[28] reported that the dune and marsh forms of *S. virginicus* were both able to adjust morphologically and physiologically to anaerobic conditions. However, adaptations to anaerobic conditions varied in the two populations, with plants found in wetter marsh habitats having a reduction in biomass from 56.6 to 28.4 g, development of more rhizome aerenchyma, and production of dwarf plants with reduced underground biomass. Dune plants, from a dry sandy habitat, when grown under anaerobic conditions, had an increase in stolon production but no statistically significant reduction in total biomass production from aerobic controls (45.7 to 38.6 g). Increased stolon development and more rhizome aerenchyma tissue in dune plants enabled them to transport oxygen to roots and maintain aerobic respiration. Donovan and Gallagher[28] found no indication of anaerobic respiration in either form under unaerated conditions, indicating that both marsh and dune plants could adapt to anaerobic conditions. The distribution of these two populations on the dune and marsh was not limited by their ability to tolerate anaerobic conditions.

A correlation was found between the lack of aeration and an increase in abscisic acid (ABA) production in the glycophyte *Nicotiana rustica*.[29] Plants treated with salt solutions maintained their turgor under unaerated conditions or when treated with exogenous ABA, while plants from no-salt treatments were wilted when aeration treatments were stopped. Mizrahi et al.[29] hypothesized that the resistance of this species to both aeration stress and salinity stress was due to an increase in endogenous ABA concentrations.

Scirpus americanus had an increase in ethylene production when plants were waterlogged in greenhouse experiments. Seliskar[30] reported that increased ethylene production was associated with a decrease in plant growth and an increase in stem aerenchyma tissue. Ethylene concentrations in plants ranged from 0.4 nl/g dry weight under drained conditions to 1.8 nl/g dry weight in the waterlogged plants. Other species occurring in the area, *Panicum amarulum* and *Spartina alterniflora*, did not exhibit an increase in ethylene production in the waterlogged treatments. Measurements of ethylene production in the field after periods of heavy precipitation indicated a significant increase for plants of *Scirpus americanus* growing at lower sites but not at higher sites. Exogenous treatments of *S. americanus* plants with ethylene produced plants with 47.7% aerenchyma tissue compared with 10.3% in air controls, while ethylene treated plants increased 13.8% in height compared with 64.9% in the air controls.[30] These data may explain the reduced growth of a number of halophytic taxa when they were exposed to anoxic conditions. Hypoxia caused a reduction in the salt tolerance threshold and a reduction in sodium ion exclusion for *Helianthus annuus*. Root-shoot ratios were reduced in unaerated plants, dry weight of both roots and shoots were reduced, leaf expansion was decreased, and plants were prone to wilting.[31] Aerenchyma development was considered to be an adaptation to hypoxic conditions, but aeration enhanced both growth and salt exclusion beyond the level that any adaptations to anaerobic conditions could produce in *H. annuus*.

Dry mass production of *Salicornia stricta*, a coastal halophyte from the Orne Estuary (France) that is usually subjected to 20 to 30 tidal inundations per month, was stimulated by once-daily artificial tidal inundation treatments in the laboratory (Figure 3).[32] However, *S. ramosissima* from an inland saline location at Lorraine (France) had its highest dry mass production in the treatment that was not exposed to inundation (Figure 4). The response of *S. stricta* to once-daily tidal inundation appears to be an evolutionary adaptation to the tidal regimen that it is exposed to on the Orne Estuary. Langlois and Ungar[32] experimentally

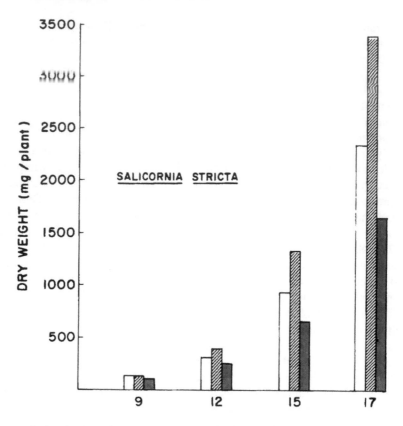

FIGURE 3.   The effect of tidal inundation (white bar = none, diagonal bar = once daily, black bar = twice daily) on the biomass production of *Salicornia stricta*. (From Langlois, J. and Ungar, I. A., *Aquatic Bot.*, 2, 43, 1976. With permission.)

compared the effect of once-daily, twice-daily, and no tidal inundation on the height growth and protein nitrogen content of *S. stricta* and *S. ramosissima*. *S. stricta* had its best height growth at once daily submergence (191.9 mm vs. 165.0 mm), while *S. ramosissima* had maximal height growth in the nontidal treatment (174.0 mm vs. 138.4 mm). Protein nitrogen content of plants followed the same pattern as the results that were obtained for dry mass production and height growth. The optimal growth pattern for both of these plant species was found to be in the tidal treatment that resembled the conditions that they were exposed to in their native habitats. *S. stricta* is exposed to 20 to 30 d per month of tidal inundation in the field, and it made optimal growth in the laboratory in the once-daily inundation treatments. An adaptation to tidal inundation by *S. stricta* improved its ability to occupy tidal locations. The inland population of *S. ramosissima* may have been isolated from coastal forms of *Salicornia* for a prolonged period of time, because it was inhibited by tidal submergence and made optimal biomass production in controls that did not receive a tidal treatment. Langlois[33] reported that a once-daily treatment with tidal inundation stimulated the growth and dry matter production in species of *Salicornia*. Binet[34] also determined that *Cochlearia anglica*, an estuarine halophyte, had better growth in the low marsh habitat, where it was exposed to regular tidal inundation, than in the high marsh environment.

Salicornia virginica did not commonly occur in the lower *Spartina foliosa* zone on the Mare Island salt marshes (California). Water depth in the *S. foliosa* zone averaged 4 cm greater than in the *Salicornia virginica* zone during tidal periods. Laboratory experiments were performed by Mahall and Park[7] to determine if artificial tidal regimens would have a differential effect on these two salt marsh dominants. Growth of *Salicornia* seedlings was

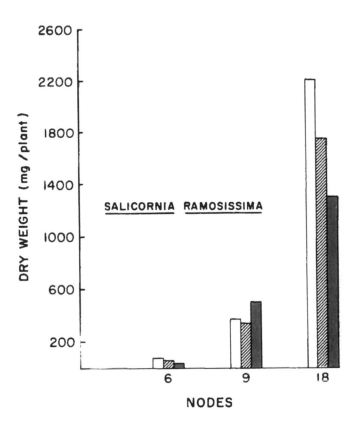

FIGURE 4.   The effect of tidal inundation regimes (white bar = none, diagonal bar = once daily, black bar = twice daily) on the biomass production of *Salicornia ramosissima*. (From Langlois, J. and Ungar, I. A., *Aquatic Bot.*, 2, 43, 1976. With permission.)

reduced by 64% of that in 1-cm shallow tides by twice daily 9-cm tides, while growth of older plants was reduced by 37%. New shoot development was reduced by about 50% in deep tides, and the inhibition of vegetative reproduction of *Salicornia* in the *Spartina* zone may be because of the inhibition of rerooting by tidal immersion. Growth of *Spartina* plants was not significantly inhibited by tidal treatments, indicating that *Salicornia* may be limited in its migration into the *Spartina* zone because of the inhibitory effect of tidal immersion.[7]

The effect of flooding, aeration, and sulfide accumulation on the growth of two coastal salt marsh species, *S. alterniflora* and *Panicum hemitomum* was investigated by Koch and Mendelssohn.[35] No significant difference was determined for dry mass production between aerated and unaerated *Panicum*, while *Spartina* was inhibited by aeration treatments. However, the greatest reduction in yield of culms, roots, and rhizomes was in the unaerated + sulfide treatment for *Panicum*, while only root biomass showed a reduction for *Spartina*. Koch and Mendelssohn[35] concluded that *Spartina* growth was inhibited in the aerated treatment because of the nitrogen limitation caused by plant uptake and loss by denitrification from the soil.

The deleterious effects of sulfide on the growth parameters of three halophytes, *Spartina alaterniflora*, *S. cynosuroides*, and *Borrichia frutescens*, was determined by Bradley and Dunn[36] (Table 4). Field *in situ* sulfide concentrations on the Sapelo Island salt marsh in Georgia ranged from 0.02 m*M* on the creek bank to 3.0 m*M* in the inner salt marsh. Laboratory experiments indicated that plant biomass production for *S. cynosuroides* was inhibited at >1.0 m*M* sulfide, with no growth at 2.0 m*M* (Table 4). For *B. frutescens* and *S. alterniflora* each increment of sulfide from 0.5 to 2.0 m*M* caused a decrease in dry mass

TABLE 4
Mean Increase in Growth Parameters after 10 Weeks for *Spartina alterniflora, S. cynosuroides, and Borrichia frutescens* Exposed to Sulfide Treatments

Species	Growth parameter	Control	1.0 mM	2.0 mM
Spartina alterniflora	Height (cm)	64.1	39.6	0.0
	Total dry mass (g)	6.0	3.0	0.5
	Leaf area (cm²)	98.7	45.2	—
S. cynosuroides	Height (cm)	49.9	18.2	0.0
	Total dry mass (g)	4.1	1.5	0.4
	Leaf area (cm²)	84.2	27.0	—
Borrichia frutescens	Height (cm)	24.7	15.0	—
	Total dry mass (g)	4.7	1.2	—
	Leaf area (cm²)	71.3	22.1	—

From Bradley, P. M. and Dunn, E. L., *Am. J. Bot.*, 76, 1707, 1989. With permission.

TABLE 5
Correlation Coefficients (r) Showing the Relationship of Plant Cover (%) for *Salicornia europaea* and *Puccinellia maritima* to Sulfide Concentration in British Salt Marshes[a]

Species	Month			
	June	July	September	October
Salicornia europaea	0.41	0.55	0.63	0.50
Puccinellia maritima	−0.37	−0.55	−0.92	−0.82

[a]   All correlations were significant at $p < 0.05$.

From Ingold, A. and Havill, D. C., *J. Ecol.*, 72, 1043, 1984. With permission.

production (Table 4). The reduction in biomass production for *S. alterniflora* was from 6.0 g/plant in controls to 0.5 g/plant in the 2.0 mM sulfide treatment. Dry mass yield for *B. frutescens* decreased from 4.7 g/plant in controls to 1.2 g/plant in the 1.0 mM sulfide treatment, and a decrease from 49.9 g/plant in controls to 18.2 g/plant in the 1.0 mM sulfide treatment was found in *S. cynosuroides*. Bradley and Dunn[36] concluded that sulfide was inhibitory to plant growth parameters and that it could be one of the factors limiting a species distribution on the Sapelo Island salt marsh. However, it was not considered to be the primary factor limiting plant distribution because of the experimental responses to sulfide by the three species investigated.

Ingold and Havill[37] determined that there was a positive correlation between the sulfide content of the soil and the distribution of *Salicornia europaea* on British coastal salt marshes (Table 5). However, the pattern for *Puccinellia maritima* was negatively correlated with soil sulfide content.[37] Growth of unaerated *Aster tripolium* plants was not inhibited by anaerobic conditions, and when compared with aerated controls there was a slight promotion of growth by the unaerated treatment.[38] Flood-intolerant species, *Hordeum vulgare* and *Agropyron pungens*, were inhibited by anaerobic conditions. The unaerated sulfide treatment was more inhibitory to growth than either aerated or unaerated treatments for the species investigated. Unaerated *Aster tripolium* plants had a greater amount of root aerenchyma tissue (17.6%) than did aerated plants (11.4%).[38] The sulfide-treated plants had increased

TABLE 6
Cytochrome Oxidase (COase) and Alcohol Dehydrogenase
(ADH) Activity in the Roots of Plants under Aerated (A),
Unaerated (U) and Sulfide + Unaerated (S) Treatments

Species/treatment	COase (μmol/min/mg protein)	ADH (μmol/min/mg protein)
Agropyron pungens		
A	0.138	0.094
U	0.029	1.823
S	0.012	1.586
Hordeum vulgare		
A	0.193	0.125
U	0.047	1.290
S	0.016	1.332
Aster tripolium		
A	0.077	0.047
U	0.047	0.165
S	0.012	0.268
Salicornia europaea		
A	0.324	0.134
U	0.294	0.113
S	0.179	0.234
S. dolichostachya		
A	0.228	0.068
U	0.292	0.386
S	0.272	0.513
S. fragilis		
A	0.137	0.114
U	0.100	0.231
S	0.167	1.630

From Pearson, J. and Havill, D. C., *J. Exp. Bot.*, 39, 431, 1988. With permission.

sulfur content in the roots, increasing from 0.5 mg/g dry mass in unaerated to 4.7 mg/g dry mass in sulfide-treated plants. The sulfate content of roots increased from 0.94 mg/g dry mass in unaerated plants to 3.14 mg/g dry mass in the sulfide-treated plants, indicating that most of the increase in sulfur was in the form of sulfate for *A. tripolium*. Pearson and Havill[38] concluded that the sulfide was being taken up by plant roots and oxidized to sulfate inside the plant, because of the 30% reduction in root ATPase activity which may be due to the oxidation of sulfide in plant roots. The ability of species to tolerate anaerobic conditions and sulfide toxicity does not appear to be related to the percentage of aerenchyma tissue in roots. *Aster tripolium*, which was the most tolerant species, had a relatively small percentage of aerenchyma tissue when compared to some of the intolerant species.[38]

The amount of aernechyma tissue present in the roots of *Salicornia europaea* reached a mean value of 7.1% under field conditions. Laboratory-grown aerated plants had 3.1%, unaerated 4.5%, and unaerated-sulfide treated plants 5.4% aerenchyma tissue in their roots by volume.[39] Other species studied, including *Agropyron pungens*, *Aster tripolium*, and *Oryza sativa*, had aerenchyma content in their roots ranging from 15.9 to 16.5%. No correlation was found between the level of aerenchyma tissue present in the roots of these plant species and their ability to withstand hypoxia or sulfide stress.[39] Changes in the activity of the enzymes alcohol dehydrogenase (ADH) and cytochrome oxidase (COase) were measured, and it was found that nonwetland species had an increase in ADH activity from 900 to 1800% and a decrease in COase activity from 80 to 92% of the aerated control (Table 6). Under aerated conditions *A. tripolium* had 64% greater cytochrome oxidase activity and

TABLE 7
The Effect of Water Depth on Growth Parameters of
Scirpus maritimus var. *paludosus*

Growth	Water depth (cm)						
	−50	−40	−30	−20	−10	0	+10
Height, fruiting culm (cm)	108.8	122.5	111.2	101.7	88.6	86.3	83.5
Mean total biomass (g/tuber)	2.1	6.9	5.9	11.3	11.3	16.8	22.3
Mean inforescence biomass (g/culm)	0.31	0.29	0.31	0.17	0.09	0.05	0.09
Mean below-ground biomass (g/stem)	0.22	0.39	0.29	0.39	0.53	0.45	0.53

Note: −50 = soil surface 50 cm below water.

From Lieffers, V. J. and Shay, J. M., *Can. J. Bot.*, 59, 118, 1981. With permission.

an 81% decrease in ADH activity. *Salicornia europaea* had 91% of the control COase activity and 84% of the ADH activity in unaerated treatments and was only slightly affected by the sulfide treatment. In the three species of *Salicornia* studied, *europaea*, *dolichostachya*, and *fragilis*, there was no significant correlation between the activities of the enzymes COase and ADH.[39]

Oxygen concentrations were determined to be much lower in shoot bases of *Spartina patens* than for *S. alterniflora*.[40] During simulated flooding in light in the laboratory, shoot base oxygen levels dropped from about 118 to 19% (% volume) for *S. alterniflora* and from 16 to 4% for *S. patens*. Gleason and Zieman[40] hypothesized that *S. patens* might not be able to invade the regularly flooded lower portions of the marsh that are occupied by *S. alterniflora*, because of its decreased ability to supply oxygen to roots and rhizomes during periods of flooding.

Burdick[41] determined that flooding caused a reduction in the specific gravity from 0.78 to 0.58 in old and new roots of the halophyte *S. patens*. Flooding was found to stimulate aerenchyma production, inhibit the growth of new roots, and increase the mortality of older roots. Aerenchyma tissue was formed primarily by cell lysis when both new and older roots were flooded. Burdick[41] concluded that the production of aerenchyma tissue by *S. patens* could be important in maintaining flooding tolerance under conditions of poor aeration when plants were exposed to waterlogging.

Lieffers and Shay[42,43] reported that *Scirpus maritimus* var. *paludosus* grows in saline-flooded soils in inland salt marshes. Height of shoots along environmental gradients ranged from 8 to 140 cm, and above-ground biomass ranged from 3 to 936 g/m². Plant shoots reached their maximum size in water that was greater than 40 cm deep and in low soil salinity. Similar results were obtained for reproductive biomass production of plants along environmental gradients (Table 7). Dry and highly saline habitats inhibited the growth of *S. maritimus* in the field sites investigated in Canada. Laboratory investigations indicated that *S. maritimus* plants fruited at a greater height in deeper water than in shallow water ($r = -0.77$), ranging from 120 cm in deep water (−50 cm depth) to 80 cm in shallow water (+10 cm depth). Inflorescence biomass decreased with a reduction in water depth from about 0.3 g/plant at −50 cm depth to 0.05 g/plant at +10 cm depth ($r = -0.92$).[44]

Bruguiera gymnorrhiza, a mangrove species from South Africa, had more negative leaf water potentials when plants were exposed to flooding conditions in $^1/_3$ seawater than did the controls.[45] Stomatal resistance was higher in the leaves of flooded plants and water potentials decreased from −2.5 MPa after 20 d of flooding to −4.0 MPa after 80 d of flooding. Termination of flooding yielded water potentials similar to controls in up to the 12-d treatment, but plants flooded for 40 and 80 d had more negative water potentials both during and after flooding than controls which were not flooded. Stomatal closure caused an increase in stomatal resistance, which was a direct response to flooding and the corresponding low oxygen levels. Naidoo[45] concluded that stomatal closure in response to flooding was probably controlled by increased ABA levels in the leaves.

IV. CONCLUSIONS

Tidal inundation changes some characteristics of coastal salt marsh soils by reducing substrate oxygen levels, lowering the soil redox potential, causing sulfide toxicity, changing the availability of ions, increasing the erosion of substrate, or deposition of sediment.[35,37,46,47] Flooding, which occurs in inland salt marshes because of rising water tables, would have a similar effect on soil chemistry. However, plants growing in the inland salt marsh habitats are not exposed to the mechanical effects of tidal action.

Abrasion and sedimentation factors related to tidal activity have a definite influence in determining the zonational patterns and cyclical vegetation dynamics in salt marshes.[11] Oxygen deficiencies have been shown to inhibit the biomass production in a number of halophytes. Anoxic conditions have been shown to reduce the level of aerobic respiration, induce high levels of anaerobic respiration, and cause the accumulation of toxic metabolic end products. However, plants may be able to increase the level of soil aeration by transferring oxygen from shoots to roots and soil. Growth of halophytes in inland marshes and upper portion of coastal marshes is more limited by waterlogging and tidal inundation than is that of lower marsh species, which have evolved under conditions of nearly daily tidal inundation.[32] Sulfide toxicity can directly limit plant growth, but oxygen deficiencies, salinity, and reduced nitrogen availability may also play a role in the inhibition of growth.[46,47] The presence of sulfides in soils can also indirectly limit plant growth by retarding nitrogen uptake by plant roots.[47]

REFERENCES

1. **Johnson, D. S. and York, H. H.**, The relation of plants to tide-levels, *Carnegie Inst. Wash. Pub.*, 206, 1, 1915.
2. **Purer, E. A.**, Plant ecology of the coastal salt marsh lands of San Diego County, *Ecol. Monogr.*, 12, 81, 1942.
3. **Chapman, V. J.**, *Salt Marshes and Salt Deserts of the World*, J. Cramer, Bremerhaven, 1974.
4. **Ranwell, D. S.**, *Ecology of Salt Marshes and Sand Dunes*, Chapman and Hall, London, 1972.
5. **Redfield, A. C.**, Development of a New England salt marsh, *Ecol. Monogr.*, 42, 201, 1972.
6. **Waisel, Y.**, *Biology of Halophytes*, Academic Press, New York, 1972.
7. **Mahall, B. E. and Park, R. B.**, The ecotone between *Spartina foliosa* and *Salicornia virginica* L. in salt marshes of northern San Francisco Bay. III. Soil aeration and tidal immersion, *J. Ecol.*, 64, 811,1976.
8. **Cooper, A.**, The effects of salinity and waterlogging on the growth and cation uptake of salt marsh plants, *N. Phytol.*, 90, 263, 1982.
9. **Olff, H., Bakker, J. P., and Fresco, L. F. M.**, The effect of fluctuations in tidal inundation frequency on a salt-marsh vegetation, *Vegetatio*, 78, 13, 1988.
10. **McKee, K. L. and Patrick, W. H.**, The relationship of smooth cordgrass (*Spartina alterniflora*) to tidal datums: a review, *Estuaries*, 11, 143, 1988.
11. **Runge, F.**, Dauerquadratbeobachtungen bei Salzwiesen-assoziationen, in *Grundfragen und Methoden in der Pflanzensoziologie*, Tuxen, R., Ed., Junk, The Hague, 1972, 419.
12. **Armstrong, W., Wright, E. J., Lythe, S., and Gaynard, T. J.**, Plant zonation and the effects of the spring-neap tidal cycle on soil aeration in a Humber salt marsh, *J. Ecol.*, 73, 323, 1985.
13. **Pount, H. and Revel, J. C.**, Influence of water table on the distribution of a saline-soil species *Limonium diffusum* (Pourret) Kuntze, Plumbaginaceae, *J. Biogeog.*, 9, 437, 1982.
14. **Thibodeau, F. R. and Nickerson, N. H.**, Differential oxidation of mangrove substrate by *Avicennia germinans* and *Rhizophora mangle*, *Am. J. Bot.*, 73, 512, 1986.

15. **Justin, S. H. F. W. and Armstrong, W.,** Oxygen transport in the salt marsh genus *Puccinellia* with particular reference to the diffusive resistance of the root-shoot junction and the use of paraffin oil as a diffusive barrier in plant studies, *J. Exp. Bot.* 34, 980, 1983.

16. **Stelzer, R. and Lauchli, A.,** Salt and flooding tolerance of *Puccinellia peisonis.* IV. Root respiration and the role of aerenchyma in providing atmospheric oxygen to the roots, *Z. Pflanzenphysiol.* 97, 171, 1980.

17. **Raskin, I. and Kende, H.,** Mechanisms of aeration in rice, *Science*, 228, 327, 1985.

18. **Wiegert, R. G., Chalmers, A. G., and Randerson, P. F.,** Productivity gradients in salt marshes. the response of *Spartina alterniflora* to experimentally manipulated soil water movement, *Oikos,* 41, 1, 1983.

19. **Mendelssohn, I. A., McKee, K. L., and Patrick, W. H.,** Oxygen deficiency in *Spartina alterniflora* roots: metabolic adaptation to anoxia, *Science,* 214, 439, 1981.

20. **Mendelssohn, I. A. and McKee, K. L.,** Determination of adenine nucleotide levels and adenylate energy charge ratio in two *Spartina* species, *Aquatic Bot.,* 11, 37, 1981.

21. **Mendelssohn, I. A. and McKee, K. L.,** *Spartina alterniflora* die-back in Louisiana: time-course investigation of soil waterlogging effects, *J. Ecol.,* 76, 509, 1988.

22. **Burdick, D. M. and Mendelssohn, I. A.,** Waterlogging responses in dune, swale and marsh population of *Spartina patens* under field conditions, *Oecologia,* 74, 321, 1987.

23. **Karimi, S. H. and Ungar, I. A.,** The effect of salinity on the ionic content and water relation of *Atriplex triangularis* Willd., in *Symp. Biology of Atriplex and Related Chenopods,* Tiedemann, A. R., McArthur, E. D., Stutz, H. C., Stevens, R., and Johnson, K. L., Eds., Gen. Tech. Rep. INT-172, Forest Service, U.S. Department of Agriculture, Ogden, UT, 1984, 124.

24. **Karimi, S. H. and Ungar, I. A.,** Development of epidermal salt hairs in *Atriplex triangularis* Willd. in response to salinity, light intensity, and aeration, *Bot. Gaz.,* 150, 68, 1989.

25. **Stelzer, R. and Lauchli, A.,** Salt- and flooding tolerance of *Puccinellia peisonis.* I. The effect of NaCl- and KCl-salinity on growth at varied oxygen supply to the root, *Z. Pflanzenphysiol.,* 83, 35, 1977.

26. **Stelzer, R. and Lauchli, A.,** Salt- and flooding tolerance of *Puccinellia peisonis.* II. Structural differentiation of the root in relation to function, *Z. Pflanzenphysiol.,* 84, 95, 1977.

27. **Stelzer, R. and Lauchli, A.,** Salt- and flooding tolerance of *Puccinellia peisonis.* III. Distribution and localization of ions in the plant, *Z. Pfanzenphysiol.,* 88, 437, 1978.

28. **Donovan, L. A. and Gallagher, J. L.,** Anaerobic substrate tolerance in *Sporobolus virginicus* (L.) Kunth., *Am. J. Bot.,* 71, 1424, 1984.

29. **Mizrahi, Y., Blumenfeld, A., and Richmond, A. E.,** The role of abscisic acid and salination in the adaptive response of plants to reduced root aeration, *Plant Cell Physiol.,* 13, 15, 1972.

30. **Seliskar, D. M.,** Waterlogging stress and ethylene production in the dune slack plant, *Scirpus americanus, J. Exp. Bot.,* 39, 1639, 1988.

31. **Kriedmann, P. E. and Sands, R.,** Salt resistance and adaptation to root-zone hypoxia in sunflower, *Aust. J. Plant Physiol.,* 11, 287, 1984.

32. **Langlois, J. and Ungar, I. A.,** A comparison of the effect of artificial tidal action on the growth and protein nitrogen content of *Salicornia stricta* Dumort. and *Salicornia ramosissima* Woods, *Aquatic Bot.,* 2, 43, 1976.

33. **Langlois, J.,** Influence du rythme d'immersion sur la croissance et le metabolisme proteique de *Salicornia stricta* Dumfort, *Oecol. Plant,* 6, 227, 1971.

34. **Binet, P.,** Etudes d'ecologie experimentale et physiologique sur *Cochlearia anglica* L. I. Etudes dans l'estuaire de l'Orne, *Oecol. Plant,* 1, 7, 1965.

35. **Koch, M. S. and Mendelssohn, I. A.,** Sulphide as a phytotoxin: differential responses in two marsh species, *J. Ecol.,* 77, 565, 1989.

36. **Bradley, P. M. and Dunn, E. L.,** Effects of sulfide on the growth of three salt marsh halophytes of the southeastern United States, *Am. J. Bot.,* 76, 1707, 1989.

37. **Ingold, A. and Havill, D. C.,** The influence of sulphide on the distribution of higher plants in salt marshes, *J. Ecol.,* 72, 1043, 1984.

38. **Pearson, J. and Havill, D. C.,** The effect of hypoxia and sulphide on culture-grown wetland and non-wetland plants. I. Growth and nutrient uptake, *J. Exp. Bot.,* 39, 363, 1988.

39. **Pearson, J. and Havill, D. C.,** The effect of hypoxia and sulphide on culture-grown wetland and non-wetland plants. II. Metabolic and physiological changes, *J. Exp. Bot.,* 39, 431, 1988.

40. **Gleason, M. L. and Zieman, J. C.,** Influence of tidal inundation on internal oxygen supply of *Spartina alterniflora* and *Spartina patens, Est. Coast. Shelf Sci.,* 13, 47, 1981.

41. **Burdick, D. M.,** Root aerenchyma development in *Spartina patens* in response to flooding, *Am. J. Bot.,* 76, 777, 1989.

42. **Lieffers, V. J. and Shay, J. M.,** Distribution and variation in growth of *Scirpus maritimus* var. *paludosus* on the Canadian prairies, *Can. J. Bot.,* 60, 1938, 1982.

43. **Lieffers, V. J. and Shay, J. M.,** Seasonal growth and standing crop of *Scirpus maritimus* var. *paludosus* in Saskatchewan, *Can. J. Bot.,* 60, 117, 1982.

44. **Lieffers, V. J. and Shay, J. M.,** The effects of water levels on the growth and reproduction of *Scirpus maritimus* var. *paludosus, Can. J. Bot.,* 59, 118, 1981.
45. **Naidoo, G.,** Effects of flooding on leaf water potential and stomatal resistance in *Bruguiera gymnorrhiza (L.) Lam., N. Phytol.,* 93, 369, 1983.
46. **DeLaune, R. D., Smith, C. J, and Patrick, W. H.,** Relationship of marsh elevation, redox potential and sulfide to *Spartina alterniflora* productivity, *Soil Sci. Soc. Am. J.,* 47, 930, 1983.
47. **Bradley, P. M. and Morris, J. T.,** Influence of oxygen and sulfide concentration on nitrogen uptake kinetics in *Spartina alterniflora, Ecology,* 71, 282, 1990.

Chapter 9

NITROGEN

I. INTRODUCTION

It has been reported in a number of investigations with species from coastal salt marsh communities that the availability of nitrogen was one of the primary environmental factors limiting plant biomass production in saline environments.[1-13] Nitrogen availability was also determined to be a factor limiting plant biomass production in inland saline areas.[14,15] The salt marsh habitat has a complex of variables controlling plant growth, including waterlogging, reducing environment, sulfide accumulation, salinity, tidal erosion, and accretion, and interactions between these factors may be limiting plant growth in some salt marsh zonal communities. It has been observed by several researchers carrying out fertilization experiments in coastal marshes either that field nitrogen levels do not vary significantly in the zonal communities or that the zone of depressed plant growth had more nitrogen than the zone containing vigorous growth.[4,5,16] However, when nitrogen fertilization of the zone with inhibited growth is carried out, a significant increase in biomass production occurs. Howes et al.[17] hypothesized that one of the other interacting factors in the soil environment was impeding nitrogen uptake by roots, even though sufficient inorganic nitrogen was present in the marsh soils.

II. NITROGEN FERTILIZATION EXPERIMENTS IN SALT MARSHES

The effect of nitrogen and phosphorus fertilization of a *Juncus gerardi*-dominated plant community from a Baltic seashore meadow was determined.[1] Shoot length of *J. gerardi* increased 30 to 40% from controls when soils were fertilized with 5 M NH_4/m^2. The nitrogen content of dry matter for *J. gerardi* averaged 790 μM/g in controls and 880 μM/g dry matter in plots treated with 5.0 M NH_4/m^2 (Table 1). Similar increases in shoot nitrogen were found in *Plantago maritima*, averaging 643 μM/g in controls and 725 μM/g in the fertilized plots. Plant biomass yield in the *J. gerardi* community treated with NH_4 was elevated by 55% (284 to 442 g/m^2) over the controls when fertilization treatments of 5.0 M nitrogen/m^2 were applied (Table 1). Nitrate fertilization did not increase yields significantly, and this could be because most of the available nitrogen for plants was in the form of ammonia in the reducing environment of salt marshes.[1] It was calculated that plants in the *J. gerardi*-dominated community took up about 2% of the 0.5 M phosphorus/m^2 used in the fertilization treatments and about 5% of the nitrogen in the 2.5 M/m^2 ammonium treatment. However, treatments with phosphorus alone increased nitrogen uptake in plants by about 3%. Phosphorus treatments alone caused an increase in biomass yields of the treated vegetation that was 28% greater than controls, indicating that both nitrogen and phosphorus were not readily available to plants growing in soils of the Baltic salt marshes.[1]

Increasing the nitrogen supply to 100 kg/ha for *Atriplex litoralis* plants growing on a driftbank along the shore of the North Sea (Sweden) produced an increase in biomass production from 5.99 g in controls to 8.90 g/plant in fertilized treatments.[18] Increase in dry mass yields per plant over controls were 42% for stems (3.23 g), 58% for leaves (0.93 g), 65% for reproductive organs (3.69 g), and 17% for roots (1.05 g) (Table 2). Fertilization treatments with NPK significantly increased yields of *A. litoralis* when compared with treatments with individual elements (Table 3). The author hypothesized that nitrogen could be a limiting factor to the biomass production of *A. litoralis* in a balanced nutrient solution

TABLE 1
Mean Production (g Dry Mass/m^2) and Mean Contents (μM/g Dry Weight) of Phosphorus (P) and Nitrogen (N) in Standing Crop and in the Shoots of *Juncus gerardi* and *Plantago maritima*

Treatment (mol/m^2)	Production (g/m^2)	Biomass P	Biomass N	Juncus P	Juncus N	Plantago P	Plantago N
Control	284	35	792	36	790	32	643
Na$_2$HPO$_4$ (0.5)	382	53	840	44	763	50	747
NaNO$_3$ (2.5)	306	35	815	44	793	35	768
NaNO$_3$ (5.0)	394	36	910	29	840	43	785
NH$_4$Cl (2.5)	375	33	887	31	880	33	747
NH$_4$Cl (5.0)	442	31	955	27	880	32	725
Na$_2$HPO$_4$ + NaNO$_3$ (2.5)	382	50	832	43	793	49	705

From Tyler, G., *Bot. Notiser*, 120, 433, 1967. With permission.

TABLE 2
Mean Weight of Different Plant Fractions of *Atriplex litoralis* with Increased Nitrogen Supply in Pot Experiments at the Driftbank in 1979 (Dry Matter g/Plant)

Treatment	Plant fraction Stems	Leaves	Flowers plus seeds	Roots	Whole plant
Without N	2.27 ± 2.04	0.59 ± 0.71	2.23 ± 1.58	0.90 ± 0.75	5.99
Single N[a]	2.65 ± 1.55	0.59 ± 0.28	3.13 ± 2.82	0.86 ± 0.49	7.23
Double N[b]	3.23 ± 2.24	0.93 ± 0.69	3.69 ± 2.61	1.05 ± 0.55	8.90

[a] Equal to 100 kg ha^{-1}.
[b] Equal to 200 kg ha^{-1}.

From Steen, E., *Oikos*, 42, 74, 1984. With permission.

TABLE 3
Mean Weight of Roots and Shoots of *Atriplex litoralis* with Separate and Combined N, P, and K Fertilization in a Pot Experiment at the Driftbank

Treatment	Mean air dry weight (g ind^{-1}) Root	Shoots
No fertilizer	0.86 ± 0.25	1.63 ± 0.42
P only	0.86 ± 0.27	1.67 ± 0.47
K only	1.08 ± 0.28	3.23 ± 0.77
N only	1.21 ± 0.27	2.35 ± 0.41
NPK	2.66 ± 0.72	5.60 ± 0.90

From Steen, E., *Oikos*, 42, 74, 1984. With permission.

TABLE 4
August Standing Crop of *Spartina foliosa* and *Salicornia virginica* in Competition Plots with and without Nitrogen (Urea) Addition[a]

Spartina biomass	+ *Salicornia*	− *Salicornia*
Urea	625	1282
Control	577	898
Salicornia biomass	+ *Spartina*	− *Spartina*
Urea	1525	1484
Control	1038	1316

[a] Data are means in g/m² (N = 2) for above-ground biomass.

From Covin, J. D. and Zedler, J. B., *Wetlands*, 8, 51, 1988.
With permission.

in saline habitats. However, Steen[18] concluded that shortages in other nutrient elements such as phosphorus and potassium could influence the effect of nitrogen by limiting biomass production.

Field experiments were established by Covin and Zedler[12] to determine the effect of urea fertilization, 11.2 g/m², on the biomass production and total nitrogen content of *Spartina foliosa* and *Salicornia virginica* on marsh sites at the Tijuana Estuary (California). Biomass production for August in enriched plots containing *Spartina foliosa* was 56% (1230 g/m²) greater than in unfertilized plots (790 g/m²). The biomass production of *Salicornia virginica* was 1484 g/m² in August in fertilized plots vs. 1316 g/m² in unfertilized plots, averaging 13% higher in the fertilized plots. The nitrogen content of *Spartina foliosa* leaf tissue in July was 1.6% of the dry mass in fertilized plots and 1.3% of the leaf dry mass in unfertilized plots.[12] The availability of nitrogen could be critical for *S. foliosa*, because *Salicornia virginica* is a better competitor for nitrogen. It was determined that the removal of *S. virginica* increased the yields of *Spartina foliosa* in fertilized plots by 2.1-fold, while the biomass production of *Salicornia virginica* did not increase when *Spartina foliosa* was removed (Table 4).[12]

The effects of fertilization of field plots on the Stiffkey salt marsh (England) were analyzed by Jefferies and Perkins.[6] It was determined that the frequency of several species increased after ammonium and nitrate fertilization, including *Spergularia marina, Halimione portulacoides,* and *Suaeda maritima.* On the other hand, the frequency of *Armeria maritima* decreased from >50 to <20% after 5 years, which was probably due to the poor competitive ability of *Armeria.*[6] The reproductive capacity of several species, as measured by inflorescence number m² per square meter, was higher for the nitrogen fertilized plots than controls for *Spergularia marina, Halimione portulacoides,* and *Limonium vulgare.* About a twofold increase in dry mass was found in nitrogen fertilized treatments compared with controls for the following species: *Puccinellia maritima, Limonium vulgare, Salicornia europaea, Halimione portulacoides,* and *Spergularia marina.* Fertilization treatments did not produce an increase in yields of *Armeria maritima* and *Triglochin maritima* in comparison with control plots. After 3 or 4 years of perturbation with nitrogen fertilization, the dominant species on the salt marsh did not change, and no species disappeared from the fertilized salt marsh plots. Therefore, Jefferies and Perkins[6] concluded that the level of primary production in the Stiffkey salt marsh system is probably not limited by a shortage of nitrogen, but by low water potentials and high soil salinities.

Nitrogen fertilization in field plots with potassium nitrate was determined by Loveland and Ungar[14] to increase the biomass production of tall *Salicornia europaea* plants by 62% in a salt marsh at Rittman, OH (Figure 1). Short *S. europaea* plants from the more stressful

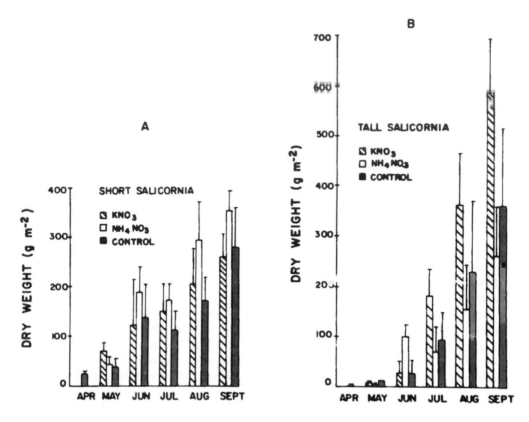

FIGURE 1.   The effect of nitrogen fertilization on seasonal aerial standing biomass for (A) short *Salicornia europaea* and (B) tall *S. europaea* on an inland salt pan. (From Loveland, D. G. and Ungar, I. A., *Am. Midl. Nat.*, 109, 346, 1983. With permission.)

zone were not stimulated by KNO_3, but their growth was stimulated by the NH_4NO_3 treatment. Shoot nitrogen content of *S. europaea* individuals was inversely related to the growth response to fertilization (Figure 2). The habitats containing the tall- and short-growth form of *S. europaea* differed from one another in two major environmental variables: the salinity was higher and the nitrogen content of soils lower in habitats containing the short form of *S. europaea*.[14] *Hordeum jubatum* did not have an increase in biomass yield when treated with nitrogen fertilization. *Atriplex triangularis* was stimulated by fertilization, but because of the high variability in yields within treatments the differences between treatments were not statistically significant.[14]

Saline grasslands in Argentina were fertilized with ammonium sulfate (762 kg/ha/year) and triple superphosphate (208 kg/ha/year) to determine the effect of treatments on species' abundance and yields. Ginzo et al.[19] reported that the relative frequency of species changed with fertilization treatments. The most frequent and abundant species in control plots in 1978 were *Distichlis scoparia*, *Hordeum euchaston*, *Sporobolus pyramidatus*, *Plantago myosurus*, *Gaudinia fragilis*, and *Distichlis spicata*. After 3 years of fertilization from 1976 to 1978, the frequency of *D. scoparia* and *S. pyramidatus* decreased markedly, while *Lepidium parodii* and *Hypochoeris* spp. had an increase in frequency in plots that were fertilized. Dry matter production of grasses and forbs increased in the ammonium sulfate + triple superphosphate treatments in all 3 years, with dry mass production of grasses being 400 g/m²/year and forbs being about 50 g/m²/year vs. about 100 g/m²/year for grasses and 20 g/m²/year for forbs in the unfertilized plots.[19]

Subarctic salt marshes on La Perouse Bay (Canada) were dominated by two graminoids,

159

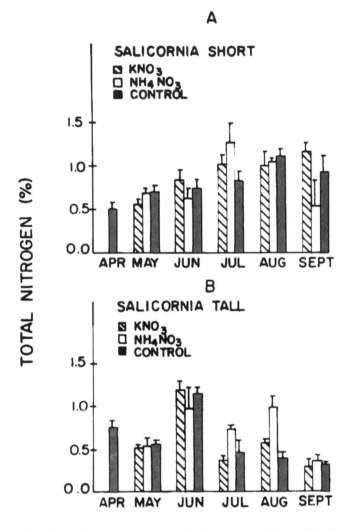

FIGURE 2. Seasonal changes in total nitrogen content of shoots for (A) short *Salicornia europaea* and (B) tall *S. europaea* from an inland salt pan. (From Loveland, D. G. and Ungar, I. A., *Am. Midl. Nat.*, 109, 346, 1983. With permission.)

Puccinellia phryganodes and *Carex subspathacea*, and their net primary production was estimated to be 97 g/m²/year by Cargill and Jefferies.[20] Phosphorous was evidently not limiting on these salt marshes, since fertilization with phosphorous did not increase the biomass production of these subarctic salt marshes. Vegetation of a mixed stand containing *Puccinellia* and *Carex* had their production increased to 200 and 182 g/m², respectively when treated with ammonium sulfate and sodium nitrate. Similar results were obtained for pure stands of these species in which controls averaged 64 and 65 g/m² while ammonium-treated plots had 176 and 160 g/m² biomass for *Puccinellia* and *Carex*, respectively.[20]

III. NITROGEN AND HALOPHYTE GROWTH

The effect of nitrogen fertilization on the growth of several strand-line species from Bergen op Zoom (Netherlands) was examined by Rozema et al.[9] using laboratory experiments. The growth of *Atriplex hastata* and *A. littoralis* was stimulated by 60 and 150 m*M*

NaCl when plants were grown in Hoagland solution containing 3.5 and 14 m*M* nitrate. The stimulation of growth for *A. hastata* was 2.5 times that of controls, and for *A. littoralis* it was 2.0 times that of controls. *Salsola kali* biomass yields increased to about 1.75 times that of controls in both the 3.5- and 14-m*M* nitrate and 150-m*M* NaCl treatment. The sand dune driftline species *Cakile maritima* was not stimulated by increased nitrogen, and increasing salinity from 60 to 300 m*M* NaCl caused reductions in biomass production. Total nitrogen content of the *Atriplex* spp. under all salinity treatments ranged between 4 and 5% of plant dry mass, and it was calculated that the quaternary ammonium compound glycinebetaine made up about 12.5% of the total plant nitrogen content when plants were grown at 300 m*M* NaCl.[9]

The effect of three levels of nitrogen fertilization (0.14, 1.4, and 14 mg N/l as ammonium chloride) and salinity (0.1, 0.3, and 0.5 *M* NaCl) on the growth of *Avicennia marina* was determined by Naidoo.[11] Growth of roots and shoots were higher in the lowest salinity treatment than in the two higher salinities. Nitrogen treatments of 14 mg/l significantly increased shoot growth of *A. marina* at 0.1 and 0.3 *M* NaCl, but not at the highest salinity. Lower levels of nitrogen had no significant effect on either root or shoot growth at any salinity tested.[11]

A comparison was made by Jefferies[5] of the response of plants from low- and high-marsh zones of the Stiffkey salt marsh (England) to nitrogen enrichment under laboratory conditions. Nitrogen content of soils at the field sites varied during the year, but at any one time they were similar on the low- and high-marsh sites. The growth of lower marsh species, such as *Aster tripolium, Halimione portulacoides*, and *Suaeda maritima*, was stimulated by fertilization with 1 m*M* nitrate, while *Armeria maritima* shoot weight was not significantly affected. Intraspecific variation was observed among populations of *Salicornia europaea, Triglochin maritima, Aster tripolium*, and *Plantago maritima*, with the low-marsh populations of each species having more rapid growth rates than did high-marsh populations.[5] These differences in growth rate were most apparent when plants were treated with 1 m*M* nitrate. Lack of tidal immersion in the upper marsh caused a drying out of the soil and the development of hypersaline conditions. The upper marsh represented a more stressful environment in the summer months than was found in the low marsh which was subjected to more frequent tidal immersion. Jefferies[5] hypothesized that the presence of low soil-water potentials and hypersaline conditions during most of the growing season resulted in the selection of plant populations with lower growth rates. Reduced growth rates may be involved in making plants more tolerant of the highly stressful conditions in the high-marsh environment. High-marsh perennials such as *Triglochin maritima, Aster tripolium*, and *Plantago maritima* did not respond as strongly as low-marsh populations to nitrogen fertilization. The pattern of slow growth in the high-marsh populations could be a genetic adaptation selected for in the restrictive habitat in which these plants were growing.[5]

Laboratory experiments with two coastal halophytes, *Armeria maritima* and *Plantago maritima*, from Irish salt marshes indicated that increases in nitrogen from 0.28 to 280 mg/l and 2.8 to 140 mg/l for each species, respectively, stimulated an increase in plant biomass at the salinities tested (−0.2 to −5.0 MPa) for both species.[13] Increases in ammonium nitrate stimulated shoot and root growth of *Armeria maritima* in the salinity regimens tested, with the least increase in yield at the highest salinity, −1.97 MPa. Similar results were obtained with *P. maritima* in up to 140 mg/l nitrogen, but higher nitrogen levels were reported to be inhibitory to this species. *Armeria* accumulated nitrogen in shoots in the form of proline, while *Plantago* stored nitrogen in its roots. Available soil nitrogen ranged from 1 to 40 mg/l, varying with zone and seasonally on the North Bull salt marsh (Ireland). Plants grown at <28 mg/l nitrogen best simulated soil nitrogen conditions in the middle salt marsh zone. Growth of *A. maritima* plants was stimulated by 28 mg/l nitrogen or higher levels, but water potentials lower than −0.9 MPa retarded growth. Lack of stimulation of growth

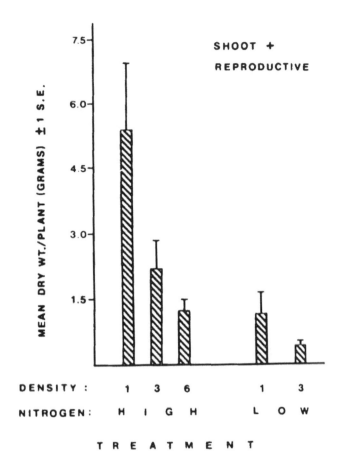

FIGURE 3. The effects of nitrogen and density treatments on the dry
mass of shoots plus reproductive structures for plants grown in the laboratory. (From Drake, D. R. and Ungar, I. A., *Am. J. Bot.*, 76, 1125,
1989. With permission.)

by N in the field was concluded to be due to the inhibitory effect of salinity.[13] *Armeria*
accumulated shoot nitrogen with increased salinity. Under field conditions, it was the only
species that demonstrated an increase in tissue nitrogen, even though growth was not stimulated. Laboratory experiments indicated that stimulation of growth with N fertilization was
most apparent at low salinities and reduced at high salinities. Therefore, Skeffington and
Jeffrey[13] concluded that the addition of nitrogen to salt marshes could increase the ability
of some halophytes to withstand salinity stress.

Laboratory investigations with *Atriplex triangularis* by Drake and Ungar[15] indicated that
high nitrogen fertilization caused a significant increase in root, shoot, and reproductive dry
mass compared with the low nitrogen treatment for all salinities tested (0 to 3% NaCl; Figure
3). In a field investigation at Constitution, OH, nitrogen was reported to stimulate growth
only in the low plant density (2 plants per 100 cm²) treatment (Figure 4).[15] *A. triangularis*
may be growing near the asymptote of its nitrogen response curve under field conditions.
Other factors such as soil salinity and competition may be overriding factors in determining
the growth responses of plants under field conditions.

The effect of two levels and two sources of nitrogen, potassium nitrate and ammonium
sulfate, on the dry mass production of *A. hastata* was determined after a 6-week growing
period.[21] Yields in potassium nitrate treatments were 204 and 247 mg/plant for the 14 and
224 mg/l nitrogen treatments, respectively, while yields were 105 mg/plant for both levels

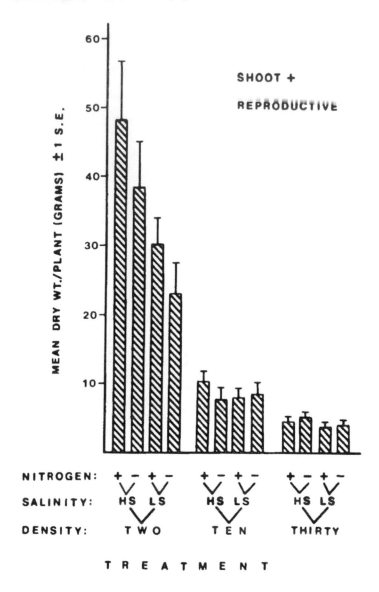

FIGURE 4. The effects of salinity, nitrogen, and density on the dry mass of shoots plus reproductive structures of field plants. +/− = with and without nitrogen fertilization; HS/LS = high/low salinity transects. (From Drake, D. R. and Ungar, I. A., *Am. J. Bot.*, 76, 1125, 1989. With permission.)

of the ammonium sulfate treatments. Biomass production was not significantly different at low- and high-nitrogen fertilization treatments, indicating that *A. hastata* is very efficient in its use of nitrogen. When both nitrogen sources were available, Weston[21] determined that *A. hastata* absorbed nitrate preferentially at low pH (4 to 6) and ammonium at high pH (7 to 8).

Casuarina obesa is a small tree which occurs on the border of salt lakes in western Australia.[22] Roots of the species contain a symbiotic nitrogen-fixing actinomycete *Frankia* spp. Reddell et al.[22] reported that plants growing in soils with 20 mg Cl/g soil contained *Frankia* nodules in their roots. Nodule growth was suppressed in experimental treatments with 1.5 mg NaCl/g soil and higher. Nitrogen fixation was reduced by 60% in *C. obesa* plants exposed to 1.5 mg NaCl/g soil and inhibited in the higher salt treatments, 15 mg

NaCl/g soil. Reductions in the production of photosynthate by the host plant may limit *Frankia* nodule production under hypersaline condition. Selection for high productivity in saline habitats would require salt-tolerant *C. obesa* plants and *Frankia* genotypes that could form symbiotic associations under saline conditions.[22] The nitrogen fixation in root nodules of *Casuarina equisetifolia* var. *incana* collected from Southport (Australia) was not limited by salinities up to 200 mM NaCl, but 500 mM NaCl caused a decrease of 40% in the rate of nitrogen fixation compared with control treatments (Table 5).[23] Nodulation in control plants increased biomass yields of plants by 4.3 times (112.03 g/plant) compared with uninoculated plants of *C. equisetifolia* grown in nutrient solution without NaCl. Biomass production of *C. equisetifolia* also decreased from a maximum of 16.43 g/plant at 50 mM NaCl to a minimum of 5.04 g/plant in the 500-mM NaCl treatment. Nodule dry mass decreased from 0.67 g/plant in the 50-mM NaCl treatment to a minimum for nodulated plants of 0.23 g/plant when plants were grown in 500 mM NaCl (Table 6). The chief limitation for nodule production was hypothesized to be related to the reduced amount of photosynthate available for translocation to the root when plants were exposed to high-salinity treatments.[23]

Aster tripolium from the low-salt marsh of the southwest coast of the Netherlands was grown to determine the effect of continuous vs. intermittent nitrogen supplies (3.7 mM calcium nitrate + 4.5 mM potassium nitrate) on plant growth.[24] The number of hours that plants were exposed to fertilization per week ranged from 4 to 168. Dry mass yields of stems and leaves increased significantly in treatments of up to 24 h. However, Stienstra[24] found that although longer periods of exposure (48 to 168 h) produced an increase in shoot dry mass, it also caused a reduction in root dry mass. Increasing the exposure time to nitrate produced an increase in the percent leaf and root nitrogen content on a dry mass basis from 0.7 to 3.63% in leaves and from 0.45 to 0.57% in roots in the 4-h and 168-h treatments, respectively.[24]

Saline areas in the Dead Sea region of Jordan had ammonium levels in the surface soils ranging from 19 μg N/g soil in April to 1 μg N/g soil in August.[25] Nitrate levels in the surface soils ranged from 15 to 60 μg N/g soil. Ammonium levels for *Arthrocnemum fruticosum* plants in this area were 25 μg N/g in shoots and 50 μg/g in roots in March, declining to 1 μg N/g in shoots and between 10 and 15 μg/g in roots during July. Plant nitrate levels followed a similar pattern, declining from 90 μg/g for roots in March to about 1 μg N/g in July. Shoot nitrate levels also declined from 30 μg N/g in March to 2 μg/g in June. Nitrate reductase activity of *A. fruticosum* plants increased in shoots from 0.2 μg N/h/g in March to 0.9 μg N/g/h in April, but decreased to 0.1 μg N/h/g in July.[25] Drying of the soil and increased soil salinity could have inhibited nitrification of ammonia in the soil. Soil and plant levels of ammonium and nitrate decreased with increasing soil salinity. Nitrate reductase activity also decreased with plant nitrate levels from May to August, indicating that nitrogen uptake and reduction in the plant was inhibited during the summer when the soil is dry and more saline.[25] Nitrate reductase activity was initially inhibited in laboratory-grown *A. fruticosum* plants when they were exposed to salinity increments from 1 to 5% NaCl.[26] However, after acclimation only the 5% NaCl treatment was inhibitory and caused a shift of nitrate reductase activity from roots to shoots. Eddin and Doddema[26] determined that lower water potentials inhibited nitrate reductase activity in roots, but in the shoot the limiting factor was the supply of nitrate from the roots.

IV. NITROGEN FERTILIZATION EFFECTS ON *SPARTINA ALTERNIFLORA*

A number of investigations have been carried out to determine whether nitrogen availability was a limiting factor determining the biomass production of *Spartina alterniflora*, the dominant species on Atlantic and Gulf coast salt marshes of North America.[3,7,8,10,16,17,27-31]

TABLE 5
Effect of NaCl Concentration and Content in Shoot, Total N Content, and N_2 Fixation of *Casuarina equisetifolia*, 24 Weeks after the Commencement of Salinity Treatment[a]

Treatment	NaCl conc. (mM)	N concentration in shoots (%)	N content of shoots (mg/plant)	Total N content (mg/plant)	Estimated N fixed (mg/mg nodule dry wt)
Uninoculated	0	1.7 ± 0.3	38.3 ± 15.3	44.7 ± 20.5	
Inoculated	0	1.9 ± 0.1	179.9 ± 13.9	232.0 ± 18.8	0.40 ± 0.07
	20	1.9 ± 0.1	146.3 ± 11.3	189.5 ± 15.1	0.43 ± 0.09
	50	1.5 ± 0.1	196.6 ± 11.3	164.6 ± 15.1	0.33 ± 0.04
	100	1.6 ± 0.1	191.7 ± 11.6	261.4 ± 15.5	0.34 ± 0.05
	200	1.8 ± 0.1	173.0 ± 13.5	216.4 ± 18.1	0.39 ± 0.07
	500	1.7 ± 0.1	65.0 ± 12.6[b]	82.7 ± 16.9[a]	0.17 ± 0.01[b]

[a]   Values are means of four replicates ± SE.
[b]   Significantly different from inoculated control at $p < 0.01$.

From Ng, B. H., *Plant Soil*, 103, 123, 1987. With permission.

TABLE 6
Effect of NaCl Concentration on Shoot, Root, and Nodule Dry Weight of *Casuarina equisetifolia*, 24 Weeks after the Commencement of Salinity Treatment

Treatment	NaCl conc. (m*M*)	Shoot dry weight (g plant^{-1})	Root dry weight (g plant^{-1})	Nodule dry weight (g plant^{-1})
Uninoculated	0	2.23 ± 1.01	0.60 ± 0.38	0
Inoculated	0	9.41 ± 0.99	2.62 ± 0.35	0.47 ± 0.05
	20	7.77 ± 0.80	2.08 ± 0.28	0.34 ± 0.04
	50	12.78 ± 0.80[a]	3.65 ± 0.28[b]	0.67 ± 0.04[a]
	100	12.15 ± 0.82[a]	3.74 ± 0.29[b]	0.64 ± 0.04[a]
	200	9.81 ± 0.96[a]	2.73 ± 0.34	0.44 ± 0.5
	500	3.95 ± 0.89[a]	1.09 ± 0.32[a]	0.23 ± 0.04[b]

[a] Significantly different from the inoculated control at $p < 0.01$.
[b] Significantly different from the inoculated control at $p < 0.05$.

From Ng, B. H., *Plant Soil*, 103, 123, 1987. With permission.

Tall- and short-growth forms have been observed throughout the range of this broadly distributed species, and a great deal of effort has been made to attempt to sort out the environmental factors which may be responsible for differences in biomass production at different sites on the salt marsh. In this complex environment a number of factors may act to limit plant growth, including salinity stress, tidal action, nutrient unavailability, lack of aeration, accumulation of sulfides, and herbivory. Nitrogen fertilization has been reported to increase yields of halophytes, affect plant form, increase seed production, change species' abundance in plant communities, and increase the nitrogen content of biomass.[4,28]

Valiela et al.[27] reported that biomass production of the low marsh dominated by *S. alterniflora*, and *S. patens* and *Distichlis spicata* the dominants of high-marsh habitats in Massachusetts, did not respond to sewage sludge (10-6-4) fertilization in the first year. However, an increase in yield was obtained in the second and third year of treatment. Low-marsh biomass production averaged 0.51 kg/m² in controls and 1.32 kg/m² in the high dosage fertilization treatment (25 g/m²/week; Table 7). On the high marsh, controls produced 0.63 kg/m², while the high dosage plots had a mean biomass production of 1.26 kg/m². Production of the two dominants of the high marsh, *S. patens* and *D. spicata*, both increased with fertilization. Fertilization of the low-marsh soils converted the community from the dwarf form *S. alterniflora* with scattered *Salicornia* spp. to the tall form *Spartina alterniflora* with the exclusion of *Salicornia* spp. In the high-marsh, *Spartina patens* increased in yield over time and replaced *D. spicata* by the third year, while *S. alterniflora* did not respond to fertilization in the high marsh.[27]

The availability of nitrate and ammonia and its effect on *S. alterniflora* plant nitrogen content was determined on the Oak Island salt marsh, North Carolina by Mendelssohn.[28] The availability of nitrate + nitrite was much lower than that of ammonium for interstitial waters in the three marsh zones (short, medium, and tall), with soil ammonium nitrogen levels of 3.4, 0.9, and 0.6 mg/l and nitrate + nitrite concentrations of 0.04, 0.04, and 0.05 mg/l in the three zones, respectively. Leaf total nitrogen content during the growing season averaged 1.5% in the short and medium *Spartina* zones and 1.7% in the tall zone. Leaves of the tall form had 124 µg N/g fresh weight, the medium 107 µg/g, and the short 114 µg/g. The tall form plants had higher amounts of ammonia and nitrate + nitrite nitrogen in the leaves and roots than the medium and short forms of *S. alterniflora*. Nitrate reductase

TABLE 7
The Effect of Sludge Fertilizer on the Net Annual Production (kg/m²) and Peak Above-Ground Biomass (kg/m²) on Low Marsh and High Marsh Vegetation in Each of 3 Years

Treatment	Year	Low Marsh		High marsh	
		Net production	Peak biomass	Net production	Peak biomass
High dosage	1971	1.53 ± 0.16	1.55 ± 0.18	1.23 ± 0.28	1.35 ± 0.36
	1972	1.13 ± 0.06	1.13 ± 0.14	1.08 ± 0.47	0.88 ± 0.14
	1973	1.31 ± 0.23	0.87 ± 0.00	1.46 ± 0.12	0.78 ± 0.00
	Mean	1.32 ± 0.12	1.18 ± 0.20	1.26 ± 0.11	1.00 ± 0.18
Low dosage	1971	0.78 ± 0.09	0.65 ± 0.03	1.41	0.88
	1972	0.87 ± 0.23	0.87 ± 0.17	0.96	1.00
	1973	0.91 ± 0.15	0.71 ± 0.03	1.77	1.02
	Mean	0.86 ± 0.04	0.74 ± 0.66	1.38 ± 0.23	0.97 ± 0.04
Control	1971	0.36 ± 0.16	0.28 ± 0.06	0.62 ± 0.11	0.44 ± 0.06
	1972	0.44 ± 0.21	0.25 ± 0.00	0.54 ± 0.17	0.35 ± 0.01
	1973	0.72 ± 0.20	0.42 ± 0.05	0.73 ± 0.25	0.54 ± 0.12
	Mean	0.51 ± 0.11	0.32 ± 0.05	0.63 ± 0.05	0.44 ± 0.05

From Valiella, I., Teal, J. M., and Sass, W. J., *J. Appl. Ecol.*, 12, 973, 1975. With permission.

activity was stimulated in both roots and leaves of the tall and short forms when plants were fertilized with nitrate (200 g/m² sodium nitrate). The increase in enzymatic activity for the short form plants was 5.3 times that of controls in the leaves and 19.4 times the control in roots, while for the tall form plants, increases were 1.2 times for leaves and 6.1 times for roots in plants receiving nitrate fertilization. These data indicate that nitrate could be used as a nitrogen source when available in the reducing environments of salt marshes.[28]

An investigation was carried out to determine the effects of 100 g dry sludge/m²/week of sewage sludge fertilization (2 g N/m²/week) on the growth of the short form of *S. alterniflora* on the Sapelo Island (Georgia) salt marsh. Haines[7] determined that the biomass production in treated plots increased (1133 g/m²) 1.5 times over that of controls (647 g/m²) live biomass. Nitrogen content of shoots in treated and untreated plots averaged about 1.4%. Shoot density in the treated plots increased to 1026/m² in January compared to 698/m² in control plots. Plants responded to fertilization within a few weeks and continued to increase production for the year. The rapid response obtained in this investigation differs from the results of Valiela et al.,[27] who did not obtain an increase in biomass yield in response to sewage sludge fertilization in the first year of their study in Massachusetts, but they determined that biomass production of *S. alterniflora* increased significantly in subsequent years of fertilization.

Dry mass production of *S. alterniflora* and *Distichlis spicata* increased when plants were exposed to nitrogen fertilization in the laboratory.[8] Phosphorus fertilization treatments were also reported to increase plant production. The ratio of below-ground to above-ground biomass decreased with an increase in nitrogen or phosphorus availability.

S. alterniflora was grown on test plots that were established on the Neuse River shoreline (North Carolina). Broome et al.[10] determined that when nitrogen or phosphorus were used alone, dry mass, plant height, and number of flowers per plant increased in both cases when nutrient treatments were compared with growth parameters of the unfertilized controls (Table 8). Biomass production without fertilization was 8 g/plant, while 112 kg/ha nitrogen yielded 30 g/plant, and 49 kg/ha phosphorus yielded 34 g/plant. However, a combination of 112 kg N + 49 kg P increased yields to a much higher level of 174 g/plant. The highest biomass production, 227 g/plant, was obtained in a fertilization treatment with 224 kg N/ha + 99

TABLE 8

Effects of Ammonium Sulfate and Superphosphate
Fertilization of *Spartina alterniflora* Transplanted June 13,
1978 and Sampled September 13, 1978 along the Neuse
River Shoreline near Oriental, NC

Fertilizer rate (kg/ha)		Dry wt (g/plant)	Height (cm)	No. flowers/plant	No. stems/plant
N[a]	P[b]				
0	0	8	103	2	5
0	49	34	134	3	12
112	0	30	108	4	14
56	12	38	122	7	15
112	12	97	154	10	24
224	12	66	108	6	24
56	25	120	160	12	25
112	25	108	147	10	25
224	25	146	169	8	24
56	49	94	159	11	22
112	49	174	163	12	28
224	49	214	173	13	32
56	99	106	147	9	24
112	99	96	122	12	27
224	99	227	175	16	35
LSD $_{05}$		69	28	7	14
(cv %)		54	17	57	42

[a] Source of N was ammonium sulfate.
[b] Source of P was concentrated superphosphate.

From Broome, S. W., Seneca, E. D., and Woodhouse, W. W., *Estuaries*, 6, 212, 1983. With permission.

kg P/ha (Table 8). The authors concluded that tidal waters were deficient in these elements and did not supply sufficient nitrogen and phosphorus for optimal plant growth.[10]

Laboratory experiments carried out by Morris[32] indicated that the dry mass production of *S. alterniflora* was highest in the high nitrate fertilization treatment of 116 mg/l nitrogen (41.6 g/plant) and lowest in the low nitrate treatment of 4.1 mg/l nitrogen (10.1 g/plant). The highest plant growth rate was for individuals in the high-nitrate treatment, averaging 0.6 g/plant/d in late August vs. 0.1 g/plant/d in the low-nitrate treatment.[32] Plants limited by nitrogen had low leaf nitrogen content, averaging 1% of the dry weight, while plants not limited by nitrogen had >2% nitrogen on a dry weight basis in leaves. In contrast to these results, Webb and Dodd[33] determined that fertilization of *S. alterniflora* transplants did not produce a significant difference in the height or density of plants. However, the time of planting was significant, with winter transplants producing a denser stand of stems by the end of the first growing season than did spring transplants.

Hopkinson and Schubauer[31] estimated that the uptake of nitrogen by *S. alterniflora* was 34.8 g/m²/year on the Sapelo Island (Georgia) salt marshes. About 54% of the above-ground nitrogen requirement of the plant was determined to be met by internally recycled nitrogen. Nitrogen conservation by *S. alterniflora* was suggested to be significant in salt marsh habitats which have been shown to be nitrogen limited. The authors concluded that *S. alterniflora* played a significant role in salt marsh habitats by converting inorganic nitrogen to an organic form and, subsequently, with the senescence of plant organs in depositing organic matter on the marsh surface.[31]

Buresh et al.[16] reported that the total nitrogen content for shoots of *S. alterniflora* decreased with distance from the tidal creek stream edge on a Louisiana salt marsh. In September, the nitrogen content of shoots ranged from about 1% at 1.5 m from the stream to 0.6% in plants from 7.6 to 17 m from the edge of the stream. Phosphorus content of plants followed a similar declining pattern along the gradient from 0.11 to 0.04% total phosphorous in shoots in September. The soil nitrogen content actually increased with distance inland from the stream, from 0.4% at 1.5 m from the stream to 0.9% at 24 m, with organic nitrogen making up 99% of the total soil nitrogen. Fertilization of *S. alterniflora* plots was done with 200-kg/ha phosphorus and 200-kg/ha nitrogen, and the results indicated that phosphorus was not limiting in this system. Above-ground biomass production in phosphorus-treated plots was 1.031 kg/m^2 and this was not significantly different from the 1.061 kg/m^2 in control plots. Nitrogen-fertilized plots had higher above-ground (1.357 kg/m^2) and below-ground biomass (6.322 kg/m^2) than the controls, which had 5.494 kg/m^2 below-ground yield.[16] These data from a streamside marsh support the data of Patrick and Delaune,[34] which indicated that nitrogen and not phosphorus was limiting the primary production in salt marshes dominated by *S. alterniflora* in Louisiana. In the Louisiana marsh, primary production of plants was increased by 15% with nitrogen fertilization compared with a 28% increase over controls in the more recent study by Buresh et al.[16] The nitrogen content of plants in the fertilized plots was always higher during the growing season than for unfertilized controls, ranging from 0.9 to 1.3% in fertilized treatments compared with 0.7 to 0.9% in controls. It was concluded that nitrogen availability was a significant factor determining the biomass production of *S. alterniflora* on Louisiana salt marshes. Both investigations on Louisiana salt marshes indicated that soil nitrogen availability was lower for populations at the inland locations than for streamside populations of *S. alterniflora*.[16,34]

Rates of nitrogen uptake were measured for *S. alterniflora* under controlled conditions in the laboratory, and Morris[29] determined that plants were probably not limited by the supply of nitrogen per se in salt marsh soils. The half-saturation constants for ammonium (0.057 mg N/l) were too low to account for nitrogen-limited growth of the short form of *S. alterniflora* in the field when compared with the inorganic ammonium levels (3 mg NH$_4$-N/l) in marsh interstitial waters. A number of environmental variables were suggested that could account for the observed restrictions on growth by nitrogen deficiencies in the short *S. alterniflora* zone, including oxygen deficiencies in the root zone, high soil salinity, high sulfide content, and competition for carriers by other ions.[29] Any of these factors might be responsible for limiting nitrogen uptake in areas away from tidal creek banks and lead toward the observed dwarf form of *S. alterniflora*.[29]

The interactions between sediment oxidation, nitrogen, salinity, and drainage on the growth of *S. alterniflora* were investigated at 21 sites on the Great Sippewissett Marsh (Massachusetts) by Howes et al.[17] The short *S. alterniflora* was 10 to 45 cm tall, and the tall form was about 2 m in height. Plant biomass production was correlated with the redox potential of the soil sediment at 2 cm (r = 0.79). However, salinity and water-soluble ammonia were not strongly correlated with plant biomass, with KCl extractable ammonia having a correlation of r = -0.40. Stepwise multiple regression indicated that the redox potential at 2-cm depth accounted for 62% of the variation in data. When the redox potential is removed as a factor, the water table depth accounted for 26% of the variation. If the plant biomass is related to salinity and nitrogen alone, available nitrogen accounted for 16% and salinity 4% of the variation. The tall form of *S. alterniflora* had biomass yields of 1700 g/m^2, compared with 270 and 1100 g/m^2 in unfertilized and fertilized short-form plots, respectively (Table 9). Howes et al.[17] concluded that the most direct limiting factor to production could be the amount of nitrogen available. Factors limiting N availability, such as reducing environments, high salinity, and sulfide accumulation, could indirectly cause a reduction in plant biomass yields. Fertilization treatments caused an increase in plant pro-

TABLE 9
Effects of Nitrogen Fertilizer Treatments on Plant Evapotranspiration and Associated Variables for *Spartina alterniflora* in a New England Salt Marsh[a]

Treatment	Plant height (cm)	Live above-ground biomass (g dry weight m^{-2})	Leaf area coverage (m^{2}m^{-2})	Evapotranspiration (1 m^{-2}d^{-1})
Fertilized	102	1100 ± 90	7.3 ± 0.7	9.1 ± 1.1
Control	30	270 ± 66	1.9 ± 0.1	4.6 ± 1.0
Quotient (fertilized/control)		4.1	3.8	2.0

[a] Values are means ± 1 SE, n = 4.

From Howes, B. L., Dacey, J. W. H., and Goehringer, D. D., *J. Ecol.*, 74, 881, 1986. With permission.

duction and plant water uptake, and increased plant growth would increase the oxygen supply to the roots and sediments (Table 9).[17]

V. CONCLUSIONS

Fertilization with nitrogen increased plant production as well as water uptake in salt marshes, and the increased root growth should serve to increase the oxygen supply available for root growth.[17] In salt marsh habitats a number of factors could influence plant growth, ranging from periodic changes in the weather to nutritional deficiencies and hypersaline edaphic conditions. Field and laboratory investigations indicated that nitrogen deficiencies might be the primary limiting factor for the growth of *S. alterniflora* and other marsh species.[1-16] Therefore, factors limiting nitrogen uptake could produce symptoms of soil nitrogen deficiencies even when soil nitrogen concentrations were adequate for growth.[4,5,16] Several environmental factors are involved in limiting plant nitrogen uptake from the soil, including hypersaline conditions, reducing environments, low soil-water potentials, and sulfide accumulation.[17] It is necessary to sort out the direct and indirect effects of environmental factors on plant production in salt marsh habitats. Experiments will have to be designed to take into account the many interactions involved in the nitrogen nutrition of halophytes and how these relate to biomass production in salt marsh habitats.

REFERENCES

1. **Tyler, G.,** On the effect of phosphorus and nitrogen, supplied to baltic shore-meadow vegetation, *Bot. Notiser*, 120, 433, 1967.
2. **Pigott, C. D.,** Influence of mineral nutrition on the zonation of flowering plants in coastal salt marshes, in *Ecological Aspects of the Mineral Nutrition of Plants*, Rorison, I. H., Ed., Blackwell Scientific, Oxford, 1969, 25.
3. **Valiela, I. and Teal, J. M.,** Nutrient limitation in salt marsh vegetation, in *Ecology of Halophytes*, Reimold, R. J. and Queen, W. H., Eds., Academic Press, New York, 1974, 547.
4. **Gallagher, J. L.,** Effects of an ammonium nitrate pulse on the growth and elemental composition of natural strands of *Spartina alterniflora* and *Juncus roemerianus*, *Am. J. Bot.*, 62, 644, 1975.
5. **Jefferies, R. L.,** Growth responses of coastal halophytes to inorganic nitrogen, *J. Ecol.*, 65, 847, 1977.
6. **Jefferies, R. L. and Perkins, N.,** The effects on the vegetation of the additions of inorganic nutrients to salt marsh soils at Stiffkey, Norfolk, *J. Ecol.*, 65, 867, 1977.
7. **Haines, E. B.,** Growth dynamics of cordgrass, *Spartina alterniflora* Loisel., on control and sewage sludge fertilized plots in a Georgia salt marsh, *Estuaries*, 2, 50, 1979.

8. **Smart, R. M. and Barko, J. W.**, Nitrogen nutrition and salinity tolerance of *Distichlis spicata* and *Spartina alterniflora*, *Ecology*, 61, 630, 1980.

9. **Rozema, J., Dueck, T., Wesselman, H., and Bijl, F.**, Nitrogen dependent growth stimulation by salt in strand-line species, *Acta Oecol. Oecol. Plant.*, 4, 41, 1983.

10. **Broome, S. W., Seneca, E. D., and Woodhouse, W. W.**, The effects of source, rate and placement of nitrogen and phosphorus fertilizers on growth of *Spartina alterniflora* in transplants in North Carolina, *Estuaries*, 6, 212, 1983.

11. **Naidoo, G.**, Effects of salinity and nitrogen on growth and water relations in the mangrove, *Avicennia marina* (Forsk.) Vierh., *N. Phytol.*, 107, 317, 1987.

12. **Covin, J. D. and Zedler, J. B.**, Nitrogen effects on *Spartina foliosa* and *Salicornia virginica* in the salt marsh at Tijuana estuary, California, *Wetlands*, 8, 51, 1988.

13. **Skeffington, M. J. S. and Jeffrey, D. W.**, Response of *Armeria maritima* (Mill.) Willd. and *Plantago maritima* L. from an Irish salt marsh to nitrogen and salinity, *N. Phytol.*, 110, 399, 1988.

14. **Loveland, D. G. and Ungar, I. A.**, The effect of nitrogen fertilization on the production of halophytes in an inland salt marsh, *Am. Midl. Nat.*, 109, 346, 1983.

15. **Drake, D. R. and Ungar, I. A.**, Effects of salinity, nitrogen, and population density on the survival, growth, and reproduction of *Atriplex triangularis* (Chenopodiaceae), *Am. J. Bot.*, 76, 1125, 1989.

16. **Buresh, R. J., DeLaune, R. D., and Patrick, W. H.**, Nitrogen and phosphorus distribution and utilization by *Spartina alterniflora* in a Louisiana Gulf Coast marsh, *Estuaries*, 3, 111, 1980.

17. **Howes, B. L., Dacey, J. W. H., and Goehringer, D. D.**, Factors controlling the growth form of *Spartina alterniflora*: feedbacks between above-ground production, sediment oxidation, nitrogen and salinity, *J. Ecol.*, 74, 881, 1986.

18. **Steen, E.**, Root and shoot growth of *Atriplex litoralis* in relation to nitrogen supply, *Oikos*, 42, 74, 1984.

19. **Ginzo, H. D., Collantes, M. B., and Caso, O. H.**, Fertilization of a halophytic natural grassland in Argentina: herbage dry matter, botanical composition, and mineral content, *Turrialba*, 36, 453, 1986.

20. **Cargill, S. M. and Jefferies, R. L.**, Nutrient limitation of primary production in a sub-arctic salt marsh, *J. Appl. Ecol.*, 21, 657, 1984.

21. **Weston, R. L.**, Nitrogen nutrition in *Atriplex hastata* L., *Plant Soil*, 20, 251, 1964.

22. **Reddell, P., Foster, R. C., and Bowen, G. D.**, The effects of sodium chloride on growth and nitrogen fixation in *Casuarina obesa* Miq., *N. Phytol.*, 102, 397, 1986.

23. **Ng, B. H.**, The effects of salinity on growth, nodulation and nitrogen fixation of *Casuarina equisetifolia*, *Plant Soil*, 103, 123, 1987.

24. **Stienstra, A. W.**, Nitrate accumulation and growth of *Aster tripolium* L. with a continuous and intermittent nitrogen supply, *Plant Cell Environ.*, 9, 307, 1986.

25. **Doddema, H., Eddin, R. S., and Mahasneh, A.**, Effects of seasonal changes of soil salinity and soil nitrogen of the N-metabolism of the halophyte *Arthrocnemum fruticosum* (L.) Moq., *Plant Soil*, 92, 279, 1986.

26. **Eddin, R. S. and Doddema, H.**, Effects of NaCl on the nitrogen metabolism of the halophyte *Arthrocnemum fruticosum* (L.) Moq., grown in a greenhouse, *Plant Soil*, 92, 373, 1986.

27. **Valiela, I., Teal, J. M., and Sass, W. J.**, Production and dynamics of salt marsh vegetation and the effects of experimental treatment with sewage sludge, *J. Appl. Ecol.*, 12, 973, 1975.

28. **Mendelssohn, I. A.**, Nitrogen metabolism in the height forms of *Spartina alterniflora* in North Carolina, *Ecology*, 60, 574, 1979.

29. **Morris, J. T.**, The nitrogen uptake kinetics of *Spartina alterniflora* in culture, *Ecology*, 61, 1114, 1980.

30. **Valiela, I.**, Nitrogen in salt marsh ecosystems, in *Nitrogen in the Marine Environment*, Academic Press, New York, 1983, 649.

31. **Hopkinson, C. S. and Schubauer, J. P.**, Static and dynamic aspects of nitrogen cycling in the salt marsh graminoid, *Spartina alterniflora*, *Ecology*, 65, 961, 1984.

32. **Morris, J. T.**, A model of growth responses by *Spartina alterniflora* to nitrogen limitation, *J. Ecol.*, 70, 25, 1982.

33. **Webb, J. W. and Dodd, J. D.**, *Spartina alterniflora* response to fertilizer, planting dates, and elevation in Galveston Bay, Texas, *Wetlands*, 9, 61, 1989.

34. **Patrick, W. H. and Delaune, R. D.**, Nitrogen and phosphorous utilization by *Spartina alterniflora* in a salt marsh in Barataria Bay, Louisiana, *Estuarine Coastal Mar. Sci.*, 4, 59, 1976.

Chapter 10

GENECOLOGY

I. INTRODUCTION

The occurrence of genetically defined races for species of halophytes in saline habitats would indicate that natural selection is taking place in these highly stressful environments. Selection for ecotypes that are adapted to local climatic and edaphic conditions should increase the fitness of species and their chances to survive in particular salt marsh habitats throughout the range of a species. Ecotypic development along salinity gradients in salt marshes would permit more salt-tolerant genotypes to invade the extreme portions of marshes, which are characterized by low plant species diversity. Halophytes are selected for in saline habitats, because they have developed some specific mechanisms to adjust their plant water status, growth, and seed germination to the high salinity and low water potentials that exist in salt marsh and salt desert habitats. These adaptations may be related to the development of specific structural modifications such as salt hairs, salt glands, and succulence, or avoiding salt stress by remaining dormant during periods of high salinity stress and making maximum growth during periods of the growing season when salinity levels are reduced.

Ecotypic differentiation among populations of halophytes has been reported by Seneca,[1] Ahmad and Wainright,[2] and Jerling,[3] for species with broad latitudinal distributions, inland vs. coastal forms, and along topographical gradients within a single marsh. Differences in morphometric characteristics, height growth, biomass yield, phenology, and salt tolerance among populations have been determined for several halophytic species.[1,4-8] Variation in salt tolerance of seeds from different populations of plants have been investigated for several species of halophytes by Seneca,[1] Workman and West,[9] Cavers and Harper,[10] Mooring et al.,[11] Bazzaz,[12] and Kingsbury et al.[13]

II. ECOTYPIC DIFFERENTIATION

An investigation into the occurrence of genetic differentiation in fitness-related characters in *Plantago maritima* was carried out along a submergence-grazing gradient in a Baltic seashore meadow. Jerling[3] determined that genetic variation occurred along the environmental gradient in that plants differed in morphology, seed size, germination behavior, growth pattern, and biomass allocation. Genotypes of *P. maritima* from locations characterized by high plant density and subjected to more intense interspecific competition were found to have higher survival percentages at high-density sites than did those transplants originating from low-density locations. Survival of upper meadow genotypes that were subjected to heavy grazing pressure was better in environments resembling their natural conditions than that for genotypes from the central and lower portions of the marsh (Table 1). In the heavily grazed upper meadow *P. maritima* genotypes had shorter and narrower leaves than genotypes from the less intensively grazed central meadow where leaves were broadest and longest.[3]

Populations of *Agrostis stolonifera* collected from coastal beach and nonsaline habitats were compared by Wu[7] to determine if they differed in salt tolerance. The *A. stolonifera* seashore population had higher root elongation at salinities ranging from 68 to 238 meq/l than did roots of seedlings originating from nonsaline habitats. An index of tolerance to seawater varied on a dry mass basis from 28.1% of controls in the bank population to 89.5% of controls in the seashore population, indicating that genetic differentiation can occur in populations within a local area (Figure 1).[7] The fact that salt-tolerant individuals occurred

TABLE 1
Survival and Mortality of Lower-Meadow, Central-Meadow, and Upper-Meadow Genotypes of *Plantago maritima* in the Grazing Simulation Experiment

	Upper meadow	Central meadow	Lower meadow
Initial number	200	200	200
Survivors	182	169	127
Mortality (%)	9.0	15.5	26.5

From Jerling, L., *Oikos*, 53, 341, 1988. With permission.

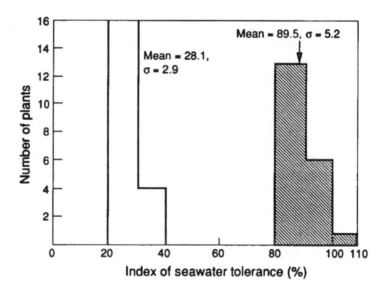

FIGURE 1.   Index of seawater tolerance for the seashore population (shaded bar) and the adjacent bank population (open bar) of *Agrostis stolonifera* from Oregon. (From Wu, L., *N. Phytol.*, 89, 471, 1981. With permission.)

in the nontolerant population supports the hypothesis that salinity tolerance could evolve rapidly in *A. stolonifera* populations when they are exposed to salinity stress. Coastal and inland ecotypes of *A. stolonifera* that were grown in a uniform garden were found to differ in the amount of salt that is retained in leaves after immersion in saltwater.[2] The maritime ecotype had more wettable leaves and retained less salt than the inland ecotype. Extracellular wax deposits were found to be significantly different between the coastal and inland plant populations, and Ahmad and Wainright[2] hypothesized that the structure and distribution of wax on the surface of leaves were significant factors in determining both the wettability and ion content of leaves of *A. tenuis*.

The genetic variability within *Puccinellia maritima* populations was examined in a uniform garden experiment. Clones were collected (a total of 56) and transplanted from six geographic regions on the British Coast.[6] Variation in tiller number, plant height, and plant width were the three most significant heritable vegetative characteristics in these populations. Date of inflorescence emergence and date of anthesis were two significant reproductive variables between biotypes. The genotypes of *P. maritima* collected from grazed salt marshes tended to be fast tillering, small, short leaved, and prostrate when compared with plants

TABLE 2
Means Analysis on Morphometric Traits of *Spartina patens* Plants Grown in a Common Garden[a]

Character	Subpopulations		
	Marsh	Swale	Dune
Vegetative culms			
Tiller number per clone	30.48 ± 3.06	26.19 ± 2.05	35.31 ± 2.60
Vegetative culm height	26.1 ± 0.91	49.36 ± 19.49	24.95 ± 0.78
Number of leaves per culm	7.70 ± 0.18	7.81 ± 0.21	7.53 ± 0.15
Reproductive culms			
Tiller number per reproductive clone	25.83 ± 2.29	27.33 ± 4.27	42.62 ± 4.64
Spikelet number per spike	35.56 ± 1.53	34.67 ± 1.30	39.75 ± 1.37
Total seed number per clone	187.1 ± 33.2	288.0 ± 48.1	400.8 ± 96.3
Flowering culms per clone	1.83 ± 0.33	2.92 ± 0.48	3.38 ± 0.64
Rhizomes			
Rhizome number per clone	6.11 ± 0.81	7.11 ± 0.93	7.35 ± 0.84
Total rhizome length (cm) per clone	545.6 ± 90.1	625.2 ± 82.0	646.3 ± 91.6
Rhizome length (cm) per vegetative culm	21.54 ± 3.08	25.60 ± 1.97	25.89 ± 2.56
Rhizome length (cm) per reproductive culm	94.71 ± 28.81	71.59 ± 7.24	161.54 ± 48.43

[a] Underlined means are not significantly different from one another.

From Silander, J. A., and Antonovics, J., *Evolution*, 33, 1114, 1979. With permission.

from ungrazed salt marsh sites. Gray and Scott[6] determined that plants collected from high-salt marshes, which may have higher soil nutrient levels, had higher total biomass than did plants originating from pioneer or midlevel salt marshes.

Microdifferentiation was found in subpopulations of *Spartina patens* from marsh, swale, and dune sites along the North Carolina coast by Silander and Antonovics.[5] Swale and dune populations were more salt tolerant than marsh populations when exposed to 3.5% sea salts. In a uniform garden experiment, it was found that dune plants had more seed production, tillering, and rhizome spread than the other two populations (Table 2). Silander and Antonovics[5] found that subpopulations had more genetic variance within them than was found among the three subpopulations studied. The dune plants were found to have the greatest phenotypic plasticity and the salt marsh plants the least. Genetic diversity based on allozyme variation decreased from salt marsh to dune population.[5,14] Responses of plants to reciprocal trans-planting of subpopulations into salt marsh, swale, and dune habitats indicated that genetic differentiation occurs among the subpopulations studied.[15] The dune habitat caused a reduction in survival and fitness values for plants from salt marsh and swale subpopulations, while transplants from all subpopulations had highest survival and fitness levels in the marsh habitat. Results from this experiment indicated that the dune habitat was most severe and imposed the greatest selection pressure (Table 3). Silander[15] concluded that the relative

TABLE 3
Proportion of Transplanted Ramets
of *Spartina patens* Surviving

Subpopulation origin	Site reared		
	Dune	Swale	Marsh
Dune	1.00	0.97	1.00
Swale	0.86	0.94	1.00
Marsh	0.83	0.92	1.00

From Silander, J. A., *Evolution*, 39, 1034, 1985.
With permission.

TABLE 4
Ratio of Relative Fitness of
Transplanted Subpopulation Genets
of *Spartina patens* to Native
Subpopulation Genets

Subpopulation origin	Site reared		
	Dune	Swale	Marsh
Dune	1.00	1.19	0.96
Swale	0.81	1.00	0.87
Marsh	0.78	1.00	1.00

From Silander, J. A., *Evolution*, 39, 1034, 1985.
With permission.

fitness of populations tended to be greater at their source site than in transplant habitats (Table 4).

III. HEIGHT FORMS IN *SPARTINA ALTERNIFLORA*

A significant amount of research has been carried out to determine if phenotypic plasticity or genetic differentiation of specialized ecotypes occur locally and in the latitudinal distribution of populations of the dominant Gulf and Atlantic coast species of the U.S., *Spartina alterniflora*. The relative significance of ecological vs. genetic factors in determining plant form, salt tolerance, and physiological responses to microedaphic conditions on the Atlantic coast of the U.S. has been reviewed by Anderson and Treshow[16] and Mendelssohn et al.[17] Morphological differentiation of a short form in the high-low marsh and a tall form along streams for the halophyte *Spartina alterniflora* has been reported by a number of researchers, including Stalter and Batson,[4] Gallagher et al.,[8] Anderson and Treshow,[16] Adams,[18] and Shea et al.[19] Differences in plant height for *S. alterniflora* along an environmental gradient have been interpreted to be because of genetic differentiation of subpopulations on a given marsh[4,8,18] and differences in microenvironmental conditions in the different zones of the marsh.[19,20]

Electrophoretic investigations by Shea et al.[19] and Valiela et al.[21] determined that there were no differences between the protein banding patterns of the tall- and short-growth forms of *S. alterniflora*. Field transplants of the tall form to the high marsh resulted in plants that were short form after three growing seasons, and short plants from the high marsh when transplanted to the low marsh resembled the tall-form plants in the low marsh. Therefore,

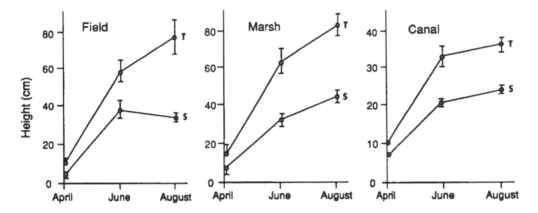

FIGURE 2.   Plant height of *Spartina alterniflora* (T = tall form, S = short form) from a Delaware salt marsh (Marsh) and for transplants (Field, Canal) into two common garden locations. (From Gallagher, J. L., Somers, G. F., Grant, D. M., and Seliskar, D., *Ecology*, 69, 1005, 1988. With permission.)

Shea et al.[19] concluded that the difference in plant size was due to phenotypic plasticity rather than the formation of genetically isolated ecotypes.

Adams[18] determined that soil salinity levels were higher in areas occupied by the short (37.7 mS/cm) than in the medium (31.2 mS/cm) and tall form (31.8 mS/cm) of *S. alterniflora* under field conditions in North Carolina. Laboratory growth experiments, using transplanted vegetative material, indicated that the mean biomass of tall-form plants was significantly greater than that of the short form. In a short-term 1-year transplant investigation carried out in a South Carolina salt marsh, Stalter and Batson[4] determined that dwarf *S. alterniflora* was inherently dwarf, but although tall-form plants were reduced in size, they were still taller than the dwarf plants growing with them after 1 year. A 9-year transplant experiment begun in 1978 was carried out in three common gardens to determine if genetic or environmental factors were responsible for tall- and short-form *S. alterniflora* in Delaware salt marshes.[8] In all the cases the tall form remained larger than the short form after the 9-year growth period (Figure 2). Also, above-ground biomass was statistically significantly higher in 1979 for the tall form, while in 1984, biomass was statistically significantly different at two of the three sites. Reproductive characteristics such as flower culm height (90.0 to 50.9 cm tall) and flower culm density (35 to 10 culms/m²) were still higher in the tall vs. the short form of *S. alterniflora* in 1986. Gallagher[8] et al. concluded that differences in plant form were not only due to phenotypic plasticity, but also to genetic differences in the two forms of *S. alterniflora*. Fertilized salt marsh plots in the Great Sippewissett Marsh (Massachusetts) produced plants in the short-form plots that resembled those of the tall form in morphology and biomass production, but after 4 years of fertilization treatment the conversion of short to tall form was not complete.[21] Since electrophoretic data indicated no genetic differences between growth forms, Valiela et al.[21] suggested that other factors such as soil salinity, iron availability, or water supply might also be involved in determining the growth pattern of short-form plants.

Field transplants were made in North Carolina from 12 populations of *S. alterniflora* plants that were originally collected from New England south to the Gulf coastal salt marshes. Seneca[1] determined that southern populations flowered later and had longer growing periods than northern populations. In this common garden field experiment, latitudinal populations differed from one another in height, flowering time, and morphology after two growing seasons (Table 5). A study was carried out by Somers and Grant[22] to determine differences in the phenology of flowering of *S. alterniflora* plants grown from seeds collected from salt marshes ranging from Maine to Virginia. The source of seeds significantly affected the time

TABLE 5

Population by Photoperiod-Thermoperiod Treatment Means for Height (cm) and Number of Culms of *Spartina alterniflora* Based on Eight Seedlings per Treatment and Flowering Response in 30—26°C Thermoperiod

| Coastline | Population | Height (cm) by photoperiod-thermoperiod | | | | No. of culms photoperiod-thermoperiod | | | | Flowering response[a] | |
| | | 18—14°C | | 30—26°C | | 18—14°C | | 30—26°C | | 30—26°C | |
		Short	Long	Short	Long	Short	Long	Short	Long	Short	Long
Atlantic	Plum Island, MA	39.2	43.5	62.0	86.6	4.5	4.0	20.2	16.1	8	8
	East Greenwich, RI	36.1	64.2	46.4	80.4	3.0	3.8	22.6	18.1	8	12
	Milford, CT	52.6	48.2	62.2	78.9	6.5	5.6	39.6	15.9	8	12
	Long Island, VA	66.4	63.0	61.2	95.9	4.4	4.4	21.2	12.4	8	11
	Assateague Island, VA	45.8	44.0	60.5	82.9	5.8	5.2	27.2	10.2	10	11
	Oregon Inlet, NC	53.5	49.5	74.0	113.9	6.0	6.5	22.0	14.2	11	12
	Okracoke Island, NC	31.6	31.8	54.9	82.5	5.4	6.6	24.5	19.1	12	—
	Snow's Cut, NC	29.1	33.0	68.0	92.1	3.1	3.9	18.6	10.8	—	—
	Sapelo Island, GA	37.8	41.2	70.1	85.4	4.1	3.8	16.0	8.5	—	—
	Merritt Island, FL	14.0	14.0	45.9	49.9	1.2	1.3	7.1	4.0	—	—
Gulf	Ocean Springs, MS	22.2	24.6	65.1	63.1	1.2	1.0	4.9	4.8	—	—
	Port Aransas, TX	53.6	38.8	89.0	96.4	1.4	1.2	7.5	7.4	—	—

[a] Number of weeks after start of experiment that emergence of an inflorescence was observed that floral primordia were present.

[b] Dashes indicate that flowering and flower initiation did not occur.

From Seneca, E. D., *Am. J. Bot.*, 61, 947, 1974. With permission.

of flowering for plants in experimental plots in Delaware; plants from Maine flowered earlier, and plants from Virginia did not flower until plants from Maine had completed their flowering process. These results indicated that genetic differentiation occurred in the latitudinal populations in regard to the flowering process. Both of the above-mentioned investigations appear to demonstrate clearly that latitudinal differences in environment have selected for specific ecoclinal variation in populations of *S. alterniflora*.[1,22]

A number of environmental factors have been hypothesized to induce the two extreme height forms in *S. alterniflora*, including soil salinity, nitrogen content of soil, sulfide levels, water-logging, and reduced aeration.[8,16] Germination response and seedling growth were determined for seeds collected from short, medium, and tall forms of *S. alterniflora*.[11] No differences in seed germination or seedling growth could be related to the parental height form of *S. alterniflora* in this investigation. Mooring et al.[11] concluded from their investigation that growth forms of *S. alterniflora* did not represent different local ecotypes, but were ecophenes resulting from exposure to different levels of soil salinity on the salt marsh. Positive correlations in plant height and biomass production with soil ammonia concentrations, and negative correlations with soil salinity were found in populations of *S. alterniflora* from Rhode Island by Nixon and Oviatt.[20] Greenhouse experiments with *S. alterniflora* indicated that maximum biomass production (52.8 g/pot) occurred in the low salt, high nitrogen-aerated treatment, while lowest biomass production (4.8 g/pot) was in the high salt (4.5% total salts), low N-unaerated treatment.[23] Plant density and mean height data followed the same pattern as the biomass data, with lowest values in the high salt, unaerated, and low N treatment. Linthurst and Seneca[23] concluded that differences between three environmental variables (salinity, nitrogen, and aeration) at different sites on the North Carolina salt marshes could explain most of the variation between the short and tall forms of *S. alterniflora*.

Delaune et al.[24] determined that both height growth and dry matter production of *S. alterniflora* decreased with distance from the stream bank in a Louisiana salt marsh, from 2185 g/m^2 in the streamside zone to 739 g/m^2 at an upland site 16 m from the stream. A significant correlation (r = 0.74) was found between soil bulk density and the dry weight of plants. Decreases in soil nutrient levels on a soil volume basis were negatively related to plant size as one moved from the tall form at the streamside to the short form of *S. alterniflora* at the more inland site. Yields of streamside plants increased 15% with nitrogen fertilization compared with untreated controls, indicating that nitrogen was a limiting element in this habitat.[25] Phosphorus increments did not increase the biomass production of *S. alterniflora*. Laboratory investigations by Haines and Dunn[26] indicated that seedlings were very plastic in their response to NaCl and NH_4-N treatments. Increases in nitrogen levels caused an increase in all growth parameters measured (height, leaf width, dry weight, rhizome length) under each of the salinity treatments, ranging from 0.5 to 4.0% NaCl. Production of both tall and short forms of *S. alterniflora* varied with the conditions of the solution cultures, supporting the view that height variation in the field is caused by environmental factors rather than by genetic selection for specific ecotypes.[26]

Gallagher[27] determined that an ammonium nitrate fertilization treatment of 200 kg/ha under field conditions would markedly enhance yields in the short form of *S. alterniflora* plants in a Sapelo Island (Georgia) salt marsh. Tissue N levels increased to about double the control values 10 weeks after treatment, but 1 year later nitrogen levels did not differ from untreated plants. After 1 year the yield of short-form plants increased 2.6 times that of untreated controls. Gallagher[27] concluded that nitrogen availability limited growth of plants in the short-form but not in the tall-form *S. alterniflora* zone. The total nitrogen content of leaves and roots of the short-form plants was lower than that of the tall form. However, Mendelssohn[28] found that both soil NH_4-N and total N were higher in the short-form zone than in the tall-form zone in a North Carolina salt marsh. He suggested that soil

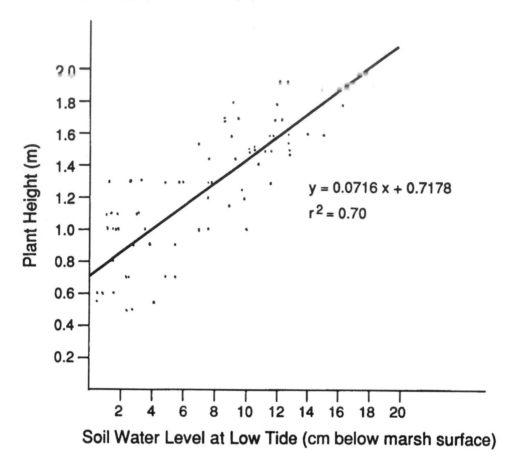

FIGURE 3.   Relationship between the height of *Spartina alterniflora* plants and soil water at low tide in a North Carolina salt marsh. (From Mendelssohn, I. A., McKee, K. L., and Postek, M. T., *Wetlands Ecol. Manage.*, 1, 223, 1982. With permission.)

waterlogging in the short-form *S. alterniflora* zone may reduce the ability of plants to take up nitrogen. The precise mechanism for the inhibition of nitrogen uptake is not known, but oxygen deficiencies, iron availability, increased sulfides, and other soil toxins may all be directly or indirectly involved in inhibiting nitrogen utilization by *S. alterniflora* in North Carolina salt marshes.[28] Ammonium nitrogen fertilization of field plots increased the aerial biomass of short-form plants up to 2.7 times of nonfertilized controls, while nitrate-N treatments caused a 2.0-times increase in biomass yield in a North Carolina salt marsh.[29] Similar fertilization treatments caused a 1.4-times increase in the tall form of *S. alterniflora*. These data also indicate that the availability of nitrogen could be a limiting factor in the short-form habitats. Mendelssohn[29] determined that short-form field plants had lower leaf and root nitrogen content than did the tall-form, and root glutamate dehydrogenase activity was twice as great in the tall compared to the short form.

Height growth of *S. alterniflora* under field conditions in North Carolina salt marshes was related ($r^2 = 0.70$) to the degree of soil drainage and aeration in stands (Figure 3).[30] When soil drainage was experimentally retarded in the field sites, total biomass production of the tall form was only 27% of the controls, medium form was 10% of the control, and short form was 36% of the control (Table 6). Similar differences were obtained for a reduction in height growth of the tall and medium, but no statistically significant height reduction was observed in the short-form plants exposed to reduced drainage. Mendelssohn and Seneca[30]

TABLE 6
The Effect of Differential Soil Drainage on the Height (cm), Density (Culms pot^{-1}), Living Aerial, Root, Rhizome, Total, and Dead Root plus Rhizome Standing Crops (g pot^{-1}), and Underground to Aerial (U/A) Ratio of the Tall-, Medium-, and Short-Height Forms of *Spartina alterniflora*

Drainage treatment by height form[a]	Height	Density	Aerial	Root	Rhizome	Total biomass	Dead root + rhizome	U/A
Tall form								
Completely drained	74b	14c	26b	6a	21c	54c	26b	1.04a
Medium drained	89a	17bc	53a	5a	27b	85b	65a	0.60b
Undrained	90a	23a	57a	5a	39a	100a	82a	0.77b
Medium form								
Completely drained	72a	24b	20b	2b	12c	38c	48b	0.70b
Medium drained	62b	24a	42a	4b	28b	73b	81a	0.76b
Undrained	73a	26a	47a	9a	40a	96a	86a	1.04a
Short form								
Completely drained	51a	22c	31ab	11a	18a	60a	130a	0.94a
Medium drained	46a	27b	26b	7b	11b	44b	130a	0.69a
Undrained	39b	40a	36a	8b	15ab	59a	124a	0.64b

[a] Values within a height form followed by different letters are significantly different at the 0.05 probability level.

From Mendelssohn, I. A. and Seneca, E. D., *Estuarine Coast. Mar. Sci.*, 11, 27, 1980. With permission.

TABLE 7
Summary of Biological Variable Mean for *Spartina alterniflora* by Plant Source and Aeration

Treatment		n	Density (culms/pot)	Aerial biomass (g/pot)	Height (cm)	Leaf width (mm)	Root biomass (g/pot)
Tall	A	4	14	44.4	138.6	16.9	45.4
Tall	U	4	3	4.5	33.2	5.4	10.6
Short	A	4	14	30.1	118.2	13.6	32.1
Short	U	4	6	4.4	41.5	6.0	8.6
Pioneer	A	4	17	31.5	108.8	13.5	42.7
Pioneer	U	4	9	8.3	72.8	8.3	12.9
LSD$_{05}$			7	14.9	35.8	4.3	23.1

From Linthurst, R. A., *Am. J. Bot.*, 67, 883, 1980. With permission.

concluded that drainage and aeration were significant factors controlling plant growth under field conditions in these salt marshes. Laboratory investigations with tall- and short-form *S. alterniflora* indicated that unaerated treatments of the tall form had about 10% of the aerial biomass of the aerated treatment, while short-form plants had yields that were 7% of the aerated treatment (Table 7).[31] Height growth for the tall form was 24% of that for aerated plants, while for the short form it was 35% of aerated treatments. A similar reduction in leaf width and root biomass was found by Linthurst[31] in both height forms of *S. alterniflora* in unaerated treatments (Table 7).

IV. ORIGIN OF MICROSPECIES

Microspecies can develop in salt marsh habitats because of geographic or temporal isolation of populations. Another factor influencing reproductive isolation is that in some genera such as *Salicornia* there is apparently selection for a breeding system that favors selfing.

Eleuterius[32] reported that a population of *Juncus roemerianus* from a Louisiana salt marsh produced a very distinct dwarfed form of the plant in a hypersaline habitat, which differed from the other six populations investigated. Reproductive isolation in *J. roemerianus* populations could be due to a number of factors, including geographic features, restricted pollen dispersal, and differences in salt tolerance of genotypes. The dwarfed salt flat population was exposed to salinity levels >6.5% total salts. Reproductive isolation of this population from the others could be because its plants flower earlier in the spring than do the other populations and have a relatively short, 1-month flowering period. A mosaic of genetically distinct populations may be selected for in salt marshes dominated by *J. roemerianus*. Genetic differences could occur in populations because of geographic or temporal isolation, inbreeding, and the selective pressure of hypersaline conditions on populations of *J. roemerianus*.[32] Reciprocal transplants in the field and common garden experiments indicated that the populations of *J. roemerianus* were genetically distinct.[33] Mean leaf length of plants from the dwarfed population decreased from 66.2 to 34.5 cm when they were transplanted at their home site, but increased to 72.2 cm when transplanted to the site with the least salt stress. Plants normally growing in favorable habitats had a mean leaf length of 215.6 cm and were not significantly inhibited by transplanting in place. However, all of the plants transplanted to the high salt habitat died, demonstrating that this population was less salt tolerant than the dwarf population. Common garden experiments in a nonsaline environment indicated that the dwarf plants remained smaller, 137.6 cm, than those plants originating from less saline habitats, 210.4 cm, even after 3 years of cultivation. Dry mass of dwarf plants increased from 44.9 to 65.5% of that for tall plants by the third year in the uniform garden plots. Eleuterius[33] concluded that ecotypic differentiation occurred in the salt marsh populations of *J. roemerianus*, as was indicated by the persistence of plant form and survival of plants in the more salt-stressed habitat.

Jefferies et al.[34] described differences in rates of growth for two populations of diploid *Salicornia europaea* from upper- and lower-marsh portions of the Stiffkey salt marsh (England). Plants from the upper-marsh population had slower growth rates and produced less biomass than did plants from the lower-marsh population. Reciprocal transplants indicated that these differences were persistent, and the transplants maintained similar growth form and biomass yields to the plants from the sites in which they originated. Similar results were obtained in a reciprocal transplant investigation by Ungar[35] with *Salicornia* populations from two locations on an inland Ohio salt pan. Growth patterns and biomass yields resembled the site of origin more closely than plants from the site into which they were transplanted (Table 8). These data from field transplant investigations indicate that genetic differences could occur between plants from the two populations of *Salicornia* growing on the same salt marsh. Electrophoretic variability was estimated for 15 populations of diploid *Salicornia* from British coastal marshes.[36] Based on morphological characteristics, the upper-marsh form could be classified as the microspecies *S. ramosissima* and the lower-marsh form as *S. europaea*. The electrophoretic mobilities for 24 of the enzymes were identical in all of the plants investigated from upland and lowland sites. Six of the isozymes had allozyme variants at the gene loci. At three sites where plants were collected from upper and lower marshes, greater than 65% of the individuals in the lowland sites had one enzymatic pattern: even though these plants occur on the same salt marsh more than 70% of the individuals at the upland sites had a second enzymatic pattern at six loci (AAT-1, ME, PGM-1, PGM-2, SKDH-1, SKDH-2; Table 9). No heterozygotes were found, and these plant microspecies represent different homozygous lineages which maintain reproductive isolation because of a cleistogamous breeding system. Lower nitrogen levels, periodic hypersalinity, and competition from perennials on the upper marsh at Stiffkey have apparently selected for plants

TABLE 8
Characteristics of Controls and Reciprocal Transplants of *Salicornia europaea* from the Edge *Salicornia* and Tall *Salicornia* Zones on October 28, 1981[a]

	Controls		Transplants (origin)	
	Edge	Tall	Edge	Tall
Sample size (N)	10	10	25	25
Height (cm)	17.1 ± 2.0	21.4 ± 1	17.2 ± 3.9	20.7 ± 2.1
No. vegetative internodes per main shoot	13.1 ± 1.3	14.0 ± 2.3	11.5 ± 2.2	15.0 ± 2.3
Branches per plant				
Nodes with ≥1 branch (%)	1.4	48.6	18.1	39.7
No. nodes with branches	0.2 ± 0.4	7.6 ± 1.1	2.1 ± 2.1	6.0 ± 2.4
Shoot dry mass (g)	0.11 ± 0.02	0.31 ± 0.03	0.13 ± 0.02	0.32 ± 0.03
No. floral branches	0.3 ± 0.5	3.8 ± 0.4	4.2 ± 3.7	10.7 ± 4.9
No. floral internodes	13.4 ± 3.6	41.2 ± 10.3	21.4 ± 15.5	45.6 ± 16.5

[a] Means ± SE.

From Ungar, I. A., *Ecology*, 68, 569, 1987. With permission.

TABLE 9
Relative Mobilities of Enzymes in *Salicornia europaea* and *S. ramosissima* When the Front Migrates (100 mm)

	S. europaea (mm)	*S. ramosissima* (mm)
Aspartate amino transferase 1	67	65
Phosphoglucomutase 1	42	38
Phosphoglucomutase 2	34	37
Malic enzyme	25	18
Shikimate dehydrogenase 1	49	39
Shikimate dehydrogenase 2	29	34

From Jefferies, R. L. and Gottlieb, L. D., *N. Phytol.*, 92, 123, 1982. With permission.

in the microspecies *S. ramosissima*, which have low fecundity and slow growth rates. In the more variable and unpredictable lower marsh, selection has been for plants which show continuous growth throughout the summer and high fecundity, *S. europaea*.[36] The diploid species of *Salicornia* on British salt marshes are difficult to distinguish morphologically because they have a great deal of phenotypic plasticity. However, the populations can be distinguished genetically by use of electrophoretic evidence. Wolff and Jefferies[37,38] reported that isozyme data were also useful in distinguishing the species of *Salicornia* in northeastern North America. Lack of genetic variability in *Salicornia* populations was suggested to be due to a number of factors: an autogamous breeding system, founder effects, and isolation. A numerical taxonomic investigation was carried out for diploid *S. europaea* populations collected from upper-, middle-, and lower-marsh locations on British salt marshes.[39] Small-scale variation in plant form could be found at sites within a location, but results between sites were not consistent regarding morphological characteristics. Ingrouille and Pearson[39] concluded that the variation in form of *S. europaea* plants at salt marsh locations was probably the result of responses of plants to microedaphic conditions. Morphological recognition of microspecies was not possible for this *S. europaea* aggregate, because of the variability of morphological characteristics between populations. They stated that the establishment of genetically defined species, which are homozygous and do not interbreed, would require more genetic analysis and a better definition of morphological and developmental distinctions between populations.

V. CONCLUSIONS

The selection for more tolerant genotypes is a critical factor in the survival of halophytes in salt marsh habitats. Resistance to a number of forms of stress, including salt stress (ionic and osmotic), waterlogging, poor aeration, sulfide accumulation, and restrictions in the availability of nutrients such as nitrogen, are necessary for plant survival, growth, and reproduction along salinity gradients. More research is needed to determine the precise physiological mechanisms which are selected for in ecotypes that are successful in saline habitats. At the seed germination stage, dormancy may be the best mechanism for avoiding hypersaline conditions,[40] because seedlings would be exposed to conditions that were unfavorable if seeds germinated at high salinity and the stressful environment would cause high mortality. At the developmental stage where most energy is put into vegetative growth, mechanisms are needed for restricting Na and Cl uptake as well as maintaining adequate rates of metabolic activity. The costs of salt resistance and the evolutionary consequences need to be considered. Although some data are available concerning the mechanisms that increase salt tolerance in some crop species, little research has been carried out regarding the costs and precise mechanisms that favor ecotypic differentiation of highly salt-tolerant halophytes.[41-43]

REFERENCES

1. **Seneca, E. D.**, Germination and seedling response of Atlantic and Gulf coasts populations of *Spartina alterniflora, Am. J. Bot.*, 61, 947, 1974.
2. **Ahmad, I. and Wainwright, S. J.**, Ecotype differences in leaf surface properties of *Agrostis stolonifera* from salt marsh, spray zone and inland habitats, *N. Phytol.*, 76, 361, 1976.
3. **Jerling, L.**, Genetic differentiation in fitness related characters in *Plantago maritima* along a distributional gradient, *Oikos*, 53, 341, 1988.
4. **Stalter, R. and Batson, W. T.**, Transplantation of salt marsh vegetation, Georgetown, South Carolina, *Ecology*, 50, 1087, 1969.
5. **Silander, J. A. and Antonovics, J.**, The genetic basis of the ecological amplitude of *Spartina patens*. I. Morphometric and physiological traits, *Evolution*, 33, 1114, 1979.
6. **Gray, A. J. and Scott, R.**, A genecological study of *Puccinellia maritima* Huds. (Parl.). I. Variation estimated from single-plant samples from British populations, *N. Phytol.*, 85, 89, 1980.
7. **Wu, L.**, The potential for evolution of salinity tolerance in *Agrostis stolonifera* L. and *Agrostis tenuis* Sibth, *N. Phytol.*, 89, 471, 1981.
8. **Gallagher, J. L., Somers, G. F., Grant, D. M., and Seliskar, D.**, Persistent differences in two forms of *Spartina alterniflora:* a common garden experiment, *Ecology*, 69, 1005, 1988.
9. **Workman, J. P. and West, N. E.**, Germination of *Eurotia lanata* in relation to temperature and salinity, *Ecology*, 48, 659, 1967.
10. **Cavers, P. B. and Harper, J. L.**, The comparative biology of closely related species living in the same area. IX. *Rumex:* the nature of adaptation to a seashore habitat, *J. Ecol.*, 55, 73, 1967.
11. **Mooring, M. T., Cooper, A. W., and Seneca, E. D.**, Seed germination response and evidence for height ecophenes in *Spartina alterniflora* from North Carolina, *Am. J. Bot.*, 58, 48, 1971.
12. **Bazzaz, F. A.**, Seed germination in relation to salt concentration in three populations of *Prosopis farcta, Oecologia*, 13, 73, 1973.
13. **Kingsbury, R. W., Radlow, A., Mudie, P. J., Rutherford, J., and Radlow, R.**, Salt stress in *Lasthenia glabrata* a winter annual composite endemic to saline soils, *Can. J. Bot.*, 54, 1377, 1976.
14. **Silander, J. A.**, The genetic basis of the ecological amplitude of *Spartina patens*. III. Allozyme variation, *Bot. Gaz.*, 145, 569, 1984.

15. **Silander, J. A.,** The genetic basis of the ecological amplitude of *Spartina patens*. II. Variance and correlation analysis, *Evolution,* 39, 1034, 1985.

16. **Anderson, C. M. and Treshow, M.,** A review of environmental and genetic factors that affect height in *Spartina alterniflora* Loisel. (salt marsh cord grass), *Estuaries,* 3, 168, 1980.

17. **Mendelssohn, I. A., McKee, K. L., and Postek, M. T.,** Sublethal stresses controlling *Spartina alterniflora,* productivity, *Wetlands Ecol. Manage.,* 1, 223, 1982.

18. **Adams, D. A.,** Factors influencing vascular plant zonation in North Carolina salt marshes, *Ecology,* 44, 445, 1963.

19. **Shea, M. L., Warren, R. S., and Niering, W. A.,** Biochemical and transplantation studies of the growth form of *Spartina alterniflora* on Connecticut salt marshes, *Ecology,* 56, 461, 1975.

20. **Nixon, S. W. and Oviatt, C. A.,** Analysis of local variation in the standing crop of *Spartina alterniflora,* *Bot. Mar.,* 16, 103, 1973.

21. **Valiela, I., Teal, J. M., and Deuser, W. G.,** The nature of growth forms in the salt marsh grass *Spartina alterniflora,* *Am. Nat.,* 112, 461, 1978.

22. **Somers, G. F. and Grant, D.,** Influence of seed source upon phenology of flowering of *Spartina alterniflora* Loisel. and the likelihood of cross pollination, *Am. J. Bot.,* 68, 6, 1981.

23. **Linthurst, R. and Seneca, E. D.,** Aeration, nitrogen, and salinity as determinants of *Spartina alterniflora* Loisel. growth response, *Estuaries,* 4, 53, 1981.

24. **DeLaune, R. D., Buresh, R. J., and Patrick, W. H.,** Relationship of soil properties to standing crop biomass of *Spartina alterniflora* in a Louisiana marsh, *Estuarine Coast. Mar. Sci.,* 8, 477, 1979.

25. **Patrick, W. H. and Delaune, R. D.,** Nitrogen and phosphorus utilization by *Spartina alterniflora* in a salt marsh in Barataria bay, Louisiana, *Estuarine Coast. Mar. Sci.,* 4, 59, 1976.

26. **Haines, B. L. and Dunn, E. L.,** Growth and resource allocation responses of *Spartina alterniflora* Loisel. to three levels of NH_4-N, Fe, and NaCl in solution culture, *Bot. Gaz.,* 137, 224, 1976.

27. **Gallagher, J. L.,** Effect of an ammonium nitrate pulse on the growth and elemental composition of natural stands of *Spartina alterniflora* and *Juncus roemerianus,* *Am. J. Bot.,* 62, 644, 1975.

28. **Mendelssohn, I. A.,** The influence of nitrogen level, form, and application method on the growth response of *Spartina alterniflora* in North Carolina, *Estuaries,* 2, 106, 1979.

29. **Mendelssohn, I. A.,** Nitrogen metabolism in the height forms of *Spartina alterniflora* in North Carolina, *Ecology,* 60, 574, 1979.

30. **Mendelssohn, I. A. and Seneca, E. D.,** The influence of soil drainage on the growth of salt marsh cordgrass *Spartina alterniflora* in North Carolina, *Estuarine Coast. Mar. Sci.,* 11, 27, 1980.

31. **Linthurst, R. A.,** A growth comparison of *Spartina alterniflora* Loisel. ecophenes under aerobic and anaerobic conditions, *Am. J. Bot.,* 67, 883, 1980.

32. **Eleuterius, L.,** Taximetric analysis of female and hermaphroditic plants among populations of *Juncus roemerianus* under different salinity regimes, *J. Coastal Res.,* 5, 29, 1989.

33. **Eleuterius, L.,** Natural selection and genetic adaptation to hypersalinity in *Juncus roemerianus* Scheele, *Aquatic Bot.,* 36, 45, 1989.

34. **Jefferies, R. L., Davy, A. J., and Rudmik, T.,** Population biology of the salt-marsh annual *Salicornia europaea* agg., *J. Ecol.,* 69, 1, 1981.

35. **Ungar, I. A.,** Population characteristics, growth, and survival of the halophyte *Salicornia europaea,* *Ecology,* 68, 569, 1987.

36. **Jefferies, R. L. and Gottlieb, L. D.,** Genetic differentiation of the microspecies *Salicornia europaea* L. (sensu stricto) and *S. ramosissima* J. Woods, *N. Phytol.,* 92, 123, 1982.

37. **Wolff, S. L. and Jefferies, R. L.,** Morphological and isozyme variation in *Salicornia europaea* (s.l.) (Chenopodiaceae) in northeastern North America, *Can. J. Bot.,* 65, 1410, 1987.

38. **Wolff, S. L. and Jefferies, R. L.,** Taxonomic status of diploid *Salicornia europaea* (s.l.) (Chenopodiaceae) in northeastern North America, *Can. J. Bot.,* 65, 1420, 1987.

39. **Ingrouille, M. J. and Pearson, J.,** The pattern of morphological variation in the *Salicornia europaea* (L.) aggregate (Chenopodiaceae), *Watsonia,* 16, 269, 1987.

40. **Ungar, I. A.,** Germination ecology of halophytes, in *Tasks for Vegetation Science,* Sen, D. N. and Rajpurohit, K. S., Eds., Junk, The Hague, 1982, Chap. 3.

41. **Waisel, Y.,** *Biology of Halophytes,* Academic Press, New York, 1972.

42. **Staples, R. C. and Toenniessen, G. H.,** *Salinity Tolerance in Plants,* John Wiley & Sons, New York, 1984.

43. **Kik, C.,** Ecological genetics of salt resistance in the clonal perennial *Agrostis stolonifera* L , *N. Phytol.,* 113, 453, 1989.

Chapter 11

COMPETITION

I. INTRODUCTION

A review of the literature on competition experiments under field conditions was compiled by Schoener[1] and indicated that interspecific competition plays a significant role in determining the distribution of species in ecosystems. The competitive capacity of species varies in different plant communities, since selection for competitiveness differs with the amount of physical stress and disturbance that is present in a particular ecosystem. Highly stressful habitats tend to have a low species diversity, since only a small number of species have acquired the adaptations necessary to cope with the most extreme environments. Grime[2] has hypothesized that the relative competitiveness of plant species varies with the habitat in which they are found. His model predicted that the most competitive species would be found in productive habitats with crowded vegetation in which physical stress and disturbance were not extreme. The zonal communities of salt marshes are characterized by low species diversity when compared with plant communities occurring in nonsaline habitats in the same climatic region. The lack of success in establishment of halophytic species in nonsaline soils is related to three factors that make them poor competitors: a slow growth rate, a low stature, and the lack of drought resistance in many of the salt marsh species in comparison with glycophytes. Halophytes have relatively high light requirements, and reductions in available illumination by taller competitors would eliminate many of the halophytes from nonsaline habitats.

The influence of competition as a factor in the development of zonational patterns in salt marsh habitats has been suggested by a number of investigators, including Purer,[3] Hinde,[4] Clarke and Hannon,[5] Pielou and Routledge,[6] Vince,[7] Ungar et al.,[8] and Bertness and Ellison.[9] More specifically, Purer,[3] Hinde,[4] and Ungar et al.[8] hypothesized that the upper limits for the distribution of halophytic species in less saline habitats is determined by competition with facultative halophytes and glycophytes. However, Ungar[10] and Barbour[11] proposed that the limit for a species distribution in the more highly saline environments was probably because most of the organisms were unable to tolerate physiologically to the physical conditions in the more stressful habitat. Other researchers concluded that mortality of plants in inland and coastal saline habitats was primarily due to abiotic factors, such as tidal abrasion or siltation, waterlogging and associated soil factors, and salt stress.[8,12-15]

Bertness and Ellison[9] determined from field observations and experiments that zonation in a Rhode Island salt marsh was controlled by a complex of factors: abiotic stress, predation pressure, disturbance of the habitat, and interspecific competition. Partridge and Wilson[16] used field transplants to determine the tolerance to environmental stress of halophytes on an Otago, New Zealand salt marsh. They concluded that competition was limiting to the distribution of halophytes in both the least and most saline habitats.

This chapter reviews the current status of field and laboratory investigations regarding the significance of interspecific and intraspecific competition in determining the survival and zonation of halophytes in saline environments.

II. INTERSPECIFIC COMPETITION

Current research on the subject of interspecific competition in saline habitats has been focused on determining if competition is a significant factor in stressful, hypersaline environments. Field investigations have used a number of techniques to determine if resources are limiting because of the impact of interspecific competition. The major problem in this research involves distinguishing competitive effects from stress caused by the physical environment. Examples of experimental procedures include transplanting species into new environments,[9] removal of species from a habitat,[17] removal of shading plants and artificial shading of target plants to determine if light is limiting,[18] and nutrient enrichment investigations.[19]

Bertness and Ellison[9] determined from transplantation of marsh perennials to low-marsh and high-marsh habitats that all of the species had their poorest growth in the low marsh and best growth on the terrestrial border of the high marsh. The lack of correlation between the position of a dominant species on the marsh and where it made optimal growth indicates that abiotic factors might not be controlling plant distribution and that interspecific competition could be playing a significant role in the zonational dominance of species in the Rhode Island marshes investigated.[9]

Other researchers have also found significant differences between the physiological limits of tolerance of plant species and their distribution in salt marsh habitats.[4,6,20] Few field experiments have been designed to test the direct effect of competition on species transplanted into habitats in which they do not normally grow. Ungar et al.[8] and Ungar[21] carried out transplant studies in cleared and uncleared plots in various zones on an inland Ohio salt pan to determine if the physiological limits of a species were reflected in the distribution of plants and the role that competition played in limiting the distribution of plant species along a salinity gradient. It was determined that both *Salicornia europaea* and *Atriplex triangularis* made optimal growth in less saline zones than those in which they were the dominant species (Figure 1), indicating that interspecific competition with less salt-tolerant species was limiting their distribution. Abiotic factors were limiting on the barren pan, and both species had high mortality and low biomass yields in zones with higher soil salinity than occurred in their original habitats (Table 1).

Based on laboratory seed germination experiments and field survival data at marsh and dune sites in the Frisian Islands (Netherlands), Bakker et al.[22] determined that glycophytic species could germinate on the lower marsh, but their seedlings were not capable of competing successfully with the halophytes. Similarly, seeds of halophytes could germinate on abandoned salt marshes, but were not successful because their seedlings were not competitive with glycophytes. It was concluded that species such as *Gallium mollugo, Cochlearia danica, Plantago lanceolata,* and *Bromus hordeaceus* of the upper salt marsh and dunes could not disperse into the lower marsh because of their low salt tolerance. On the other hand, lower and midmarsh species such as *A. prostrata, Limonium vulgare,* and *S. europaea* could invade open areas of the upper marsh and dunes when the plant canopy was sufficiently open to permit the penetration of light to the soil surface.[22]

Partridge and Wilson[16] found that halophyte transplants grew well in nonsaline habitats and concluded that their absence from less saline environments was due to the fact that they were poor competitors in the presence of glycophytes. At the lower boundary of the salt marsh, the limiting factors controlling plant distribution were either specific seedling establishment requirements or the tolerance limits of adult plants. Most of these transplant investigations involved the movement of adult plants from one marsh zone to another. It could be that the germination and seedling establishment phases of a species' life history are more critical in determining plant zonation than responses of well-established adult plants. Further investigations should be carried out by using transplants from different developmental stages to determine the tolerance limits and competitive ability of halophytes in different zones of the salt marsh.

Vince and Snow[23] concluded that the distribution of species in zonal communities may indicate that each plant species can find optimal growth situations in different portions of salt marshes (Table 2). They caution that experiments must be conducted to determine the significance of both biotic and abiotic factors in the development of zonational patterns.

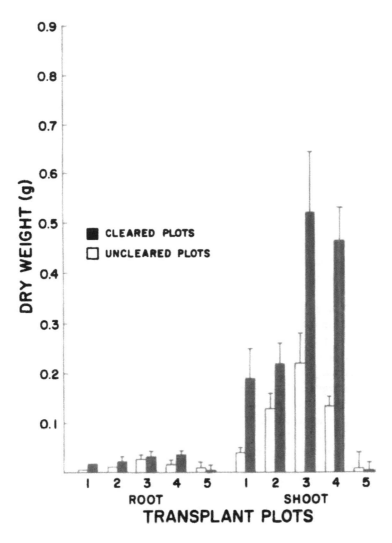

FIGURE 1.   Root and shoot dry mass of *Salicornia europaea* plants in five zones
along a gradient from low (5) to high (1) salinity. (From Ungar, I. A., Benner,
D. K., and McGraw, D. C., *Ecology*, 60, 329, 1979. With permission.)

Laboratory experiments with marsh species, which were characteristic of the different zones
in the subarctic Susitana Flats salt marsh (Alaska), demonstrated that all species made optimal
growth in the least-saline waterlogged treatment.[20] Reciprocal transplants in the field demonstrated that several species found in less saline habitats, *Carex lyngbyaei*, *C. ramenskii*,
and *Poa emurens*, were inhibited in the mudflat zone where salinities averaged 3.5% total
soluble salts. *Triglochin maritima* showed no significant difference in biomass yields at any
of the transplant sites along the marsh gradient. *Puccinellia nutkanensis* had higher biomass
production in three other marsh zones than in the zonational community it dominated; its
growth was inhibited by edaphic conditions in the portion of the marsh which it dominated.
Snow and Vince[20] concluded that physical gradients could limit a species distribution, because
the physiological limits of tolerance for populations to edaphic conditions were reached at
some portion of the gradient. Competition might be the limiting factor at less saline sites
where many species could potentially survive and reproduce. In terms of niche theory, the
fundamental niche of marsh species appeared to be broader than realized, indicating that

TABLE 1
Survival Percentages of *Salicornia europaea* and *Atriplex triangularis* Transferred to Cleared and Uncleared Plots in Five Salt Marsh Zones

Zone	Soil salinity (MS/cm)	*Salicornia europaea*		*Atriplex triangularis*	
		Cleared	Uncleared	Cleared	Uncleared
Salt pan	>40	12 ± 12	14 ± 6	0 ± 0	0 ± 0
Salicornia	33	70 ± 11	59 ± 9	0 ± 0	10 ± 5
Atriplex	13	90 ± 4	82 ± 8	26 ± 8	74 ± 9
Hordeum	9	66 ± 8	62 ± 6	72 ± 13	76 ± 14
Meadow	6	0 ± 0	14 ± 7	42 ± 16	54 ± 15

From Ungar, I. A., Benner, D. K., and McGraw, D. C., *Ecology*, 60, 329, 1979; and Ungar, I. A., in *Symp. Biology of Atriplex and Related Chenopods*, McArthur, E. D., Stutz, H. C., Stevens, K. R., and Johnson, K. L., Eds., Gen. Tech. Rep. INT-172, Forest Service, U.S. Department of Agriculture, Provo, UT, 1984, 40.

biotic interference could play a significant role in determining plant distribution when edaphic factors are well within a species' tolerance limits.[6,9-11,19-22]

A number of laboratory investigations have been conducted to determine the competitive ability of halophytes with glycophytes under various environmental conditions. Cords[24] reported that perennial grasses such as tall fescue and tall wheatgrass outcompete the more salt-tolerant *H. jubatum* in less saline conditions. *Hordeum jubatum* was found to make optimal growth when grown alone under high soil moisture, high fertility, and low soil salinity.[25] A glycophyte, *Dactylis glomerata* was found to restrict *H. jubatum* growth in nonsaline soils, but it offered little competition when grown in wet saline soils. Suehiro and Ogawa[26] reported that the glycophyte *Chenopodium album* produced more biomass than *Atriplex gmelini* when plants were grown in mixed stands and at lower salinities. However, at salinities of 0.25- to full-strength seawater *A. gmelini* produced greater biomass than *C. album* when the two species were grown together. Results of a replacement series experiment indicated that *A. gmelini* had its highest biomass production at 0.25-strength seawater in pure culture, but this shifted to an optimum production at 0.45-strength seawater in mixed culture; while *C. album* had an optimum biomass yield at 0.25-strength seawater in pure culture, this shifted to a lower salinity for optimum biomass production at 0.12-strength seawater in mixed culture. The halophyte, *A. gmelini*, was the better competitor at all salinities ranging from 0.14- to full-strength seawater, while *C. album* had higher biomass production than *A. gmelini* only at the lowest range of seawater concentration below 0.14-strength seawater. Suehiro and Ogawa[26] concluded from their investigations of interspecific competition between a glycophyte and halophyte that the behavior of species in mixed stands cannot always be predicted from a species' performance in pure stands.

The growth of an herbaceous perennial halophyte, *Jaumea carnosa*, decreased 52% in comparison with controls, when it was grown with the glycophyte *Lolium perenne* at low salinity (400 mg/l). Under high-salinity (11,600 mg/l) treatments, the inhibitory effect of *L. perenne* was not apparent (Table 3).[27] *J. carnosa* made its best growth in the nonsaline controls, which indicates that its physiological optimum for growth was under conditions in which it was not found growing under natural field conditions. The fact that the physiological limits of tolerance of halophytic species are often broader than natural plant distributions indicate supports the hypotheses of Ungar[10] and Barbour,[11] which stated that halophytes could be restricted to salt marsh habitats because of their inability to compete with glycophytes in nonsaline habitats. Therefore, the success of halophytes in saline habitats is probably due to the fact that soil salinity levels are beyond the tolerance limits for

TABLE 2
Above-Ground Biomass (g/m²) at Study Sites on Susitna Flats, AK

Species	Outer mudflats		Outer sedge marsh	Inner mudflats		Riverbank levee	Inner sedge marsh	
	1	2	3	4	5	6	7	8
Triglochin maritimum	68 ± 16	—	40 ± 19	285 ± 26	173 ± 30	—	42 ± 37	—
Puccinellia phryganodes	41 ± 6	4 ± 1	—	—	—	—	—	—
P. nutkaensis	3 ± 1	43 ± 8	—	—	—	—	—	—
Suaeda depressa	—	3 ± 2	—	—	—	—	—	—
Carex ramenskii	—	—	267 ± 12	19 ± 3	6 ± 1	—	—	—
Potentilla egedii	—	—	19 ± 6	115 ± 20	21 ± 1	69 ± 6	—	—
Plantago maritima	—	—	—	—	149 ± 27	—	—	—
Glaux maritima	—	—	—	—	33 ± 5	—	—	—
Poa eminens	—	—	—	—	28 ± 5	185 ± 36	—	—
Festuca rubra	—	—	—	—	—	65 ± 13	—	—
Ligusticum scoticum	—	—	—	—	—	11 ± 8	—	—
Lathyrus palustris	—	—	—	—	—	6 ± 3	—	—
Carex lyngbyaei	—	—	—	—	—	—	375 ± 6	1 ± 1
Carex spp.	—	—	—	—	—	—	45 ± 20	—
C. pluriflora	—	—	—	—	—	—	4 ± 0	217 ± 31
Total	112 ± 19	50 ± 10	326 ± 21	419 ± 17	410 ± 20	336 ± 47	466 ± 57	218 ± 30

From Vince, S. W. and Snow, A. A., *J. Ecol.*, 72, 651, 1984. With permission.

TABLE 3
The Effect of Competition on Dry Mass of *Jaumea*
carnosa and *Lolium Perenne*

NaCl (mg/l)	Planting ratio (J:L)	*Jaumea* dry mass/flat (g)	*Lolium* dry mass/flat (g)
400	30:0	0.21	—
	Mix	0.09	10.28
	0:300	—	12.46
4,000	30:0	0.13	—
	Mix	0.11	7.37
	0:300	—	7.41
11,600	30:0	0.11	—
	Mix	0.11	4.82
	0:300	—	4.94

Note: Mix planting is a pool of ratios of 20:100, 15:150, and 10:200 (J:L).

From Barbour, M. G., *Oecologia*, 37, 93, 1978. With permission.

glycophytes;[10,11,28] because of salt stress, interspecific competition with intolerant species is no longer a significant factor in determining a species' growth and distribution.

The salt-tolerant species *Hordeum marinum* and *Chloris gayana* were inhibited under nonsaline conditions by the glycophyte *Triticum vulgare* cv. *Florence*.[29] Competitive inhibition by wheat in nonsaline media increased with fertilization, indicating that competition was for nutrients and not light in these experimental treatments. Szwarcbaum and Waisel[29] determined that the competitive inhibition of the two more salt-tolerant species by wheat was eliminated under saline conditions.

Species distributions in the zonational communities of the Tijuana Estuary salt marshes, California were reported to be broadly overlapping by Zedler.[30] Only *Spartina foliosa* was an exception, and a statistical treatment of distribution data indicated that it had no positive associations but several negative associations with eight of the other marsh species. The lower limits of distribution for *Salicornia virginica* were previously suggested to be controlled by competitive inhibition from *Spartina foliosa*.[3,4] Zedler[30] hypothesized that the limits of distribution for *Salicornia virginica* in the middle of its range are determined by interspecific competition with *Batis maritima* and *Salicornia bigelovii*. Covin and Zedler[19] studied the effect of nitrogen fertilization on the interaction between *Spartina foliosa* and *Salicornia virginica* on the Tijuana Estuary salt marshes. Enrichment of the salt marsh with urea increased the biomass yields of both species in monocultures (13% for *S. virginica*, 43% for *Spartina foliosa*). In competition studies, *Salicornia* biomass increased 47%, while *Spartina* increased only 8% with fertilization. It was found that *Salicornia virginica* significantly increased in biomass production when urea fertilization was used with or without the presence of *Spartina foliosa*. In contrast, the removal of *S. foliosa* from fertilization-treated plots did not stimulate an increase in *Salicornia virginica* production. However, the presence of *S. virginica* in enriched treatments caused a 49% decrease in the yield of *Spartina foliosa*, demonstrating that it used nitrogen more efficiently than *S. foliosa* (Table 4, Chapter 9).

Gray and Scott[31] carried out a de Wit[32] replacement series experiment in the laboratory to determine the effect of competition on the zonational pattern on the Morecambe Bay salt marshes, England. They determined that the competitive ability of *Puccinellia maritima*, *Festuca rubra*, and *Agrostis stolonifera* varied with soil water level (Figure 2) and seawater concentration of the medium. The halophyte *P. maritima* was more competitive than

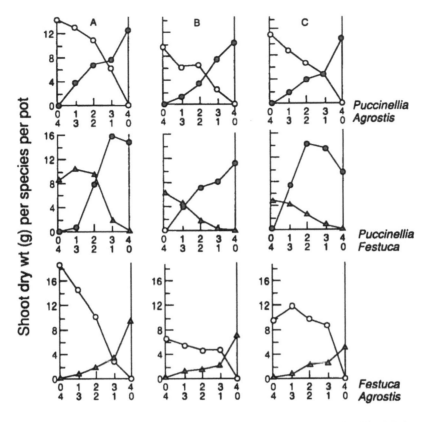

FIGURE 2.   Effects of soil water level on competition between *Puccinellia* (filled circle), *Festuca* (triangle), and *Agrostis* (open circle) at various plant ratios and water levels: (A) pot base, (B) 5 cm below soil surface, (C) at soil surface. (From Gray, A. J. and Scott, R., *J. Appl. Ecol.*, 14, 229, 1977. With permission.)

the other two species under seawater treatments. *Agrostis* was the best competitor in freshwater conditions, and *Festuca* was a poor competitor with the other two species when grown at high water tables. Small-scale distribution of the three species on the salt marshes was concluded to be caused by differences in their competitive abilities under various waterlogging and salinity conditions.[31]

Perturbation studies were carried out by Silander and Antonovics[17] to determine if removal of species on a coastal barrier island in North Carolina affected the niche breadth of the target species. In the majority of the cases studied, coexisting species increased in importance after the removal of competitors (Table 4). Removal of *Fimbristylis spadiceae* on the high marsh led to the increased importance of *Spartina patens*, *Aster* sp., and *Limonium carolinianum*. The low marsh demonstrated an increase in importance for *S. patens*, *Aster* sp., and *L. carolinianum* when *Spartina alterniflora* was removed from plots. Interactions between species were found to be diffuse as in the examples given above, but in other instances they were specific and ranged from reciprocal to nonreciprocal in terms of the effect of removal. Silander and Antonovics[17] concluded that competitive interactions between species played a significant role in the structuring of these salt marsh communities. Russel et al.[33] determined niche breadths and niche overlaps of plant species in two salt marshes in Sussex, England. They concluded that interspecific competition affected resource partitioning between species on these salt marshes; higher species richness on the Hayling marsh led to significantly lower mean species' niche overlap and lower niche breadth than were found in the East Head marsh, which had a less diverse species composition.

TABLE 4
Species Removal Response Coefficients for Plants in the High Marsh and Low Marsh along the Core Banks, North Carolina[a]

Species removed	Species retained	Control coefficient	Removal coefficient
High marsh			
Fimbristylis spadiceae	*Spartina patens*	5.54	7.57
	Limonium carolinianum	0.18	0.99
	Sabatia stellaris	0.48	0.48
Spartina patens	*Fimbristylis spadiceae*	6.20	7.33
	Limonium carolinianum	0.18	0.24
	Sabatia stellaris	0.48	0.68
Low marsh			
Spartina alterniflora	*Spartina patens*	4.73	5.24
	Fimbristylis spadiceae	2.05	1.86
	Limonium carolinianum	1.07	1.59
S. patens	*Spartina alterniflora*	5.97	5.63
	Fimbristylis spadiceae	0.91	0.83
	Limonium carolinianum	1.07	1.09

[a] C_{17} = Nij/Ni where Nij is change in the percent cover of species J following the removal of species i, and Ni is the relative cover species i.

From Silander, J. A. and Antonovics, J., *Nature (London)*, 298, 557, 1982. With permission.

Early emergence of sprouts and seedlings were observed to give a species an advantage in competition for light on Dutch salt marshes.[34] Removal of competitors affected biomass yields differently at low- and high-marsh sites. *Puccinellia maritima* yields increased 2.6 times in lower sites, but not significantly (1.3 times) when *Spartina anglica* was removed from plots at upper sites, while *Spartina anglica* biomass increased more significantly at higher sites when *P. maritima* biomass was removed (4.4 times) than at lower sites where there was an insignificant (1.2 times) increase in yield. Laboratory experiments indicated that *Aster tripolium* suppressed *P. maritima* growth in low-nitrogen treatments, while *P. maritima* had little effect on *A. tripolium*. Scholten et al.[34] concluded that competition for nutrients might be significant in salt marshes at early stages of plant development, while at later developmental stages the availability of light was probably the most critical factor.

In an investigation of the effect of competition on the establishment of *Salicornia europaea* in a coastal salt marsh at Rumstick Cove, RI, Ellison[18] reported that the growth of *S. europaea* was inhibited by the encroaching perennial plant canopy. Biomass yields of *S. europaea* (1.11 g/m^2 in the open) beneath the perennial canopy were reduced 91% by *Spartina patens*, 10% by *Distichlis spicata*, and 77% by *Juncus gerardi*. Removal of perennial species led to an increase in biomass production and fecundity of *Salicornia europaea* in comparison with control plants that were growing beneath the perennial graminoid canopy. The effect of all of the perennial species was similar, indicating that shading might be the critical factor inhibiting the growth of *S. europaea*. Artificial shading treatments indicated that at 50% of full sunlight, biomass production of *S. europaea* decreased from 682 to 36 mg/plant (5.3% of plants growing in full sunlight), and with 10% of full sunlight the production decreased further to 11 mg/plant (1.6% of plants growing in full sunlight). Other characteristics were also affected by shading: branch number, root length, seeds produced per plant, and survival were reduced by shading (Table 5). These results corroborate the hypothesis of Goldberg and Werner,[35] which stated that interspecific competition could be a diffuse phenomenon and not species specific. The general effect of shading by graminoids on the Rumstick Cove salt marsh was inhibitory; the data collected indicated that the inhibition of *S. europaea* growth by perennials was not a species-specific phenomenon.[18]

TABLE 5
The Effect of Artificial Shading on the Growth Characteristics of *Salicornia europaea* on the Rumstick Cove Salt Marsh, Massachusetts

Character	Light (%)		
	100	50	10
Biomass (mg)	682	36	11
Branch (no.)	150.3	2.0	0.6
Root depth (mm)	25	12	11
Seeds/plant (no.)	1522	0	0
Survival (%)	100	7	0

From Ellison, A. M., *Ecology*, 68, 576, 1987. With permission.

TABLE 6
The Effect of Plant Density on Survival and Dry Mass Production of Species from Salt Marsh Habitats[a]

Species	Plants (m²)		Survival (%)		Dry mass (mg/plant)		Ref. (site)
	H.D.	L.D.	H.D.	L.D.	H.D.	L.D.	
Atriplex triangularis	10,000	1,000	33.0	40.0	82.3	162.0	38 (Ohio)
Salicornia europaea	29,500	1,000	62.8	70.0	47.2	203.8	37 (Ohio)
S. europaea	32,000	4,000	30.0	80.0	20.0	70.0	41 (England)
Spergularia marina	100,000	5,000	25.0	75.0	—	—	48 (Sweden)
Salicornia patula	15,300	500	4.6	8.4	29.0	2,700.0	43 (Poland)
Plantago maritima	2,500	100	48.8	47.6	—	—	42 (Sweden)

[a] H.D. = high density; L.D. = low density.

III. INTRASPECIFIC COMPETITION

Intraspecific competition may influence the survival, growth, and fecundity of populations in saline habitats. A number of researchers have reported that halophytes may be morphologically plastic in their response to increased plant density, but that mortality was not directly correlated with a density-dependent factor (Table 6).[3,15,36-39] Ungar[15] and Jefferies et al.[36] reported that mortality of halophytes growing in inland and coastal salt marshes was probably due to physical factors in the highly stressful salt marsh environment, such as flooding and high salinity, rather than competition.

An analysis of three *S. europaea* populations in an inland salt marsh at Rittman, OH indicated that peak density of plants in a population was not significantly ($r^2 = 0.02$, $p > 0.05$) related to the percent survival of plants, with only 2% of the variance being explained by the regression formula $y = 13.1306 + 0.0179 X$.[15] Abiotic physical stress accounted for about 85% of the mortality over the 3-year investigation period. These data corroborate earlier results by McGraw and Ungar,[40] which indicated that pan populations of *S. europaea* that had the lowest density and were exposed to the highest soil salinities always had the highest mortality, averaging >95%. Similar results were reported, demonstrating density-

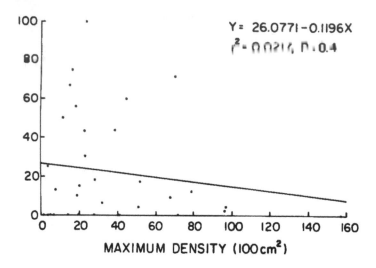

FIGURE 3.   The relationship between peak plant density in 100-cm² quadrats and survival in populations of *Atriplex triangularis*. (From Ungar, I. A., in *Symp. Biology of Atriplex and Related Chenopods*, McArthur, E. D., Stutz, H. C., Stevens, K. R., and Johnson, K. L., Eds., Gen. Tech. Rep. INT-172, Forest Service, U.S. Department of Agriculture, Provo, UT, 1984, 40.)

independent mortality as a primary factor for a population of *Atriplex triangularis* under field conditions in Ohio (Figure 3).[21] Peak densities ranging from <10 to >600 plants/100 cm² did not significantly ($r^2 = 0.02$, $p > 0.05$) affect the number of individuals surviving in permanent quadrats monitored over a growing season (Y = 26.0771 − 0.1196 X). Ellison[39] investigated the dynamics of a monospecific stand of *S. europaea* on a Rhode Island salt marsh. Treatments to determine density-dependent effects on survival, production, and plant form ranged from 100 to >10,000 plants/m². It was determined that intraspecific competition had a statistically significant effect on biomass production, leading to a decrease in biomass production from 3560 mg/plant (10 plants/m²) in low-density plots to 50 mg/plant (>10,000 plants/m²) in high-density plots. Mortality was not found to be density dependent in these *S. europaea* monocultures, with 91% of the plants surviving at the highest densities. *Salicornia* was plastic in its morphological response, and the amount of branching was strongly reduced at high densities. Other field investigations have also determined that density-dependent mortality was not common for halophytes growing in highly saline salt marsh habitats.[14,41-43] An exception to these general findings for annual halophytes was reported from a field experiment by Drake and Ungar.[44] *Atriplex triangularis* growing under optimal conditions at experimental densities ranging from 200 to 3000 plants/m² showed an increase in mortality, ranging from 5.3 to 48.0%, based on the initial density of the plants in the quadrats. In this experimental case where conditions were optimal for *A. triangularis*, growth and salt stress was not sufficient to induce mortality, and thinning occurred in the higher density plots. Therefore, these data suggest that annual halophytes could be subjected to density-dependent mortality under moderate salinity conditions.[44]

The effect of intraspecific competition on survival of halophytes was reported to be closely correlated with the growth form of halophytic species.[39,45] Three species of plants in the family Chenopodiaceae were studied by Ellison:[45] *Salicornia europaea*, a leafless stem succulent; *Atriplex triangularis*, a broad-leafed species; and *Suaeda maritima*, with linear leaves. Initial densities were 10,000 plants/m² for this investigation, and plants were grown for 18 weeks to determine the effect of density on survival in these experimental populations. No density-dependent mortality was reported for *Salicornia europaea*, while 10% of the *Suaeda maritima* population died, and >40% of the *A. triangularis* population died. The thinning slope for *S. maritima* was -2.5, and the slope for *A. triangularis* was -1.8, closer to the -1.5 predicted in the investigation by Yoda et al.[46] for thinning in natural plant populations. Ellison[45] concluded that plant geometry was the most critical factor in determining the percent survival and the slope of the thinning line in these halophyte populations. The species with the greater amount of branching and plant height, *A. triangularis*, had the highest density-dependent mortality and the closest fit to the predicted -1.5 slope for density-dependent induction of mortality.[45]

Investigations with halophytes on Ohio salt marshes had not been able to determine any significant correlation between density and survival in populations of *Salicornia europaea* and *A. triangularis*.[15,47] It was concluded from these earlier investigations that salt stress was an overriding factor in determining production in these hypersaline environments and that although density did affect plant yield, it was not significantly correlated with survival in these highly saline habitats.[37,38] Field and laboratory experiments were carried out to determine the interaction between intraspecific competition, nitrogen fertilization, and soil salinity on the survival, growth, and reproductive output of *A. triangularis*.[44] Under the moderate salinities (mean values at low- and high-salt sites were 0.4 and 1.2% total salts) present at the Constitution, OH location, Drake and Ungar[44] were able to demonstrate a significant correlation between density and salinity and a number of plant variables, including survival and shoot and reproductive dry mass. Similar results were obtained in laboratory experiments with *A. triangularis*, where salinity and nitrogen were also determined to effect significantly both shoot and reproductive dry mass production.[44] In the low- and high-salt field sites, nitrogen significantly influenced plant yields in the low-density plots (2 plants/100 m²), but not in the higher-density plots where individual shoot yield was below 10 g/plant (initial density of 10 plants/100 cm²) or <5 g/plant (initial density of 30 plants/100 cm²) compared with a range from >20 to 48 g/plant in the low-density plots (Figure 3). As plant density in the field plots increased, the size inequality of plants increased, indicating that competition was asymmetrical.[44] Inequality led to the formation of a few large dominants and many small suppressed individuals, with the dominants producing a disproportionate amount of the total seed production in a population (Figure 4). Reproductive effort was not significantly affected by density in the field experiment. The yield of *A. triangularis* seeds per plot was not significantly affected by density, since solitary plants grew larger and produced more seed per individual than did plants in the more dense plots. Drake and Ungar[44] concluded that persistence of the population was unlikely to be influenced by variations in the level of seed production among individuals.

Riehl and Ungar[38] determined that *A. triangularis* field populations, which were initially thinned to 1, 10, 25, and 100 plants/100 cm², significantly declined in biomass production when the density of plants was increased. Biomass yields ranged from 319 mg/plant (single plant) to 82.3 mg/plant (100 plants/100 cm²). Similar results were obtained with field populations of *Salicornia europaea* (Figure 5), with a reduction in yield of 89% in plots containing 295 plants/100 cm² when compared with controls yielding 416 mg/plant.[37] The occurrence of decrease yield, which was caused by density-dependent factors, has been reported for other coastal salt marsh habitats studied by Jefferies et al.,[41] Jerling,[42] Wilkon-Michalska,[43] and Torstensson (Table 6).[48]

Replacement-series experiments were carried out between plants collected from two populations of *Cynodon dactylon*, one from alkaline soils and the other from normal soils.[49] The normal soil population of *C. dactylon* was more competitive and had higher plant production in normal soils, while plants from the alkaline soil population had higher yields and were more competitive in soils treated with salts. These results indicate that ecotypic differentiation has occurred in these two populations of *C. dactylon*, and selection has increased their competitive ability under soil conditions similar to those in their natural habitats. Similar results were obtained in an investigation by Smith[50] with two mangrove subspecies of *Ceriops tagal*, with ssp. *australis* being associated with high-salinity regions

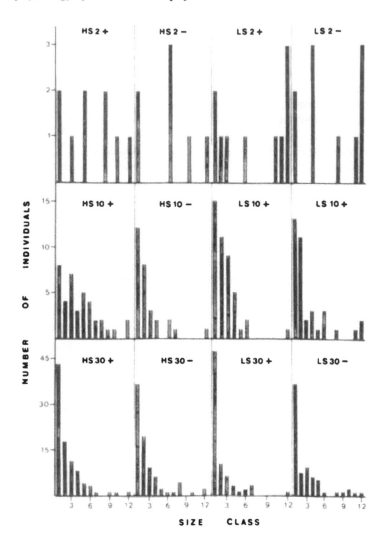

FIGURE 4.   Size class distributions for *Atriplex triangularis* plants in each of
12 experimental treatments under field conditions. HS/LS = high salt/low salt;
2, 10, 30 = plant density; +/− = with and without nitrogen fertilization. (From
Drake, D. R. and Ungar, I. A., *Am. J. Bot.*, 76, 1125, 1989. With permission.)

and ssp. *tagal* with lower-salinity portions of the salinity gradient.[50] The ssp. *tagal* was
more competitive at lower salinities and ssp. *australis* was the better competitor at higher
salinities. Differential success of these subspecies at different ends of the gradient played a
significant role in their distribution, since both subspecies made optimal growth at 1.5%
NaCl when competition was not a factor.[50]

Seed size and plant density affected the competitive ability of *Atriplex triangularis*.[51]
Large seeds germinated earlier than small seeds, and this factor may have a significant
influence on plant biomass production and survival percentages in populations. The fecundity
of individuals was strongly affected by plant density, with seed production being reduced
from a high value of 251 seeds per plant when grown singly to 24 seeds per plant for plants
at densities of 100 plants per treatment. Ellison[51] concluded that germination date had a
more direct effect than seed size in determining plant mortality in high density populations
of *A. triangularis*. Khan and Ungar[52] demonstrated that the seedlings of *A. triangularis*
produced earlier in the growing season had a better chance of surviving to reproductive
maturity than did seedlings that were produced in late spring and summer.

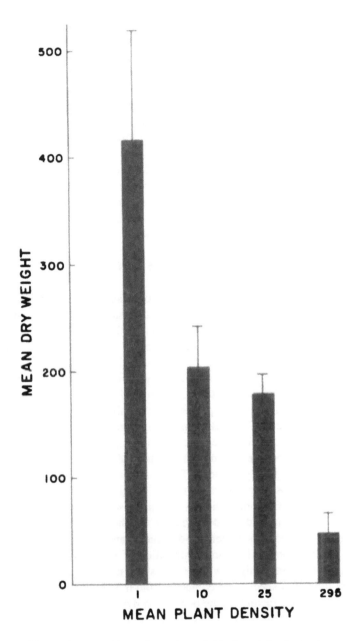

FIGURE 5. The effects of experimental thinning on the dry mass production (mg/plant) of a population of *Salicornia europaea* from an inland salt pan.

Khan[53] was able to demonstrate, using both field and laboratory experiments, that density-dependent factors played a significant role in determining the total biomass yield, reproductive output, and fecundity of *A. triangularis* plants. In low-salt plots (0.5% total salts) biomass production was reduced from 7.8 to 0.2 g/plant for 100 cm² plots containing one plant (low density) and 25 plants (high density), respectively. Reproductive output was reduced from 27.1 to 12.9% of the total dry mass production in the low- and high-density treatments. Laboratory experiments indicated similar decreases in biomass and reproductive output with increases in plant density. In the high-salt habitat (2.0% total salts) at Rittman, OH, biomass yields ranged from 0.6 to 0.1 g/plant in low-density (1 plant/100 cm²) and high-density (25 plants/100 cm²) field plots, indicating that both increased density and high soil salinity could cause a decrease in the biomass production of plants. Reproductive output in the high-salt plots was not significantly affected by increased plant density and ranged from 28.1 to 22.2% for the low- and high-density treatments, respectively. Although increases in density caused a reduction in reproductive biomass production in the high salt plots, they did not significantly change the percentage of resources allocated by plants to seed production. Khan[53] concluded that both density-independent soil salinity stress and density-dependent factors were involved in regulating the biomass production in populations of *A. triangularis*.

Goldsmith[54] also reported that a density-dependent reduction in growth occurred when *Festuca rubra* was grown in tap water but not in seawater treatments. The more salt-tolerant *Armeria maritima* had density-dependent inhibition of growth only in the seawater treatment. The less salt-tolerant *F. rubra* had a competitive advantage over *A. maritima* in the tap water treatments, while in seawater *A. maritima* was the better competitor.

IV. CONCLUSIONS

Interspecific competition probably plays a very significant role in determining the species diversity and relative cover in less saline portions of salt marshes and in ecotone areas between the zonal communities on salt marshes. Data collected up to this point indicate that seeds of most halophytes germinate best at low salinities or in nonsaline conditions and that salt stimulation of growth occurs at salinity levels that are usually less than 30% of seawater concentrations. Because most salt marsh species can grow to reproductive maturity in portions of the salt marsh with reduced salinity stress, it is hypothesized that competition for space, nutrients, and light become limiting to species in these habitats. Studies of ecotones indicate that removal of species generally caused an increase in production of the remaining species, indicating that competition for some resource is limiting in these transition zones on salt marshes. In the most hypersaline portions of coastal and inland salt marshes the diversity and density of species are greatly reduced. Edaphic conditions are most probably beyond the tolerance limits for most angiosperms in these highly stressful portions of the marsh. Intraspecific and interspecific competition is low in these hypersaline sites, and mortality tends to be driven by density-independent factors. Further investigations are necessary to sort out the relative significance of density-dependent and density-independent factors in determining the species composition and distribution of salt marsh habitats. Field experiments, using a multifactorial design, will be necessary to determine specifically the environmental factors for which species are competing and the relative importance of the effects of edaphic factors vs. biotic factors on the population dynamics of plants in these salt marsh habitats.

REFERENCES

1. **Schoener, T. W.**, Field experiments in interspecific competition, *Am. Nat.*, 122, 240, 1983.
2. **Grime, J. P.**, *Plant Strategies and Vegetation Processes*, John Wiley & Sons, New York, 1979, 8.
3. **Purer, E. A.**, Plant ecology of the coastal salt marshlands of San Diego County, California, *Ecol. Monogr.*, 12, 81, 1942.
4. **Hinde, H. P.**, Vertical distribution of salt marsh phanerograms in relation to tide levels, *Ecol. Monogr.*, 24, 209, 1954.
5. **Clarke, L. D. and Hannon, N. J.**, The mangrove swamp and salt marsh communities of the Sydney district. IV. The significance of species interactions, *J. Ecol.*, 59, 535, 1971.
6. **Pielou, E. C. and Routledge, R. D.**, Salt marsh vegetation: latitudinal gradients in the zonation patterns, *Oecologia*, 24, 311, 1976.

7. **Vince, S. W.**, Vegetation patterns in an Alaskan salt marsh, *Univ. Michigan School Nat. Resour. News*, 22, 14, 1981.
8. **Ungar, I. A., Benner, D. K., and McGraw, D. C.**, The distribution and growth of *Salicornia europaea* on an inland salt pan, *Ecology*, 60, 329, 1979.
9. **Bertness, M. D. and Ellison, A. M.**, Determinants of pattern in a New England salt marsh plant community, *Ecol. Monogr.*, 57, 129, 1987.
10. **Ungar, I. A.**, Influence of salinity on seed germination of succulent halophytes, *Ecology*, 43, 763, 1962.
11. **Barbour, M. G.**, Is any angiosperm an obligate halophyte?, *Am. Midl. Nat.*, 84, 105, 1970.
12. **Weaver, J. E.**, The quadrat method in teaching ecology, *Plant World*, 21, 267, 1918.
13. **Wendelberger, G.**, Zur Soziologie der kontinentalen Halophytenvegetation Mitteleuropas, *Akad. Wiss. Wien Math. Naturwiss. Kl.-Denkschr.*, 108, 1, 1950.
14. **Jefferies, R. L., Davy, A. J., and Rudmck, J.**, Population biology of the salt-marsh annual *Salicornia europaea* agg., *J. Ecol.*, 69, 17, 1981.
15. **Ungar, I. A.**, Population characteristics, growth, and survival of the halophyte *Salicornia europaea*, *Ecology*, 68, 569, 1987.
16. **Partridge, T. R. and Wilson, J. B.**, The use of field transplants in determining environmental tolerance in salt marshes of Otago, New Zealand, *N. Z. J. Bot.*, 26, 183, 1988.
17. **Silander, J. A. and Antonovics, J.**, Analysis of interspecific interactions in a coastal plant community — a perturbation approach, *Nature (London)*, 298, 557, 1982.
18. **Ellison, A. M.**, Effects of competition, disturbance, and herbivory on *Salicornia europaea*, *Ecology*, 68, 576, 1987.
19. **Covin, J. D. and Zedler, J. B.**, Nitrogen effects on *Spartina foliosa* and *Salicornia virginica* in the salt marsh at Tijuana Estuary, California, *Wetlands*, 8, 51, 1988.
20. **Snow, A. A. and Vince, S. W.**, Plant zonation in an Alaskan salt marsh. II. An experimental study of the role of edaphic conditions, *J. Ecol.*, 72, 669, 1984.
21. **Ungar, I. A.**, Autecological studies with *Atriplex triangularis*, in *Symp. Biology of Atriplex and Related Chenopods*, McArthur, E. D., Stutz, H. C., Stevens, K. R., and Johnson, K. L., Eds., Gen. Tech. Rep. INT-172, Forest Service, U.S. Department of Agriculture, Ogden, Utah, 1984, 40.
22. **Bakker, J. P., Dijkstra, M., and Russchen, P. T.**, Dispersal, germination and early establishment of halophytes on a grazed and abandoned salt-marsh gradient, *N. Phytol.*, 101, 291, 1985.
23. **Vince, S. W. and Snow, A. A.**, Plant zonation in an Alaskan salt marsh. I. Distribution, abundance and environmental factors, *J. Ecol.*, 72, 651, 1984.
24. **Cords, H. P.**, Factors affecting the competitive ability of foxtail barley *(Hordeum jubatum)*, *Weeds*, 8, 636.
25. **Wilson, D. B.**, Growth of *Hordeum jubatum* under various soil conditions and degrees of plant competition, *Can. J. Plant Sci.*, 47, 405, 1967.
26. **Suehiro, K. and Ogawa, H.**, Competition between two annual herbs, *Atriplex gmelini* C. A. Mey and *Chenopodium album* L., in mixed cultures irrigated with seawater of various concentrations, *Oecologia*, 45, 167, 1980.
27. **Barbour, M. G.**, The effect of competition and salinity on the growth of salt marsh species, *Oecologia*, 37, 93, 1978.
28. **Waisel, Y.**, *The Biology of Halophytes*, Academic Press, New York, 1972.
29. **Szwarcbaum, I. S. and Waisel, Y.**, Inter-relationships between halophytes and glycophytes grown on saline and nonsaline media, *J. Ecol.*, 61, 775, 1973.
30. **Zedler, J. B.**, Salt marsh community structure in the Tijuana Estuary, California, *Estuar. Coast. Mar. Sci.*, 5, 39, 1977.
31. **Gray, A. J. and Scott, R.**, The ecology of Morecambe Bay. VII. The distribution of *Puccinellia maritima*, *Festuca rubra*, and *Agrostis stolonifera* in the salt marshes, *J. Appl. Ecol.*, 14, 229, 1977.
32. **De Wit, C. T.**, On competition, *Versl. Landbouwk. Onderzoek*, 66.8, 1, 1960.
33. **Russell, P. J., Flowers, T. J., and Hutchings, M. J.**, Comparison of niche breadths and overlaps of halophytes on salt marshes of differing diversity, *Vegetatio*, 61, 171, 1985.
34. **Scholten, M., Blaauw, P. A., Stroetenga, M., and Rozema, J.**, The impact of competitive interactions on the growth and distribution of plant species in salt marshes, in *Vegetation Between Land and Sea*, Huiskes, A. H. L., Blom, C. W. P. M., and Rozema, J., Eds., Junk, Dordecht, Netherlands, 1987, Chap. 21.
35. **Goldberg, D. E. and Werner, P. A.**, Equivalence of competitors in plant communities: a null hypothesis and a field experimental approach, *Am. J. Bot.*, 70, 1098, 1983.
36. **Jefferies, R. L., Jensen, A., and Bazely, D.**, The biology of annual *Salicornia europaea* agg. at the limits of its range in Hudson Bay, *Can. J. Bot.*, 61, 762, 1983.
37. **Riehl, T. E. and Ungar, I. A.**, Growth and ion accumulation in *Salicornia europaea* under saline field conditions, *Oecologia*, 54, 193, 1982.

38. Riehl, T. E. and Ungar, I. A., Growth, water potential and ion accumulation in the inland halophyte *Atriplex triangularis* under saline field conditions, *Acta Oecol. Oecol. Plant.*, 4, 27, 1983.
39. Ellison, A. M., Density-dependent dynamics of *Salicornia europaea* monocultures, *Ecology*, 68, 737, 1987.
40. McGraw, D. C. and Ungar, I. A., Growth and survival of the halophyte *Salicornia europaea* L. under saline field conditions, *Ohio J. Sci.*, 81, 109, 1981.
41. Jefferies, R. L., Davy, A. J., and Rudmik, T., The growth strategies of coastal halophytes, in *Ecological Processes in Coastal Environments*, Jefferies, R. L. and Davy, A. J., Eds., Blackwell Scientific, Oxford, 1979, Chap. 15.
42. Jerling, L., Effects of microtopography on the summer survival of *Plantago maritima* seedlings, *Holarctic Ecol.*, 4, 120, 1981.
43. Wilkon-Michalska, J., Structure and dynamics of the inland populations of *Salicornia patula*, *Vegetatio*, 61, 145, 1985.
44. Drake, D. R. and Ungar, I. A., Effects of salinity, nitrogen, and population density on the survival, growth, and reproduction of *Atriplex triangularis* (Chenopodiaceae), *Am. J. Bot.*, 76, 1125, 1989.
45. Ellison, A. M., Morphological determinants of self-thinning in plant monocultures and a proposal concerning the role of self-thinning in plant evolution, *Oikos*, 54, 287, 1989.
46. Yoda, K., Kira, T., Ogawa, H., and Hozumi, K., Intraspecific competition among higher plants. XI. Self-thinning in overcrowded pure stands under cultivated and natural conditions, *J. Biol. Osaka Cty. Univ.*, 14, 107, 1963.
47. Wertis, B. A. and Ungar, I. A., Seed demography and seedling survival in a population of *Atriplex triangularis* Willd, *Am. Midl. Nat.*, 116, 152, 1986.
48. Torstensson, P., The demography of an annual halophyte *Spergularia marina* on a Baltic seashore meadow, *Vegetatio*, 68, 157, 1987.
49. Gupta, U. and Ramakrishnan, P. S., The effect of added salt on competition between two ecotypes of *Cynodon dactylon* (L.) Pers., *Proc. Indian Acad. Sci.*, 86B, 275, 1977.
50. Smith, T. J., Differential distribution between sub-species of the mangrove *Ceriops tagal:* competitive interactions along a salinity gradient, *Aquatic Bot.*, 32, 79, 1988.
51. Ellison, A. M., Effect of seed dimorphism on the density-dependent dynamics of experimental populations of *Atriplex triangularis* (Chenopodiaceae), *Am. J. Bot.*, 74, 1280, 1987.
52. Khan, M. A. and Ungar, I. A., Life history and population dynamics of *Atriplex triangularis*, *Vegetatio*, 66, 17, 1986.
53. Khan, M. A., Salinity and density effects on demography of *Atriplex triangularis* Willd., *Pak. J. Bot.*, 19, 123, 1987.
54. Goldsmith, F. B., The vegetation of exposed sea cliffs at South Stack, Anglesey. II. Experimental studies, *J. Ecol.*, 61, 819, 1973.

INDEX

A

ABA growth regulator, 90, 92
Abronia maritima, photosynthesis in, 74, 76
Acaena anserinifolia, growth response in, 63
ACC, see 1-Aminocyclopropane-1-carboxylic acid
Aegialitis annulata, water status in, 96
Aegiceras corniculatum
  carbon isotope ratio for, 81—82
  growth response in, 49
  photosynthesis in, 76, 81—82
  water status in, 96
Aeluropus littoralis
  cytokinins and, 90—91
  ionic content of, 109
  photosynthesis in, 80—81
Agropyron pungens, waterlogging and, 148—149
Agrostis alba, mycorrhizal fungi in, 65
Agrostis spp., germination of, 32
Agrostis stolonifera
  ecotypic differentiation and, 171—172
  growth response in, 62—63
  germination of, 10, 37
  interspecific competition and, 190—191
  ionic content of, 109
  waterlogging and, 139—140
Agrostis tenuis, ecotypic differentiation and, 172
Aizoon canariense, ionic content of, 110—111
Alcohol dehydrogenase, waterlogging and, 149—150
Allenrolfea occidentalis, 90—91, 108, 111
Alternanthera philoxeroides, 74, 77, 104
1-Aminocyclopropane-1-carboxylic acid (ACC), in ethylene production, 90
Ammonium sulfate, fertilization experiments with in salt marshes, 167
Ammophila arenaria, growth response in, 63
Anabasis setifera, ionic content of, 110
Anions, in seed germination, 9
Anthemis leucanthemifolia, ionic content of, 110
Arctotheca populifolia, germination of, 24
Armeria maritima
  germination of, 10, 18, 32, 37
  growth response in, 59, 61
  intraspecific competition and, 198
  ionic content of, 123
  mycorrhizal fungi in, 64—65
  nitrogen fertilization and, 157, 160—161
  waterlogging and, 139, 142—143
Artemisia monosperma, ionic content of, 110
Arthrocnemum fruticosum, 110, 164—165
Arthrocnemum glaucum, ionic content of, 107—108
Arthrocnemum halocnemoides, germination of, 18
Arthrocnemum macrostachyum, ionic content of, 110—111
Arthrocnemum perenne, ionic content of, 107—108
Arundinaria gigantea, ionic content of, 113

Aspartate amino transferase 1, relative mobility of in *Salicornia* spp., 181
Assimilation, salinity effects on, 73—79, 84
Aster spp., interspecific competition and, 191
Aster tenuifolius, germination of, 36
Aster tripolium
  germination of, 10, 13, 32
  growth response in, 51, 54, 59, 61
  interspecific competition and, 192
  ionic content of, 108, 110, 122—123
  mycorrhizal fungi in, 64—65
  nitrogen fertilization and, 160, 164
  phenological growth pattern of, 66
  photosynthesis in, 80
  waterlogging and, 142—143, 148—149
Atractylis flava, ionic content of, 110
Atriplex amnicola, 52—53, 126
Atriplex anthocarpa, ionic content of, 118
Atriplex californica, photosynthesis in, 74, 76
Atriplex calotheca, growth response in, 52—53
Atriplex canescens
  germination of, 23—24
  gibberellins and, 89
  ionic content of, 113, 118—119, 128
Atriplex confertifolia
  germination of, 24
  ionic content of, 108—109, 111—112
  mycorrhizal fungi in, 64
Atriplex cuneata, ionic content of, 128
Atriplex dimorphostegia, germination of, 26
Atriplex gardneri, ionic content of, 112
Atriplex glabriuscula, germination of, 24—25
Atriplex gmelini
  growth response in, 51—52
  interspecific competition and, 188
  ionic content of, 126, 131
Atriplex halimus
  cytokinins and, 90
  germination of, 11
  growth response in, 52—53
  ionic content of, 110, 128—129
  photosynthesis in, 74, 76
  water status in, 98
Atriplex hastata
  germination of, 19—20
  growth response in, 51—52, 61
  ionic content of, 108—109, 118—119, 127—128, 132
  mycorrhizal fungi in, 64—65
  nitrogen fertilization and, 159—162
Atriplex hortensis
  growth response in, 53, 61
  ionic content of, 128
Atriplex hymenelytra, ionic content of, 116
Atriplex inflata
  germination of, 25

growth response in, 51—52
ionic content of, 112—113
Atriplex laciniata, germination in, 24—25
Atriplex hymenelytra, ionic content of, 112—113
Atriplex leucophila, photosynthesis in, 74, 76
Atriplex littoralis
  growth response in, 61
  nitrogen fertilization and, 155—156, 159—160
Atriplex nitens, growth response in, 52—53
Atriplex nummularia
  germination of, 10—11, 25, 40
  growth response in, 51—53
  ionic content of, 108, 112—113, 117, 125, 128
  photosynthesis in, 76
Atriplex nuttallii, ionic content of, 112
Atriplex patula
  carbon isotope ratio for, 81
  germination of, 9, 13, 33, 36—37
  ionic content of, 107, 112
  photosynthesis in, 76, 81
Atriplex polycarpa
  germination of, 26
  ionic content of, 112
Atriplex prostrata
  germination of, 20, 36
  interspecific competition and, 186
  waterlogging and, 140
Atriplex repanda, germination of, 26
Atriplex rhagodioides, 10, 52—53
Atriplex rosea, germination of, 26
Atriplex semibaccata, germination of, 25
Atriplex spongiosa
  germination of, 25
  growth response in, 53
  ionic content of, 129, 132—133
Atriplex spp.
  growth response in, 49, 51—53
  ionic content of, 109
  phenological growth pattern of, 66
Atriplex tornabeni, 26, 65
Atriplex triangularis
  growth response in, 51
  germination of, 9—10, 18, 24, 29, 41—42
  interspecific competition and, 186, 188
  intraspecific competition and, 193—198
  ionic content of, 115—117, 119—121, 132
  nitrogen fertilization and, 158, 161
  phenological growth pattern of, 66—67
  seed banks reported for, 32—35
  seed budget of, 34
  waterlogging and, 142—144
  water status in, 96—99, 101—102
Atriplex turcomanica, ionic content of, 108—109
Atriplex vesicaria
  germination of, 25
  growth response in, 51—52
  ionic content of, 108, 112—113, 117
Austrofestuca littoralis, growth response in, 63
Auxin, salinity and, 90
Avicennia germinans, waterlogging and, 140
Avicennia marina

carbon isotope ratio for, 81—82
growth response in, 49
nitrogen fertilization and, 160
photosynthesis in, 74, 76, 80—82
water status in, 96, 98

B

Batis maritima
  carbon isotope ratio for, 81—82
  interspecific competition and, 190
  ionic content of, 113
  photosynthesis in, 74—75, 81—82
  water status in, 100—102
Benzyladenine, salinity and, 89
Beta vulgaris, photosynthesis in, 79
Biochemical inhibition, in photosynthesis, 79—80
Bolboschoenus maritimus, ionic content of, 109
Borrichia frutescens
  carbon isotope ratio for, 81—82
  ionic content of, 113
  photosynthesis in, 81—82
  waterlogging and, 147—148
  water status in, 100, 102
Bouteloua curtipendula, ionic content of, 130
Branches, in avoidance of excessive ion accumulation, 95
Brassicaceae, ionic content of, 107
Bromus diandrus, growth response in, 63
Bromus hederaceus, germination of, 13
Bromus hordeaceus, interspecific competition and, 186
Bromus mollis, ionic content of, 131
Bruguiera exaristata, water status in, 96
Bruguiera gymnorrhiza, 95, 150—151

C

Cakile edentula, 62, 102
Cakile maritima
  germination of, 24—25
  growth response in, 61
  ionic content of, 111
  nitrogen fertilization and, 160
Canavalia obtusifolia, growth response in, 49—50
Carbonate, seed germination and, 9
Carbon isotope ratios, stable, in photosynthesis, 80—82
Carboxylation pathway, in photosynthesis, induction of change in, 80, 82, 84
Carex atrubae, ionic content of, 110
Carex distans, ionic content of, 109
Carex lyngbyei, 15, 187, 189
Carex palaeea, water status in, 102—103
Carex pluriflora, interspecific competition and, 189
Carex ramenskii, interspecific competition and, 187, 189
Carex spp., interspecific competition and, 189
Carex subspathacea, nitrogen fertilization and, 159
Carex ursina, water status in, 95
Carpobrotus edulis, 80, 84, 110

O

Organic solute, production of, in salt tolerance, 107
Oryza sativa, 23, 149
Osmosis, 18—21, 42, 107

P

Pancratium maritimum, ionic content of, 110
Panicum amarulum, waterlogging and, 145
Panicum hemitomum, waterlogging and, 147
Panicum virgatum, ionic content of, 130
Parental effects, in seed germination, 38—43
Parnassia palustris, germination of, 10, 18
Pennisetum clandestinum, ionic content of, 130
Pennisetum typhoides, gibberellins and, 90
Phaseolus vulgaris
  carbon isotope ratio for, 82
  cytokinins and, 90
  gibberellins and, 89
  photosynthesis in, 78, 82
Phenology, in plant growth, 65—68
Phosphoglucomutases, relative mobility of in
      Salicornia spp., 181
Photosynthesis
  biochemical inhibition and, 79—80
  carbon isotope ratios and, 80—82
  carboxylation pathway and, 80, 82, 84
  field investigations and, 82—83
  salinity effects on assimilation and, 73—79, 84
Phragmites australis, 65, 109
Phragmites communis, 13, 51, 58—59
Pisum sativum, photosynthesis in, 80
Plagianthus divaricatus, germination of, 20
Plantaginaceae, ionic content of, 107
Plantago coronopus, 62, 65
Plantago crassifolia, ionic content of, 110—111
Plantago lanceolata, 13, 104, 186
Plantago maritima
  ecotypic differentiation and, 171—172
  germination of, 10, 13, 18, 27, 36—37
  growth response in, 59, 61—62
  interspecific competition and, 189
  intraspecific competition and, 193
  ionic content of, 108, 122—123
  mycorrhizal fungi in, 64—65
  nitrogen fertilization and, 155—156, 160
  phenological growth pattern of, 65
  photosynthesis in, 74, 77—78
  waterlogging and, 142—143
  water status in, 102—104
Plantago media, growth response in, 62
Plantago myosurus, nitrogen fertilization and, 158
Plantago spp., ionic content of, 107
Pluchea purpurascens, germination of, 35
Poa cookii, water status in, 102
Poa eminens, interspecific competition and, 189
Poa emurens, interspecific competition and, 187
Polypogon monspeliensis, germination of, 20, 36
Population variation, seed germination and, 14—18
Posidonia australis, water status in, 102

Potamogeton pectinatus, germination of, 37
Potassium, in seed germination, 9
Potentilla egedii, interspecific competition and, 189
Predation, effect of on seed production, 42
Prosopis farcta, germination of, 16—17
Pthirusa maritima, ionic content of, 114
Puccinellia distans, 13, 109, 128—129
Puccinellia festucaeformis, germination of, 13, 20—
      21
Puccinellia limoni, germination of, 13
Puccinellia maritima
  ecotypic differentiation and, 172—173
  germination of, 32, 37
  growth response in, 59, 61
  interspecific competition and, 190, 192
  ionic content of, 108, 122—123
  mycorrhizal fungi in, 64
  nitrogen fertilization and, 157
  phenological growth pattern of, 66
  waterlogging and, 139, 142—143, 148
Puccinellia nutkanensis, interspecific competition
      and, 187, 189
Puccinellia nuttalliana
  carbon isotope ratio for, 81—82
  germination of, 9—10, 13, 18, 21
  ionic content of, 109, 113
  photosynthesis in, 74, 81—82
Puccinellia peisonis
  ionic content of, 108, 122, 124
  waterlogging and, 140, 143—145
Puccinellia phryganodes
  interspecific competition and, 189
  nitrogen fertilization and, 159
  water status in, 95, 102—103

R

Reaumeria hirtella, ionic content of, 111
Reaumuria palaestina, water status in, 98
Rhizophora mangle, germination of, 27
Rhizophora mucronata, water status in, 95—96
Rhizophora stylosa, growth response in, 49
Ricinus communis, germination of, 25
Root exclusion, in avoidance of excessive ion
      accumulation, 95
Rumex crispus, germination of, 13, 17—18, 36
Ruppia maritima, 14, 113
Ruppia polycarpa, germination of, 14
Rytidosperma linkii, ionic content of, 130

S

Sabatia stellaris, interspecific competition and, 192
Salicornia bigelovii
  germination of, 20
  interspecific competition and, 190
  ionic content of, 112—113, 127
  seed banks reported for, 36
Salicornia brachiata, 23, 115, 119
Salicornia brachystachya
  germination of, 11

Milton Keynes UK
Ingram Content Group UK Ltd.
UKHW051952071024
449327UK00026B/2278